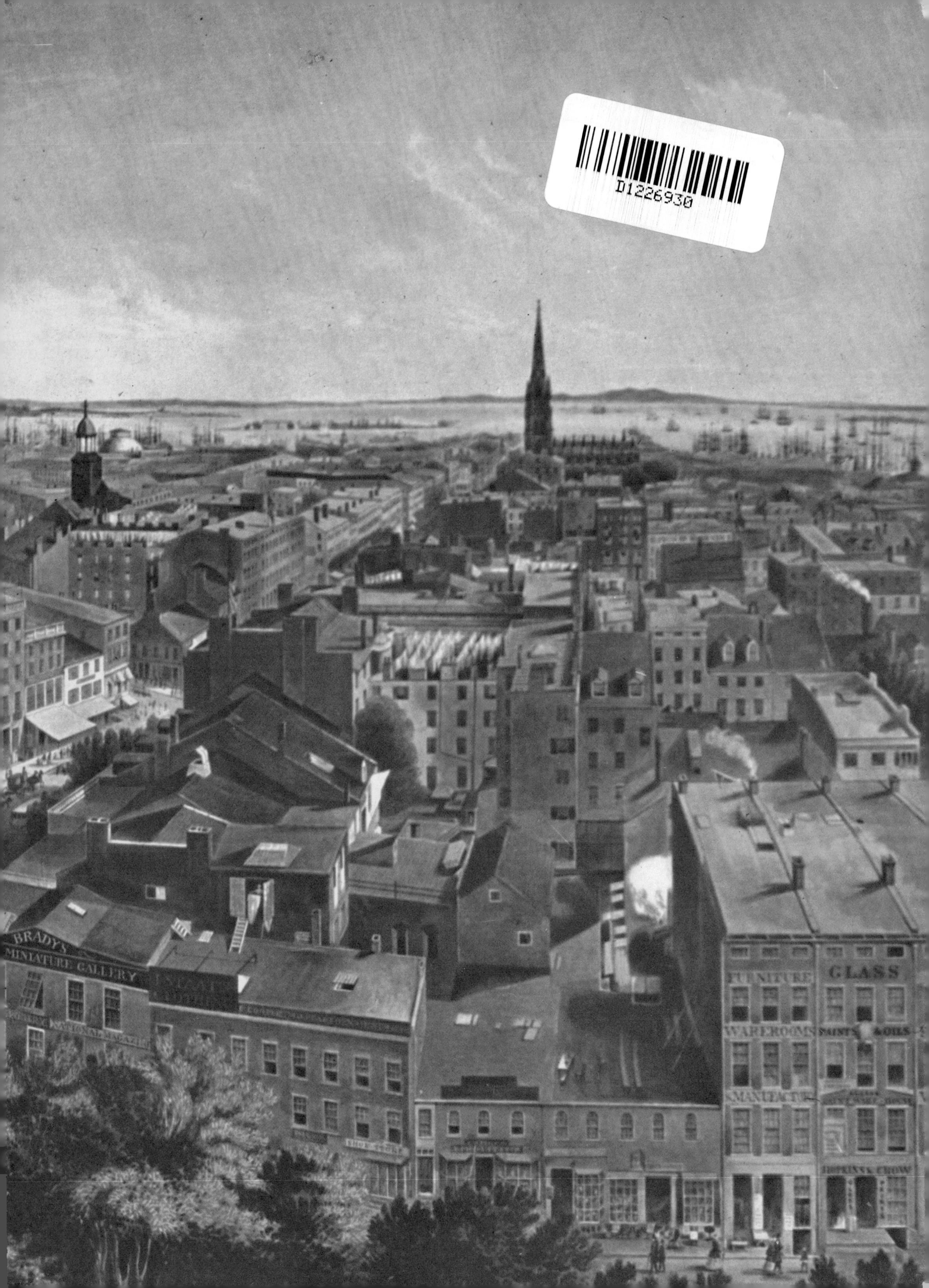
BRADY'S
MINIATURE GALLERY
FURNITURE
GLASS
WAREROOMS
PAINTS & OILS

NEW YORK FROM THE STEEPLE OF ST. PAUL'S CHURCH

Looking East, South and West

The Papprill View from St. Paul's Chapel

Aquatint, Colored
Second State

Drawing by J. W. Hill • Engraved by Henry Papprill

PUBLISHED IN 1848 BY H. I. MEGAREY

COURTESY OF DAVE DE CAMP

One of the most comprehensive panoramas of lower Manhattan in the middle of the nineteenth century, this view shows, beginning at the left, the steeples of the city's earliest churches —St. George's, the North Dutch Church, the Middle Dutch Church and Trinity Church. Barnum's Museum is at the left on the corner of Ann and Broadway, and Brady's Daguerrean Miniature Gallery is in the center foreground, on the southwest corner of Fulton Street. In the background are Bay Ridge, Governors Island, the hills of Staten Island and Bedloe's Island—not yet the home of Miss Liberty.

As You Pass By

THE USE OF THE MAPS

Most of the maps reproduced in conjunction with the Chrystie charcoal drawings indicate the point of view, direction and scope of the scenes depicted, and can be used to compare the same sites as they appear today.

To

ALEXANDRA HALSTED JOHNSTON

May she see the ever-brightening flame of Liberty gleaming over Old Father Knickerbocker's head.

K. H. D.

Oil painting on wooden panel, artist unknown, c. 1828. Notable for its contemporary view of Castle Clinton, early Manhattan fort which later became a place of amusement (Castle Garden), an immigration depot and long the city's beloved aquarium. It is now being restored as Castle or Fort Clinton. (See pages 72-73-74.)

As You Pass By

By

KENNETH HOLCOMB DUNSHEE

The City Slumbers

O'er its Silent Walls

HASTINGS HOUSE *Publishers* NEW YORK

Printed in the United States of America

INTRODUCTION

Events change a city's skyline. Each succeeding generation brings in new personalities and new accomplishments that stamp their impressions upon the physical outlines of a community. This has been the outstanding case in New York. For its size and age, New York probably has fewer landmarks or physical signs of its early history than any other major city. Nevertheless, lying beneath your feet, are the quaint neighborhoods, villages, roads, streams and the foundations or sites of buildings long since removed in the cause of expansion and progress.

It is probably not an exaggeration to state that most native New Yorkers have never seen or are not conscious of many of the present points of interest or even the marked historic sites in their own city. These are oftentimes better known to visitors and the "New Yorkers" from elsewhere. Yet, as in any other town, the citizens have a deep sentiment for their city's storied past.

To a great extent, the demands of commerce and industry have molded the city's essential character. In many ways this great metropolis has been the vortex of the commercial and industrial development of the entire nation. Except for the Spanish settlements, which were founded on dreams of conquest and treasure, Manhattan was the only major colonial settlement in the United States established purely for business reasons. The human problems of convicts, paupers or sufferers from religious persecution, fugitives from the stench of Europe, were of perfunctory interest to the Dutch West India Company. The thrifty, persistent Dutch founded and established Manhattan Island for commercial and economic development and the impression of those early Dutchmen is still upon the character of New Yorkers many generations and nationalities later.

There is probably no other city in the world of equal age which is so voluminously—but confusedly—documented and so little known. Perhaps that is why many have been discouraged in telling such a tale as that which is presented here. History, one of the most difficult studies, is often glossed over as a series of facts too undefined to make any great personal difference to us. However, anyone who has the skill to make the past, with its traditions and lessons, more real; and our forebears, who built this great country, more alive, gives us a better understanding of ourselves and our fellow men.

This volume is the result of an intense civic pride and a faith in the simple truths which are the milestones of human progress. Cynicism seems silly in an atomic age. There has never been a time when tolerance for others and love for our country, our city, our homes, has been more important.

To know something of events and the characters, exploits, even foibles, of a few of our forebears, is to give us at least a cross section of history as it has been lived. In this book the author has endeavored to focus your interest on the exact spots where the events of history and "life, liberty and the pursuit of happiness" occurred. While the plan of this work is simple, the amount of preparation which has gone into it has been prodigious. The remarkable display of illustrations, most of which have never before appeared in print, the exactness and clarity of the specially prepared maps, will give you an heretofore unknown thrill of personal acquaintanceship with the past in relation to the present.

The unique method of restoring to us many of Manhattan's earlier scenes is one that I do not believe has ever appeared before. Many years of research have gone into the project. In looking over some of the author's notes, I was amazed at the scope of his efforts and cannot think offhand of a source of history, literature and art that has not been carefully investigated to substantiate the scenes and to give each of the structures shown an honest, solid foundation. It was astounding to me to learn that in each scene research had included the contemporary street paving, surface drainage, location of pumps or wells, street lighting, as well as other objects, animate or inanimate, which were authenticated in a conscientious way.

The idea of this book originated in a section of Manhattan that has long played a part in the city's history. In old Maiden Lane, in a building near where a town pump once stood, there sits at his daily task a man of energy whose breadth of vision has had much to do with the inspiration of the present work.

I speak of Harold V. Smith, a man whom I have known for many years. And often, sitting in council with him, I have witnessed not only his many contributions to the "business that protects other business" while building up and conducting the largest fire insurance business in the world, but also his tenacity in the promotion of many cultural and educational projects. Only the good Lord knows how he has found the time.

The collecting of a museum of fire fighting relics has been one of Mr. Smith's most important accomplishments. This is housed in the headquarters of The Home Insurance Company at 59 Maiden Lane, New York. Here was both the point of inspiration and the source of a great deal of the substantiating material used in presenting many of the facts and anecdotes here related.

Who is there to doubt that New York's early firemen, running with their machines over the unpaved or cobbled streets, through mires and over the unleveled hills, knew their city better than any other group of her citizens?

No one has been more familiar as a class with the highways and byways of New York than its firemen. The author, Ken Dunshee, first became interested in the lore and history of New York through the activities and endeavors of the firemen in the preservation of life and property beginning with the days of the Dutch bucket lines.

Mr. Dunshee, who has been associated with the H. V. Smith Museum for the past twelve years, has done a work which is a fitting completion of the inspiration he received. Perhaps it is a natural heritage for him to do this. His Manhattan antecedents, I believe, go back to a lovely little farm (where Gansevoort Street and the old Greenwich Road later met) the site of Sappanikan, the Indian village. From a farmhouse on a gently sloping knoll his ancestors, the Mandevilles, de Lamars and De Camps, looked out over the majestic Hudson. Also, I believe, it was his great-uncle, Henry Webb Dunshee, who wrote "The Knickerbocker's Address to the Stuyvesant Pear Tree," as well as the "History of the Collegiate Reformed Church School," Manhattan's oldest institution of learning, of which he was headmaster from 1842 to 1883.

The team to produce this work has been a good one, combining the collecting instinct of Mr. Smith with the unflagging enthusiasm of the author and his ability to interpret the past into a living record. The result, I am confident, will be of unique and lasting interest to the reader. That this work has been produced over the years, during the course of intense business activity which has had first call on the time of the author, should entitle him to additional credit.

Here, indeed, is the best restoration of many pleasant places which have been placed on the altar of industrial progress, and the acquaintance of many interesting people whose memory should "ne'er have been forgot." You will be surprised and charmed at the folks you will meet and the scenes you will see in these pages.

George McAneny

President, American Scenic and Historic Preservation Society

CONTENTS

ILLUSTRATIONS

COLOR PLATES

CHARCOAL DRAWINGS

FOREWORD

Look about you—the marvels of the city stand before your eyes—the spires of worship, art and learning; the towers of industry, commerce and business—all proclaiming the enterprise, courage and accomplishments of free men in free competition. New York is ever-changing, ever-new—but beneath your feet as you pass by are the old landmarks and locations of scenes which were once the settings on the stage of time.

While we live in the present with our hopes and faith in the future, the deep roots of our character lie in the past. As any steamboat or ferry pilot knows, it is just as important to line up the marks on the stern as it is to point the bow of a vessel to its destination. Otherwise, the force of the current will surely lead into dangerous waters. This is one of the values of history.

Patrick Henry once said: "I have but one lamp by which my feet are guided and that is the lamp of experience. I know of no way of judging the future but by the past."

New York can best be understood by knowing some of the forces which have fashioned and developed it. The dignity and mutual respect by which Americans of every conceivable racial, religious and national origin have learned to live together is the outcome of an ideal nurtured early on Manhattan Island. The promise of its cosmopolitan beginnings in the early days of Dutch rule has been fulfilled a thousand times over. New York has become in many ways the greatest metropolis in the world, and certainly it is the greatest melting pot of peoples that the world has ever known.

If there is one single characteristic of its people, it is that of tolerance—a tolerance sometimes bewildering to other people, as it often assumes a form of indifference; but a tolerance that is bounteous in its welcome to all. New York belongs to all the people of America and of the world.

Today, in this densely populated area, where more different kinds of people are drawn into closer contact than anywhere else in the world, men often do not know their next-door neighbors and are not particularly concerned with what happens on the block next to theirs or in other sections of the city. But an investigation into the past shows that although New Yorkers have always considered their city the largest and best, nevertheless its inhabitants once lived in a neighborly home-town atmosphere, such as one sees in smaller communities throughout the country today.

Perhaps one of the best ways of recapturing the quality of those long-vanished days in the life of the great metropolis is through the eyes of those who, in their times, knew their city best—the volunteer firemen. Many people do not realize the importance of early fire-fighting in the political and social life of the city during the period of the Volunteers. For more than one hundred years these adventurous and self-sacrificing fellows, who counted among themselves many of New York's leading citizens and businessmen, hauled their machines by hand through roads, avenues, streets, alleys and shortcuts in every district of the city.

Nine mayors of New York City were active volunteer firemen or were elected largely because of the efforts of the firemen: Walter Bowne, Cornelius W. Lawrence, Stephen Allen, Isaac L. Varian, Daniel F. Tremance, C. Godfrey Gunther, William H. Wichham, Mayor Paulding and Philip Hone. While the "Little Flower," Mayor Fiorello La Guardia, did not owe his election to membership in the present Fire Department, he will forever be associated with it in people's memories.

No one can better learn the true character and makeup of their city, or fully appreciate the real love that New Yorkers had for it, than by following the firemen and fire runners through the byways of old New York and standing shoulder to shoulder with them in their fight against the demon fire to save life and property. This story is a modest attempt to bring those early scenes to life again.

This book can record little more than a small part of the history of New York with its vastness and its more than three hundred years of development. It cannot even be considered a complete guide to all of New York's monuments, markers of historical sites, etc. But the scenes depicted, mapped and described are typical examples of the atmosphere, character and appearance of the different locations at various times in history, mostly in lower Manhattan.

History must always be a compromise between many opinions and varying points of view. In reaching the compromise, however, much of the excitement may be lost as the human emotions are filtered out. For this reason, it is sometimes more rewarding to look at an event through the eyes of a single competent eye-witness. Then the scene comes to life, has warmth and dimension; the heroes and heroines, victims and villains once again are real people—as indeed they were. For this reason, the author has called upon several such competent witnesses to describe in their own words the events of New York's history as they saw them. He is indebted to the remarkable diary of Philip Hone, "The Diary Of A Little Girl" by Catherine Haven, published by Henry Collins Brown, to George Templeton Strong for his unique diary (soon to be published in a popular edition by Columbia University, which graciously permitted the author to draw upon it), and to many others.

The dates of the charcoal views in the book were chosen both for general interest at each location, and to demonstrate the changes which took place even in earlier times. In this we have had the competent and devoted help of E. P. Chrystie, architect and artist of great talent, who helped by his drawings in this medium to recapture those scenes in an authoritative and realistic manner. Each of the drawings is a careful representation of the original scene, created from the best sources of information available.

A great deal of research has been applied to the costumes of the people at different periods, the transportation and fire equipment then in use, and the general appearance of the fire laddies. Most of the photographs reproduced are rare and many of them have never before been published. The color plates include several unpublished or new-found subjects of unusual interest.

The ravages of time and the demolition due to the march of industry have etched their marks on the face of the city. Many of the lovely old structures of a bygone day have been irretrievably lost; some have been defaced or obscured by the necessities of commercial expansion or housing development; others have been replaced by modern edifices of a sometimes questionable beauty. Industrial enterprise, foreign trade and manufacturing, the never-ending growth of new services, the constant stream of immigration, the dynamism of successful business activity, all had a hand in the ruthless alteration.

Even during the time this book is being written, vast changes are going on, affecting the shape and style of the city. Whole neighborhoods and communities on the lower eastern side of the island have been completely removed to make way for huge modern housing developments, as in several other parts of the city. Only recently, the Collegiate Church of St. Nicholas, at Fifth Avenue and 48th Street, was abandoned and dismantled to make way for a modern building. Its passing was marked by a sincere sorrow in the hearts of its congregation and neighbors. The following note, accompanied by a single, wilted red rose, was dropped on its doomed doorstep by an unknown passerby on the morning of the day the wreckers began their work:

> "*To you and to me the beloved Church of St. Nicholas imparts a last message. Before Him an awful blow is struck.*
>
> "*The place where you are standing is holy ground. Every stone and grain of sand is hallowed in His service, hoary with the passing of time.*
>
> "*Farewell, beloved Church of St. Nicholas.*"

The bell which hung in the St. Nicholas tower was carefully saved and moved to a new home in the Middle Collegiate Church at Second Avenue and 7th Street. If you listen in the vicinity of the ancient crossroads of old Stuyvesant Village you will still hear the melodious tones of this clarion of liberty and independence. This bell, which originally hung in the Middle Dutch Church on Nassau at Liberty Street, two hundred and twenty years ago, is one of New York's most interesting relics, having tolled at the reading of the Declaration of Independence to the Continental troops then stationed in New York in 1776. After the Revolution, George Washington and his soldiers heard its acclaim of victory on Evacuation Day. The father of his country again heard it beat in tune with his heart when he was inaugurated as the first President of the new nation.

The building of the Brooklyn Battery Tunnel has also changed the lower west side of the city, and nearly caused the loss of Castle Clinton, over the walls of which Castle Garden, later the old Aquarium, was built. What is left of little Edgar Street now is merely a facade facing the tunnel plaza.

There is a great deal of agitation under way for a complete renovation of the lower Bowery as a follow-up to those slum clearance projects already accomplished.

Only recently, workmen removed the beautiful stone watering trough from the inside court of the New York Public Library at Fifth Avenue and 42nd Street. This was used in the days when horse-drawn vehicles were parked in the court while their owners were browsing inside the library. The area is now used for delivery trucks and for the traveling libraries which load and unload there.

The preservation of historic buildings and sites has been sacrificed to the necessities of progress. Unlike Boston, Philadelphia, Baltimore or New Orleans, New York retains but few physical reminders of its ancient past. Aside from St. Paul's Chapel, the John Street Meeting House, Fraunces Tavern, the Old Chophouse on Cedar Street, and the present City Hall, there is very little physical evidence of early Manhattan's history.

The absence of these historical treasures has imbued the author with the desire to recreate, at least in part, the living spirit of which there is so little material evidence and to bring to light the scenes of the past. It is not so much our city's relics that we revere as it is the inspiration of their association.

Kenneth H. Dunshee
New York City

DENY HIM NEVER

But he denied, saying, I know not, neither understand I what thou sayest. And he went out into the porch; and the cock crew. And the second time the cock crew. And Peter called to mind the word that Jesus said unto him, Before the cock crow twice, thou shall deny me thrice. And when he thought thereon, he wept.

St. Mark 14:68,72

This weathervane, a remarkable relic of New York's past, was presented to Washington Irving in 1836 by a prominent Manhattan resident. Although claimed to have been used on the Stadt Huys, the tavern which became New Amsterdam's first city hall, it is the author's belief that it equally could have graced either the "church in the Fort," or the second Dutch Church built at Harlem in 1686. Washington Irving used it at his homestead, Sunnyside, before he gave it to the St. Nicholas Society in 1848. It is now exhibited by the New York Historical Society.

The rooster symbol, "*aen een ketel tot de haen van de toorn*," (a copper weathercock on the steeple), once seen on a number of old Manhattan buildings, served to remind all viewers of the spiritual pitfalls which might result from any denial of Christ.

The Eagle's Nest

PROLOGUE

IT all started, really, in the Continental Eagle's Firehouse in old Maiden Lane. Originally built on the site of the old abandoned tanpits before the War, in 1790 it nestled close to Tom Stevenson's blacksmith shop on the corner of Gold Street. Old Harmanus Rutgers' brewery stood but a few hundred feet up Gold Street, a fact her boys undoubtedly had cause to remember.

It was on a certain day in May that a number of members of the company had gathered, as they often did, in the Eagle's sacred eyrie. George and Charlie were preening the feathers of that pride and joy of the company, old Eagle engine herself. One of the boys sitting at the desk in the corner put down the minute book he had been studying and idly began scribbling on a sheet of foolscap. Robert Reb sat by the fireplace rumbling on about the exciting events which had stirred the neighbors in Maiden Lane during the past few days. He was opining "folks around here will never see the likes of it again" when the chief strolled in, gave the place a critical glance and tossed an S-shaped ax handle onto the work bench. As the piece lop-lollied across the top and came to rest against a row of square axle wrenches neatly hung behind the bench, he said, "Gus finished it last night. Rob, you set it in that head so it stays."

Walking over to the corner, he glanced down for a moment at the scribbler, then peering more closely, he read:

"Tip and Tilt,
High and Low,
Rain and Frost,
Quick and Slow."

Taking a second look at the occupant of the corner, he said in a voice as restrained as he could manage, "Now, what in the name of all conscience is that?"

"Names," came the reply.

"Names of what?" he demanded.

"Could be names of all conscience, I guess! You see, Chief," he continued, "these were names in the list that John Mitchell of New Jersey sent us."

This list contained names of soldiers who had been mustered out of the Continentals following the war with Great Britain. The literary doodle brought forth no great response of enthusiasm. As the Chief sat down on the bench near the fireplace talk returned to fire engines, and to the aforesaid events in Maiden Lane.

The unexpected muster of machines took place right where Tom Jefferson used to live up the Lane beyond Smith Street. When the call came, 13 was the first to arrive with their little goose-neck engine. It gave the Eagle's fledglings full opportunity to observe the approach of the other machines which came running nearly abreast down the hill on Smith Street from Ann Street. The rumble of the heavy machines warned pedestrians of their approach. Glittering with shine and polish they made an exciting picture as the steersmen and men on the ropes paced along. Big Chick and his Good Willers made a spurt for the lead in crossing Fair or Fulton Street. In making this maneuver the Good Will machine crowded John Dee and the rival No. 4 engine so close that the right front wheel of the latter struck the curbstone. The tongue in the machine tore loose from John's and the other steersman's hands and yawing like a battering ram nearly took the back tapestries from a couple of inquisitive young ladies of quality who had stopped to see the fun. Regaining control and righting the engine, No. 4's lads got her under way, and sprinting fast, reached the rendezvous only a few seconds later than their rivals.

* * * * *

Now, what may seem to be the relating of an occurrence that took place long ago, is in reality a report, with a few liberties, it is true, of an event which happened in the 1940's in the heart of the insurance district in Maiden Lane. It was in preparation for the inauguration of the new H. V. Smith Museum on the twelfth floor of The Home Insurance Company's building at 59 Maiden Lane* that these genuine old fire engines of an earlier era, which had been in a warehouse in Ann Street, were run by employees of The Home down William Street into Maiden Lane where, to the delight and astonishment of several thousand people, they formed their peculiar muster.

On that noon hour, Bill Casey and his organization of super-efficient riggers hoisted these fully assembled machines up twelve stories for all the world to see. The novelty of the scene, together with the enthusiasm of the news cameramen and reporters, made it a lively if somewhat traffic-stopping event. Thanks to Casey's skill and an alert police department, everything went off smoothly except the traffic, and the machines were shunted in through the large windows into the twelfth floor Museum, where they may be seen to this day, in constant tribute to New York's past, present and future fire laddies.

The firehouse part of the story is not a too-inaccurate account, either, because a replica of Engine 13's firehouse of 1790 in Maiden Lane had already been built in the H. V. Smith Museum. Both the exterior and interior of this restoration were made of old pine more than 150 years old. Every detail of its construction was reproduced as nearly like the original as

**See appendix*

"Good Willers" making a spurt for the lead. The engine is a double-deck, Philadelphia-style machine

No. 4's squirrel-tail engine crowding the curb in an effort to catch the Good Will lads

humanly possible. This firehouse has become a showplace to people from all parts of the world. Its highly polished goose-neck engine is an original James Smith machine of the early 1800 period. It is decorated with painted panel, polished metal and inlaid leather, just as the machine appeared new and in her prime.

The fireplace is a reproduction of the oldest one known in a New York firehouse. Several of the relics such as brass torches, hat fronts and trumpets, were originally owned by members of Engine 13 and were found in antique collections, private homes, and out of the way places from Philadelphia to Maine. Additional relics to complete the equipment of a New York Engine Company during this period were taken from other companies of old New York.

To Harold V. Smith, the untiring collector, the preserver and guardian of untold pages of history which, except for his interest might otherwise have been lost, will go the ever-increasing appreciation of his fellow men. His efforts in assembling the nation's most complete collection of equipment, art and history of fire-fighting and fire insurance have made it possible for future generations to share in his dreams and accomplishments. This he has managed to do in spite of the fact that he has been, for many years, at the helm of the largest and most active fire insurance organization in the world.

The museum is one of the most complete in the world, important not only because of its completeness, but because of its high standard of quality, the excellent condition of the various relics and their great value in recording phases of American history and American popular culture. It represents a lifetime of careful, persevering and devoted collecting, wide knowledge and experience with material in the field, and, above all, a real love of the subject. The H. V. Smith Museum is especially noted for its display of

After the race, Good Will Engine rests up before the hoisting

BUF
BUFFET
SCHRAFFT'S

Casey and his crew readying the Eagle for her flight

Entrance to the house of Lucky 13

American skills of the handicraft era which were the matrix of American industry's present know-how. It embodies such special crafts as carriage-making and painting, wood carving, the art of the metal worker and silversmith, leather working, ceramics, etc.; local history; the history of fire-fighting and fire insurance; American folk art as it evolved. At some future date the collection very likely will become a public museum in the full sense of the term, for the material includes so much that is of intrinsic value in the American story that its preservation for study and exhibition is a public service of unique and permanent importance.

The fictionalized conversation in the museum firehouse is typical of many real conversations which took place after the business day was over and a small group would gather to relax and talk of many things in that sanctorum of peace and quiet, the firehouse of old Eagle Engine Co. 13 in Maiden Lane. Many minute books and other records which, even yet, have never been published, make those who are privileged to study them conscious in a more intimate sense of the events of the past.

Firehouse in the H. V. Smith Museum, replica of Engine 13's house at Maiden Lane and Gold Street, 1790

Eagle 13 at rest in her eyrie

Eagle ng home oost

The city slumbers; o'er its silent walls
Night's dusky mantle, soft and silent falls;
Sleep o'er the world slow waves its wand of lead,
And ready torpors wrap each sinking head;
Still'd is the stir of labor and of life,
Hush'd is the hum, and tranquill'd is the strife;
Man is at rest, with all his hopes and fears,
The young forget their sports, the old their cares,
The grave or careless, those who joy or weep,
All rest contented on the arm of sleep.

Sweet is the pillow'd rest of beauty now,
And slumber smiles upon her tranquil brow;
Bright are her dreams—yes, bright as heaven's own blue,
Pure are its joys, and gentle as its dew;
They lead her forth along the moonlit tide,
Her heart's own partner wand'ring by her side;
'Tis summer's eve; the soft gales scarcely rouse
The low-voic'd ripple and the rustling boughs,
And, faint and far, some melting minstrel's tone
Breathes to her heart a music like its own.

Composed and read for a firemen's benefit performance at the old Chatham Theatre in 1840 by a member of the Tripler family, builders of Tripler Hall. A number of the Triplers, including Thomas E. Tripler, were members of New York's Volunteer Fire Department.

The Montanus View of New Amsterdam, circa 1650. The engraver is thought to be Augustine Herrman.

A Pleasant Land To See

On that long distant day when Hendrick Hudson sailed into the lower bay of what is now New York, one of the greatest stories of human enterprise began. Although there is no conclusive evidence that the master of the Half Moon ever set foot on the island of Manhattan, his comment, "we have raised a very good land to fall in with and a pleasant land to see," and his reports to his backers in Holland were enough to convince them of the favorable prospects of trade to come in this new land.

What the eyes of Hendrick Hudson would have beheld if he had wandered over the island was described by Secretary de Rasieres before 1630. "The island of the Manhatas," he wrote, "extends in length along the Mauritse (Hudson) River, from the point where the fort, 'New Amsterdam,' is building. It is about seven leagues in circumference, full of trees, and in the middle rocky to the extent of about two leagues in circuit. The north side has good land in two places, where two farmers, each with four horses, would have enough to do without much clearing at first. The grass is good in the forest and valleys, but when made into hay is not so nutritious for the cattle as here (Holland), in consequence of its wild state, but it yearly improves by cultivation. On the east side there rises a large level field, of from 70 to 80 morgens of land, through which runs a very fine, fresh stream; so that land can be ploughed without much clearing. It appears to be good . . ."

New York was, in primitive days, the "city of hills." At the extreme south end of the island, where the ancient fort later stood, was an elevated mount, quite as high as Broadway is now at Trinity Church. The hills were sometimes precipitous, such as Beekman's and Peck's Hills, and sometimes gradually sloping, as on other hills along the line of the water coursing along the region of Maiden Lane. Between many of the hills flowed "several invasions of water," such as that which lay in the present course of Broad Street, a stream dug out to form a canal in the true Dutch tradition, to gratify Dutch recollections.

We find a note in the records of early Manhattan of the presence of sturgeon, whales and seals "playing off the rocks at the Fort." Mention is even made of pelicans! A number of the early views of the island from several directions executed by independent artists show the presence of palm trees. It is strange that they should appear in such an unlikely place and yet the draftsmanship in various early prints seems unmistakable.

The first recorded settlement of Manhattan Island was that of Adrian Block and Hendrick Christenson, commanders of the ships Tiger and Fortune respectively. According to tradition, they "caused to be built" several board and bark structures in the neighborhood of 39 Broadway, some say in Beaver Street. The first serious blow to their embryonic settlement was by fire, when Captain Block's ship, the Tiger, burned in the winter of 1613. The fact that the Fortune had returned to Holland left the tiny settlement completely isolated from the civilized world. Block determined to use the timber cut from the magnificent forest upon the island to construct a new ship and by the spring of 1614, launched a small vessel which he dubbed the Onrust, a Dutch word for Restless.

The Fortune returned in 1615, and was soon followed by other ships bearing the names of Eagle and Love, and later the Union, which in returning to Holland in 1632, was delayed by the English, who were eventually forced to release it.

One could almost foretell the future of New York by the names of the ships which were connected with its early history. The Tiger (even before Chief Tammany's time) which burned, and the return of the good Fortune were prophetic; The Restless, the island's own, was certainly indicative of the most restless city on earth; the ship Eagle stood for the future national emblem; The Nightingale, symbol of New York's traditional welcome to all, brings to mind the Swedish singer who was later to become the toast of all New York and a favorite of the New York Fire Department and the ship Eendracht, or Union, whose name and career foreshadowed the events of the Civil War.

The natural pristine beauty of Manhattan's topography has undergone more rapid and relentless changes than that of any other similar area in the world. Buried in the past, beneath the hard crust of our modern city are the hills and dales, the rocks and rills, the rivulets and creeks, the lakes and ponds, which early travelers described as having a beauty seldom surpassed in nature. If one takes the time to observe, there still may be detected some of the more general patterns of the old original surface. Wherever you go today in the city you are bound to cross the course of brooks and streams and hidden bodies of water, as well as traverse on or over the paths and roads of the old town.

A southeast view of the city about 1763 from a point midway between Ranelagh and Lispenard's. The buildings, from east to west, are: St. George's Church, the Jail, New Dutch Church, French Church, South Dutch Church, City Hall, Presbyterian Church on Wall Street (erroneously shown with a spire), King's College and the Trinity Church.

In Old Dutch Days

MANHATTAN was occupied and administered by the Dutch until 1674 though it was characterized even then by its cosmopolitan atmosphere. Reverend Isaac Joques says that as early as 1643 the island was inhabited by a score of different nationalities and sects with at least eighteen different languages spoken.

Though beset by fever, plagues and the rigors and hardships of life in a new and hostile environment, the Dutch settlers continued to pursue the traditions and customs of their fatherland. They celebrated five festivals in the year—Kerstydt (Christmas), Nieuw Jaar (New Year), a great day of cake, Pass (The Passover), Pinxter (Whitsuntide), and San Claas (Saint Nicholas or Kris Kinkle Day) celebrated on December 5. The jovial Saint Nicholas was the titular divinity of the town. If we can believe Washington Irving, his likeness graced the bowsprit of the first Dutch immigrant ship, the "Goede Vrouw," and the first church in the old fort was named for him. It was said that the old figurehead was removed from the ship and mounted upon that church.

The Stadt Huys, Coenties Slip and Pearl Street.

On Kris Kinkle Eve, even as now, the biggest thrill came to the children, many generations of whom sang the old hymn of which this is part:

Dutch rooftop

Sint Nicholaas, myn goden vriend,
Ik heb u altyd wel gediend;
Als gy my nu wot wilt geben,
Zal ik dinen als myn leven.

(Saint Nicholas, my dear, good friend,
To serve you ever was my end;
If you me now something will give,
Serve you I will, as long as I live.)

Even until later times the sons of "Oranje Boven" held their festivals in robust simplicity, reviving the recollection of their ancestors by crowning their festive boards with the very diet which was once so prized—such as Suppawn and Malk, Hoof Kaas, Zult, Hokkies, en Poetyes, Kool Slaa, Roltetje, Worst, Gofruyt Pens, etc.*

Dutch dances were very frequent; the supper on such occasions was bread and a pot of chocolate. The Rev. Dr. Laidlie, who arrived in 1764, tried to force the abandonment of dancing. He was very exact in his piety and was the first minister of the Dutch Reformed Church who was called upon or even allowed to preach in the English language.

Dutch families enjoyed sitting out on the "stoopes" in the shade of the evening, saluting passing friends and talking across the narrow streets with neighbors. Here and there, an old Knickerbocker sat with his long pipe, fuming away his cares, and ready to offer another pipe for the use of any passing friend who would sit and join him.

Dutch justice in early times was prompt and severe. An example may be found in the following excerpts from the annals of John F. Watson:

"For Scandalizing the Governor, one Hendrick Jansen, in 1638, is sentenced to stand at the fort door, at the ringing of the bell, and ask the Governor's pardon."

"For drawing his Knife upon a person, one Guysbert Van Regerslard, was sentenced, in 1638, to throw himself three times from the sail-yard of the yacht, The Hope, and to receive from each sailor there three lashes."

"The Wooden Horse punishment is inflicted, in Dec. 1638, upon two soldiers: they to sit thereon for two hours," (a military punishment used in Holland. The victim strode a sharp backed wooden horse and his body was forced down on it by a chain and iron stirrup or a weight fastened to his legs.)

At about the time Governor Peter Minuit had purchased Manhattan Island from the Indians, May 6, 1626, the first male white child born in Manhattan, Jean Vigne, saw the light of the new day, in a little farmhouse at the cornerof Wall and Pearl Streets.

The Lime-kiln man.

Back in the days of the Dutch West India Company, the gentle waters of a canal ran along what is now Broad Street, and its sides were sheathed with planks to keep the banks from caving in. Three bridges crossed the canal, at Bridge, Stone and Beaver Streets. The road above bore the name HEERE GRAFT after a similar street in Holland. The house nearest, in the sketch, was built by Peter van Couwenhoven and included most of the frontage covered by Nos. 78-86. It is at present the site of the Maritime Exchange Building, Broad Street looking North from Stone. After the English occupation of New York (1676), the canal was ordered filled in, from which resulted an unusually broad street (see page 28).

**Corn mush and milk; pork brawn or head cheese; sausage or bologna; taffy-like candy; pod peas, cabbage salad, cole slaw; small roll; sausage; fried tripe.*

(Sarah Rapelje was the first girl born in New York State, near Albany, in 1625).

Minuit was succeeded by Governor Wouter Van Twiller. This overstuffed Governor indulged himself and everyone else to the point where, were it not for Dominie Bogardus, who arrived in New Amsterdam about this time, the infant colony might never have survived the rigors of its dangers and hardships. Bogardus abided by the law and used his thunder to see that others lived up to the Divine commands as well. As has been known with others of unimpeachable character, the vigor of his spiritual leadership often led him to some very rough tussles. He was, however, well able to take care of himself with Van Twiller or any other who might cross him.

Once when he was slandered by a Dutch scold he had her hailed to court to prove her allegations and, when she failed, compelled her to stand at the gate to the Fort and acclaim by the same loose tongue that the Dominie was a good man and that she had lied about him.

It was natural that Dominie Bogardus and the fiery Governor Kieft, successor to Van Twiller, should come into conflict, as they often did. Kieft was the civic head and Bogardus the spiritual head of the young city. Each was at constant odds with the other, but the bitterest cause of enmity between them was the Indian War which the blundering Kieft brought about. It was upon his order that some of the most disgraceful massacres of Indians in American history were perpetrated. The members of the colony, especially in the outlying districts amongst the boweries and farms, suffered terribly as a result of his foolishness.

Bogardus opposed him relentlessly in this decision for war and poured out from his pulpit such diatribes that Kieft had to leave the church in order to save himself from the personal abuse. In view of his official position the Governor was at last forced to issue one of the most formidable summons ever served on a preacher.

Bogardus was accused of using the pulpit from which to make personal threats in addition to using violence and bad taste. He was accused of intoxication after partaking of the Lord's Supper; indifference to the safety of others; indifference to public authority; and embracing the cause of criminals and traitors to the authority of the King's Governor.

After long drawn-out threats, suits, evasive answers, and legal squabbles, the combatants decided they had enough and a truce was effected. In 1647 they sailed for Holland together in the ship "Princess," and were shipwrecked and drowned off the coast of England.

Along the Indian path which led to Peter Stuyvesant's Bowerie were two ponds, Buttermilk and Sweetmilk. Below, looking south along the lane, Buttermilk Pond is on the left and Sweetmilk on the right, with Nicholas Bayard's windmill in the background. The ponds were eventually filled in and disappeared on maps after 1757.

PUBLISHED BY T. W. STRONG 98 NASSAU ST. N.Y

W. C. COTTRELL & Co. 64 CORN-HILL BOSTON.

THE SCHOOL BOY.

The germ of chivalry in the heart of the boy is aroused by the Active Fireman's recital of his exploits.—

THE SCHOOL BOY

The Fireman in Four Plates, No. 1

The germ of chivalry in the heart of the boy is aroused by the Fireman's recital of his exploits.

PUBLISHED BY T. W. STRONG CIRCA 1850 COURTESY OF HAROLD V. SMITH, ESQ.

The print depicts a familiar firehouse scene in the early days of the Volunteer Fire Department, the duties of the men around the firehouse and the almost loving care taken with their engines, tools and decorations. In those times it was the dream of nearly every boy that he would someday be able to "run with the engines" and all the leisure time of many lads was spent around their neighborhood firehouse, listening to the salty dialogue of the colorful vamps and making a most appreciative audience to the tales, tall and otherwise, spun by the firemen. In a later period, during the Paid Department days, Alfred E. Smith was a good example of this youthful devotion. For a number of years, including those during which he worked as a runner for a carting concern in the Fulton Market, he "attached" himself to Engine 32 which was housed on the south side of John Street opposite Cliff. Al often took sandwiches and coffee to "his" boys during fires.

The fireman at the left is in the process of cleaning and slushing (greasing) the leather hose, to prevent it from cracking, while the other two apply "elbow-grease" to the brass of the signal lamp and the foreman's trumpet.

"Coentie's slip is showing"—rare and unusual view of the city from old Coenties Slip, circa 1880. Note "family wash," early conveyances and Third Avenue El.

Fires in the old town were frequent and sometimes disastrous, even though the colony was confined to the southern tip of the island within easy reach of the waters of the bay. As the city grew, the fire hazard increased and the number of property losses began to mount. There developed a serious need for fire-fighters and tools to work with. Buckets, hooks and ladders were provided and all able-bodied members of the town called upon to help. When fire occurred, the general method of operation seemed to be sensible enough. You took a bucket in hand, dipped it in the canal, took it out, ran as close to the fire as possible, placed the other hand on the base of the bucket, flung. The major difficulty was that, with the increase of fires and the number of amateur firemen, this fairly simple operation became a madcap adventure which resulted too often in a thoroughly drenched, if happy, milling mob and a comparatively dry and spectacular fire. Even the schout-fiscals, schepens, outwouters and krinklewosts could see the inefficiency of it.

The schout-fiscals, recognizing the danger, held one stormy meeting after another until the light of inspiration finally gleamed. Sparked by the political possibilities of organizing the people, their solution was organization—and the idea of forming bucket lines was born. Two lines of folk were formed with a good water dipper at one end and a good water flinger at the other. One line passed the loaded buckets to the fire and another one of equal length returned the empties to the water source. Gentlemen to the first line, the ladies and girls to the second—all hands around. In a manner of speaking, such a roundelay of action might even be considered as Manhattan's first Debutante Cotillion, for the social aspects of the affair became rather formalized. You could always be sure that Antje would get into line opposite Willem and that Maria would squirm in across from Jacobus, and that the Remsen widow would always push her dainty little daughter into line as close as she could to that young Roosevelt kid. Clap hands—what fun! But as the gaiety mounted, flirting got more attention than fire-fighting. The initial verve of the male line soon gave way to an uneasiness that palsied the hands and tipped the buckets. When the flingers were obliged to call "no water," more often than it was deemed necessary, the schouts, always alert to signs of slackness and irregularity, ordered that the two lines be turned back-to-back, resulting in less merriment but a decidedly more efficient organization.

In 1648, by the appointment of four fire wardens, Governor Peter Stuyvesant laid the basic foundation of New York's and the nation's volunteer fire-fighting system. It was the duty of these wardens to inspect the wooden chimneys of New Amsterdam's thatched-roofed wooden houses and to exact a penalty of four guilders for each chimney that was insufficiently swept. The money collected was devoted to the purchase of leather buckets, hooks and ladders. The buckets were of particular value in fighting fires and for many years were used to convey water until suctions for the engines and leather hose came into practical use.

The original four wardens were Martin Cregier, who kept a famous tavern opposite the Bowling Green and was one of the first two burgomasters; Adrian Keyser, member of the Dutch West India Company; and two Englishmen, Thomas Hall and George Woolsey. Hall, who had been taken prisoner by the Dutch and released on parole, was a man of wealth and influence, and Woolsey was also prominent. Other later wardens were Hendrick Hendrickson Kip, Govert Loockerman and Christian Barents.

Apparently the wardens sometimes took their inspecting authority too literally, causing some resentment among the people. The story is told of Madaleen Dircks "who presumed to insult one of the worshipful fire wardens of the city and thus to make a street riot." The young maid, in passing the door of F. W. Litschoe with her sister remarked that "There stands a chimneysweep in the door and his chimney is well swept," a classic stab which resulted in a rather stiff fine for the skeptical maiden.

Later an organization was formed, called the "Prowlers," a group consisting of eight men furnished with two hundred and fifty buckets (made by the "Knights of St. Crispin"—the town's foremost cobblers who were located in Shoemaker's Pasture) as well as many hooks and ladders. This company, also known as the "rattle-watch," patrolled the streets from nine in the evening until the drum-beat at dawn, on the alert for fire while the town slept.

Broadway and Prince Street, the winter of 1857. Red-shirted firemen and their double-deck, end-stroke engine in the foreground. Machines of this type were usually equipped with runners in wintertime to make their passage over the snow easier and swifter while racing to fires.

Lead Kindly Light

NEW YORK never has had so large a public institution as that of the Volunteer Fire Department, in which the esprit de corps remained at such a high pitch over so long a period. From the time Stormy Petrus appointed the eight "Prowlers" of 1648 until the final days of 1865, when the old department numbered more than four thousand, many generations of fire laddies gave freely of their time, health and of their lives when confronting in the public cause their greatest enemy—fire.

Serving with no other pay than the thrill of adventure and the pleasures of companionship, they faced danger with an elan only exceeded by their recklessness. The record of the sacrifices made in the line of duty by these heroes is a long one. Sometimes called roughnecks, and there is no doubt that a number of them were that and worse at various times in their career, nevertheless, their ranks included throughout the long period of their service many of the city's and the nation's finest citizens.

The combination of their social activities, the chowder parties, target shoots, collations, and annual balls, the first of which took place in 1829 at the Bowery Theatre, 46-48 Bowery Lane; their pride in parades and their daring and hardiness in working the "masheens" held particular appeal to the masculine heart. The Minute books kept by the secretaries to record fires and other events often revealed the true spirit and energy of the companies. The following excerpts from two of these old journals, while written in ink long since faded, clearly reveal an outstanding devotion to duty, the memory of which is best expressed in the old company motto of Engine 26, "True Blue Never Fades."

VOLUNTARY AID

Co. No. 5

THESE are to certify that Stephen N. Stilwell is pursuant to LAW nominated and appointed one of the Firemen of the City of New York.

By the Com. Council

June 16th 1817.

New York Volunteer Fireman's certificate, dated June 16, 1817.

Target used by Neptune Engine Company No. 2.

LADY WASHINGTON ENGINE CO. 40

Monday July 19th 2PM 7th district

Fire in south William Street 16 J. Bosch in command Number of men out 16 No officer out. Boy Burnt to death No 2 Engine stood for us we came out and ran away from them as usual

Thursday July 29 2AM 7th district

Fire corner Duane & Chatham Wm. Racey in Command Both officers absent Number of men out 25 Coming home we got as far as Howard Street and Broadway when the bell rung 8th district for 57 Vesey Street Assistant (Canfields) took Command Chased by 54 hose turned in to Murray Street and so to the fire

Thursday August 5th 3AM

Received Engine Company No 29 on their return from their Visit to Paterson New Jersey and gave them a Collation in Central Hall

Wednesday August 19th 1AM 7th district

Fire City Hall Tower Canfield in Command Pipemen J. Bosch and Wm. Racey Both officers out Number of men out 32 The plunges of our Engine got Broke

Monday August 24th 8½PM 8th district

Fire Vesey near Greenwich Foreman in Command Pipemen D. Scully and P. Canavan Both officers out Number of men out 33 J. Hutton fell through an awning

Wednesday Nov. 10th 11PM 7th district

Fire Barclay near Greenwich Canfield in Command Number of men out 30 passed by 50 Hose Cart one of 4 Engine members got run over on our rope

Monday Nov. 15th 6PM 8th district

Bell rung for the smoke of the steam engine in the park stopped cor Cedar and Greenwich Canfield in Command Number of men out 16 Foreman absent – day fine (Steamer of Exempt's Fire Co. No. 42, kept in City Hall Park)

Tuesday November 30th 6PM 8th district

Fire in Reade Near Church Wm. Duryea in Command Number of men out 34 Both officers out Chased By 19 hose Carriage We passed the Elephant (No. 8's steamer)

MINUTE BOOK—1859

CHATHAM ENGINE CO. 27

Thursday Jan. 27th 6¼PM 7th district

Fire Cor Beekman & Wm. Hutton in Command Number of men out 41 Both officers out Attacked by Engine Co 30 and Engine Co 21 J. Hutton severely Hurt sent to hospital no one else our Engine taken away by the chief

Tuesday Oct 11th 10¼PM 7th district

Fire Pearl & Wm Street W. Rohan in Command Number of men out 39 Both officers out We met the Sons of Malta on our return

Sunday Dec. 25th 9½PM 7th District

Fire Duane & Wm. W. Walsh in Command Number of men out 38 Both officers out On the next day we took our large engine to Reillys pole and Beat every apparratus playing that was there

Sunday Dec 25th 5½ 7th District

Fire Pine Near Wm Jas. C White in Command Number of men out 33 Foreman out assist. absent We passed 15 Truck Christmas day

Thursday Dec 29 5AM 7th District

Fire Beekman & Gold St. Dan Reilly in Command Number of men out 36 Both officers out Pipemen Duryea & Dalton The Last Fire in 1859 and the largest since 1835

Saturday November the 3rd at 11¼AM 7th District

Fire Water & Ro(o)sevelt St. T. F. Murry In Command number of men out 11 Rained like Hell

Thursday Feb 2nd 7PM

a large fire occurred in the tenement house 144 Elm street By which several lives were lost at which Dan'l Scully saved 5 lives Phillip Dalton saved one life that of a young girl W. Walsh had Command to the fire Pipeman P. Dalton E. Loughlin & c Both officers Racey & Canfield were out

That the volunteer firemen fared exceedingly well in the otherwise frequently critical considerations of America and American institutions by a number of peregrinating foreign observers might be taken as fair evidence of their deserving qualities. Moreau de St. Mery* (1798), Mrs. Frances Trollope (1832), Fanny Kemble (1835), William Thackeray and Clarence Mackay (1855), and Charles Dickens and Anthony Trollope (1861), all had their say. Of these, Charles

Moreau de St. Mery's American Journey, *1793-1798. Translated and edited by Kenneth Roberts and Anna M. Roberts.*

Christmas scene in old New York with firemen joining in the festivities. The engine, decked in holly, is a goose-neck or "Old New York" type machine.

Dickens probably wrote the liveliest account of the old-time fireman that has ever been published. When the English literary publication "All the Year Round," conducted by Dickens, appeared on March 16, 1861, it contained an essay, *American Volunteer Firemen*, which in this writer's opinion was undoubtedly written by "Boz" himself. In addition to the unmistakable style, it contains too many intimate, accurate observations to have been other than the accomplishment of a first-hand witness.

"The firemen of America are all volunteers. It is the law of the land that every citizen at a certain age, must come and serve for a certain specified duration of time, as either a militiaman or a volunteer. Now, as I believe the militiaman's term of service lasts five years, and a fireman's only three, you may easily imagine, among an itinerant and feverishly restless democratic youth, which is preferred.

"Besides, there are many other reasons which I have no doubt contribute to make the fireman's service more popular in America than the militiaman's. In the first place, the former service, though vexatiously frequent in its calls upon its members, is not so restrained and monotonous as that of the militiaman's; and the Americans, as self-conscious freemen, are very jealous of even the smallest and least galling restraint. Secondly, the dress is not so much of the character of a livery—which a true American always detests as a badge of serfdom; it is more loose, careless, and picturesque. Thirdly, the work is at night, when shops are shut and counting-houses closed; lastly, the service is one of stirring danger, and full of that passionate excitement that the American, whose Anglo-Saxon blood the suns of a new continent have long since fired to almost the volcanic warmth of the Indian he displaced, loves, and must have.

"I will give my first impressions of the appearance of these volunteer firemen . . . (I) was working my way from the Battery and the vast world of warehouses thereunto adjoining, into Broadway. The new region, of which I was not quite the Columbus, lay before me, with its thin wiry merchants, its sallow-faced and pale dyspeptic clerks, its hairy rowdies, its Californian itinerants, and its staring, woebegone emigrants. A party of these last (palpably Irish) had just jolted past me, seated on their sea chests, and packed in a slight-built waggon . . . Away they bolted into a new world; . . . the training-ground of nations yet unborn.

"Here, glide along the huge crimson omnibus carriages of the street-railroad; those fluttering flags over the conductor's platform, announce a great election-meeting tonight in the City Park. Here come some cotton bales, and here a cart full of oysters—sea fruit new gathered; but now a stir and oscillation in the street crowds. Now rises to the immaculate blue sky that ever smiles on New York, a bray of brass, a clamp of cymbals, and the piercing supplication of fifes, and bomb tom cannonades the drum, with expostulating groan.

Firemen parading past Niblo's in 1858 in honor of the laying of the Atlantic Cable. Probably the most impressive demonstration in the city, the firemen would parade down Broadway at the drop of a hat and the whole city would turn out to watch them.

"Ha! there breaks through the black-panted crowd (even the seediest American wears evening dress) gleams of warm scarlet! It is the rifle company of one of the New York Volunteer Firemen Societies. Here they come, four abreast. 'Fours,' with no very severe military air of stiff order and mathematical regularity, but with light, gay, swinging step, jaunty, careless, rather defiant freemen, a little self-conscious of display, but braving it out in a manly game-cock way. They are trailing rifles now, the officers swinging round in the wheels with them, glittering sword in hand.

"They wear a rude sort of shako covered with oilskin, red flannel shirts, with black silk handkerchiefs, blowing gaily (as to the ends), tied round their throats in jaunty sailor's knots; they are all young men, some quite boys. It is evidently the manner with them to affect recklessness, so as not to appear to be drilled or drummed about to the detriment of their brave democratic freedom uniform. No, they would as soon wear flamingo-plush and bell-hanging shoulder-knots.

"They have been over on what the Americans call 'a target excursion' to Brooklyn, and have been summoned together by advertisement in the New York Herald. To-morrow, there will be a paragraph about their excellent shooting, the number of bull's-eyes they made, the 'clam chowder' they partook of afterwards, and the 'good time' they had generally.

"Observe, too, a special American characteristic, the big laughing (Negro) . . . who carries the target riddled into a collander with bullet-holes.

"The song writer compares him to Pompus Caesar, whom the coloured girls peculiarly admire, and the

rand torchlight parade of the New York firemen in honor of the Prince of Wales, ctober 13, 1860. The scene shown is directly in front of the Fifth Avenue Hotel.

chorus is, I remember:

They come together
With sword and feather
Loud trumpets, drums, and hooting,
And with the mark
Bring up the dark
When they go out a shooting.

"These street processions are incessant in New York, and contribute much to the gayness of the street. Whether firemen, or volunteers, or political torch-bearers, they are very arbitrary in their march. They allow no omnibus, or van, or barouche, to break their ranks; and I have often seen all the immense traffic of Broadway (a street that is a mixture of Cheapside and Regent-street) stand still, benumbed, while a band of men, enclosed in a square of rope, dragged by a shining brass gun or a brand new gleaming fire-engine.

Fireman's flambeaux

"But, after all, it is at night-time that the fireman is really himself, and means something. He lays down the worn-out pen, and shuts up the red-lined ledger. He hurries home . . . slips on his red shirt and black dress-trousers, dons his solid japanned leather helmet bound with brass, and hurries to the guard-room, or the station, if he be on duty.

"A gleam of red, just a blush in the sky, eastward —William-street way—among the warehouses; and presently the telegraph begins to work. For, every fire station has its telegraph, and every street has its line of wires, like metallic washing-lines. Jig-jag, tat-tat, goes the indicator:

"'Fire in William-street, No. 3, Messrs. Hardcastle and Co.'

"Presently the enormous bell, slung for the purpose in a wooden shed in the City Park just at the end of Broadway, begins to swing and roll backward.

"In dash the volunteers in their red shirts and helmets—from oyster cellars and half-finished clam soup, from newly begun games of billiards, from the theatre, from Boucicault, from Booth, from the mad drollery of the Christy minstrels, from stiff quadrille parties, from gin-slings, from bar-rooms, from sulphurous pistol galleries, from studios, from dissecting rooms, from half-shuttered shops, from conversazioni and lectures—from everywhere—north, south, east, and west—breathless, hot, eager, daring, shouting, mad. Open fly the folding doors, out glides the new engine—the special pride of the company—the engine whose excellence many lives have been lost to maintain; 'A reg'lar high-bred little stepper' as ever smith's hammer forged. It shines like a new set of cutlery, and is as light as a 'spider waggon' or a trotting-gig. It is not the great Juggernaut car of our Sun and Phoenix offices—the enormous house on wheels, made as if purposely cumbrous and eternal—but is a mere light musical snuff-box of steel rods and brass supports, with axes and coils of leather, brass-socketed tubing fastened beneath, and all ready for instant and alert use.

"Now, the supernumeraries—the haulers and draggers, who lend a hand at the ropes—pour in from the neighboring dram-shops or low dancing-rooms, where they remain waiting to earn some dimes by such casualties. A shout—a tiger!

"'Hei! hei!! hei!!! hei!!!! (crescendo), and out at lightning speed dashes the engine, in the direction of the red gleam now widening and sending up the fan-like radiance of a volcano.

"Now, a roar and crackle, as the quick-tongued flames leap out, red and eager, or lick the black blistered beams—now, hot belches of smoke from shivering windows—now, snaps and smashes of red-hot beams, as the floors fall in—now, down burning stairs, like frightened martyrs running from the stake, rush poor women and children in white trailing night-gowns—now, the mob, like a great exulting many-headed monster, shouts with delight and sympathy—now, race up the fire-engines, the men defying each other in rivalry, as they plant the ladders and fire-escapes. The fire-trumpets roar out stentorian orders —the red shirts fall into line—rock, rock, go the steel bars that force up the water—up leap the men with the hooks and axes—crash, crash, lop, chop, go the axes at the partitions, where the fire smoulders. Now, spurt up in fluid arches the blue white jets of water, that hiss and splash, and blacken out the spasms of

Illuminated parade of the New York Fire Department in 1859.

The fire at Jenning's Clothing Store, 231 Broadway, the evening of April 25, 1854. At this fire—one of the most disastrous in the Department's history—eleven firemen were killed and twenty seriously injured.

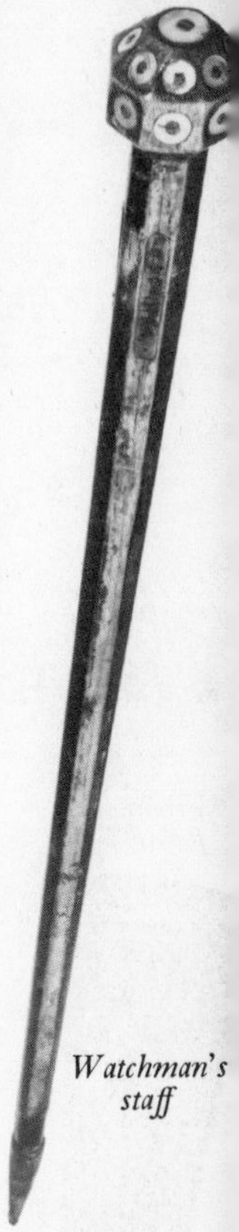

fire; and as every new engine dashes up, the thousands of upturned faces turn to some new shade of reflected crimson, and the half-broken beams give way at the thunder of their cheers.

"The fire lowers, and is all but subdued, though still every now and then a floor gives way with an earthquake crash, and into the still lurid dark air rises a storm of sparks like a hurricane of fire-flies. But suddenly there is a crowding together and whispering of helmeted heads. Brave Seth Johnson is missing; all the hook men and axe men are back but he; all the pumpers are there, and all the loafers are there. He alone is missing.

"Caleb Fisher saw him last, shouts the captain to the eager red faces; he was then breaking a third floor back window with his axe. He thinks he is under the last wall that fell. Is there a lad there will not risk his life for Seth? No! or he would be no American, I dare swear.

"Hei! hei!! hei!!! hei!!!!

"Click-shough go the shovels, chick-chick the pickaxes. A shout, a scream of

"'Seth!'

"He is there, pale and silent, with heaving chest, his breast-bone smashed in, a cold dew oozing from his forehead. Now they bear him to the roaring multitude, their eyes aching and watering with the suffocating gusts of smoke. They lay him pale, in his red shirt, amid the hushed voiceless men in the bruised and scorched helmets. The grave doctor breaks through the crowd. He stoops and feels Seth's pulse. All eyes turn to him. He shakes his head, and makes no other answer. Then the young men take off their helmets and bear home Seth, and some weep, because of his betrothed, and the young men think of her.

"Such are the scenes that occur nightly in New York. The special disgrace of the city is the incessant occurrence of incendiary fires. Yet accidental fires are exceedingly numerous, for wood is still (even in New York) the predominant building material, in consequence of the extraordinary cheapness of wood fit for building. The roofs, too, are generally of tin, and not tile or slate, and this burns through very quickly. Moreover, the universal stove (derived from the Dutch, I suppose) occasions a great use of flue pipes, and these are buried among wood, and are, even when embedded in stone, dangerous."

It is obvious from the above that much can be learned of the progress and history of Manhattan by following the signal lamps, so to speak, of many generations of the volunteer fire laddies, through the streets of the city...Lead Kindly Light!

Watchman's staff

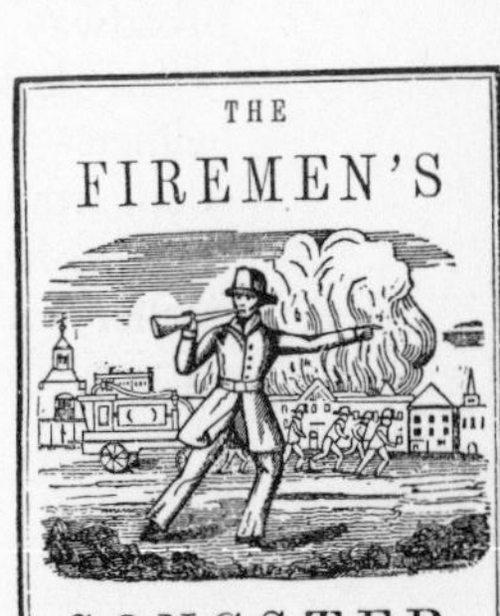

Many of the popular ballads of old New York honored the volunteer firemen.

Sheds to House the Engines from London

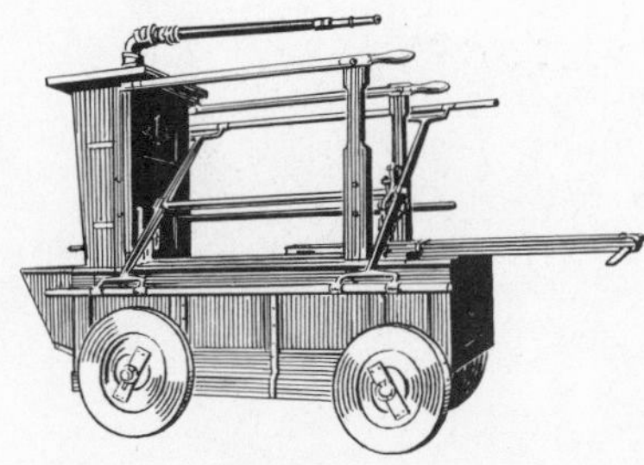

In December, 1731, New York's volunteer forces, which had been fighting fires with "buckets, hooks, chains and ladders," saw the city's first fire engines unloaded from the good ship Beaver. These machines were conveyed from the harbor to their newly-made sheds in the rear of the City Hall on Wall Street amid general rejoicing and much ceremony. For weeks, people swarmed to the spot from all over the city, to admire the new "enjines from London."

New Yorkers had been exposed to the danger of fire since the settlement of the city and a good blaze on a windy day could easily have destroyed a large part of town. The first constructive step towards creating a mechanized fire-fighting system, the acquisition of the engines, was a wise and popular move.

Built by Richard Newsham, the most successful and expert fire engine maker of his day, the engines were side-stroke, 2-cylinder machines. They were mounted on solid wooden block wheels and were awkward and difficult to maneuver, as there was no traveller or fifth wheel for steering, and it was necessary to lift the machines when turning a corner. The larger of the two (which is preserved in excellent condition at the Museum of the Fireman's Home at Hudson, New York) was equipped with auxiliary foot treadles and was operated by eight to ten men. It became the property of Engine Company No. 1.

Red flannel fireman's shirt.

The city's wealthiest and most prominent citizens composed the first engine companies. The first two overseers of Engine Co. 2 were brothers with an old and famous New York name, John and Nicholas Roosevelt, and among the leaders in No. 1 Company were Peter Rutgers, a brewer, and Johannes Vriedenburgh who was made foreman of the company.

Until 1736, both the British engines were housed in back of the City Hall, in separate sheds, No. 1 on the east, and No. 2 on the west facing Kip Street, now Nassau.

The view looks east from Kip Street. The gable of Bayard's sugar house, built in 1728, is shown at the left, the present site of the Bank of Manhattan. There were no other buildings on the north side of Wall Street between this and Gabriel Thomson's tavern at the corner of William. The lots adjoining the City Hall yard were unoccupied as late as 1743.

Because the original sheds proved unsatisfactory, the corporation, in 1736, ordered the construction of a more convenient house for the new "enjines," at the head of Broad Street, in front of the City Hall.

The entire roster of both companies served as a home guard under General Washington and few of the original members survived the Revolutionary War. In 1783, when the Department was re-organized, both companies were enlarged. No. 1 moved to a lane running down to the North River, now Barclay Street, and became known as the Hudson Company. At the Erie Canal celebration on November 4, 1825, it made a magnificent display and the members won much acclaim for their banner on which was painted a view of New York in flames, and the Genius of America pointing to direct the firemen, while overhead an eagle hovered, bearing the motto: "Where duty calls, there you will find us."

Engine Company No. 2 moved to the old Bowery or Post Road at the head of Chatham Square. It became known as the Chatham Engine Company, paying honor to William Pitt, the Earl of Chatham.

No. 2 became famous for its bitter rivalries with Engine Company 19 and Engine Company 26, and like the Hudson Company, was finally disbanded for fighting. Later re-organized, its long and colorful career came to an end in 1865, when its steam engine was transferred to Engine Co. 11 of the Paid Department.

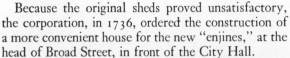

Lantern of Engine Company No. 1.

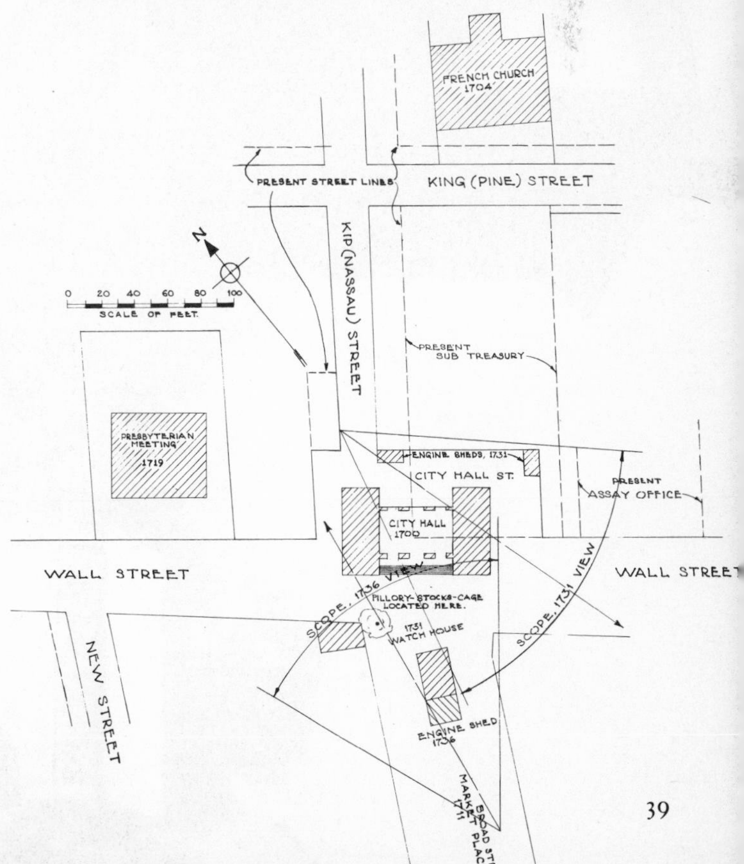

Early fireman's helmet.

At the Head of the Broad Street

THE second house built for New York's first engines was located in front of the City Hall, at the head of Broad Street. Behind it loomed the cupola of the Hall, with its huge clock of "two dyal plates of red cedar, each six foot square," and a bell (added in later times), loud enough to be heard all over the city, which announced the news of a fire and by the number of its strokes, the location.

The City Hall, built in 1700 and extensively remodelled in 1763, contained the municipal offices, the Provincial Assembly, the prison, the Supreme Court and the Admiralty Courts. In 1788, the building was extended, repaired and redecorated to make room for the Continental Congress. When the Congress moved to Philadelphia, it was again remodelled.

Directly contiguous to the engine house were the watch house, the whipping post and the pillory where criminals were punished in full sight of the populace. The hangman of the city in early times was a man named Van Johnson who was himself convicted of robbery and would have been sentenced to be hanged except that no one could be found to hang him and he couldn't or wouldn't hang himself. His punishment therefore was "lightened" to thirty-nine lashes at the whipping post, the severing of his ear and banishment from the island. Early records also reveal the punishment of two men for a most unusual crime when William Smith and Daniel Martin were given fifteen lashes at the whipping post "for stealing fiddle strings." Another story salvaged from records of this time tells the candid tale of an ambitious young man, "a Mr. Wright by name, a one-eyed man and a muff-maker by trade, who drinking hard upon a rum one evening . . . began a health of a whole halfe pint at a draught, which he had no sooner done but downee fell and never rose more."

Fireman's hat front.

Later we find a contemporary writer commenting on "two women named Fuller and Knight . . . who . . . were placed one hour in the pillory for keeping bawdy houses. If this were again enforced would not

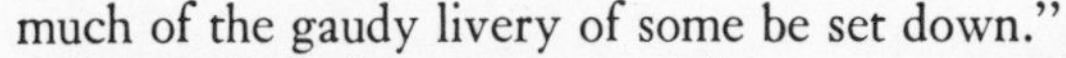

much of the gaudy livery of some be set down."

But perhaps the severest punishment was meted out to two thieves named Wallace and Willson. These two gentlemen made a grand tour of the City in a unique manner. They were placed in a cart, "so as to be publikly seen," and carried to the whipping posts of Manhattan, Kings County, Flatbush, Jamaica, and Westchester. At each place they stood in the pillory and then were lashed. In addition, after the "Grand Tour," they were fined and sentenced to prison. A heavy sentence!

The erection of the City Hall was a notable event and one which established Wall Street as the hub of the city's progress and administration.

In 1735, one of the most important public trials in America was held here, as well as the first great battle for the right of freedom of speech. John Peter Zenger*, publisher of *The New York Weekly Journal*, had dared to criticize William Cosby, the royal colonial governor. Cosby threw him into prison and brought suit for libel, and all of Zenger's papers were ordered burned near the pillory by the hangman.

Although New York was still a small town, its proud and independent citizens staunchly defended their fearless publisher. Andrew Hamilton, the most famous lawyer in Philadelphia, came quietly to defend the editor. Swayed by Hamilton's eloquence, the jury returned a verdict of "not guilty," and after nine months in prison, Peter Zenger was freed.

Hamilton was the hero of the hour. Mayor Paul Richard** presented him with the freedom of the city and the gift of a magnificent gold snuffbox, purchased by private subscription. With booming cannon and waving banners, the entire city turned out to escort him to the barge on which he was to return to Philadelphia.

**In December 1727 Zenger was the organ-blower for the new organ presented by Governor Burnet to the old South Dutch Reform Church in Garden Street.*

***Appointed July 3, 1735 upon the death of Mayor Robert Lurting who also had been a friend to Zenger.*

Federal Hall Memorial (formerly Sub Treasury Building). Site of old Engine Company No. 2.

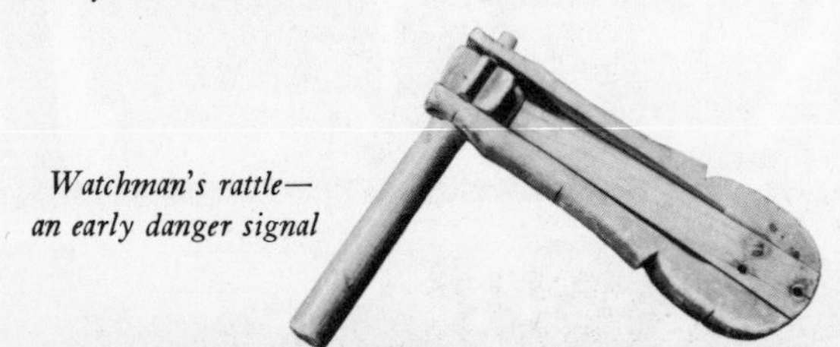

Watchman's rattle—an early danger signal

The Wreath of Roses

As the City Hall became the center of Manhattan's civic life, Wall Street grew in importance, gradually attaining its present eminence as the nation's financial and economic mecca.

The New York Stock Exchange is the direct descendant of the open air market "under the Buttonwood Tree." The stock traders and commission men of the city met here daily to transact their business, and the wide-spreading branches of Wall Street's famous old tree provided their first shelter.

In the midst of this group one could always discern a few members of Engine Company No. 15, known as the "Wreath of Roses," because of the design painted on the back panel of their machine.

When Federal Hall was dismantled, in 1813, "Wreath of Roses" was moved to Chatham Square, where she took the name of "Peterson," in honor of her foreman, who was killed in the Great Fire of 1811. The artist, David Johnson, who was a member of the company, painted his portrait on the back of the engine, and the rosewood machine, when polished, was considered the finest in the department.

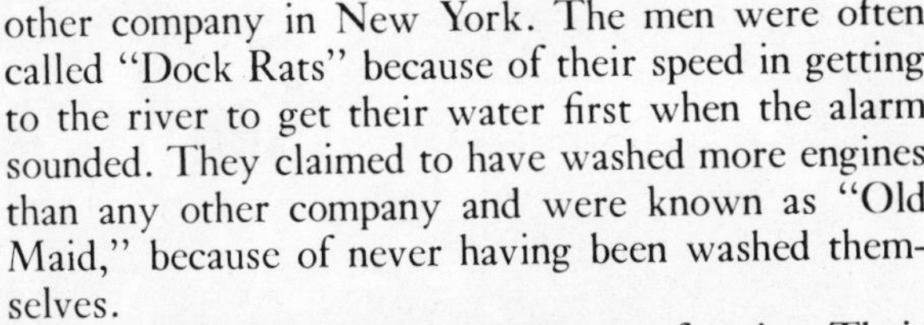

Under the Buttonwood tree

In its prime, No. 15 had as large an array of distinctions and nicknames as any fire company could boast. It had one of the largest volunteer rolls in the city, and its runners, divided into "The Fly-By-Nights" and the "Old Maid's Boys," were said to be able to "lick" any other company in New York. The men were often called "Dock Rats" because of their speed in getting to the river to get their water first when the alarm sounded. They claimed to have washed more engines than any other company and were known as "Old Maid," because of never having been washed themselves.

"Old Maid" ran a goose-neck type of engine. Their last machine, built by Harry Ludlam, was taken to Philadelphia in 1836 for a playing match, and was the only machine to succeed in throwing its stream over the cupola of the Exchange Building.

William Brandon, who was a member of No. 15, won national fame in the development of fire-fighting. After service with No. 15, he became successively, foreman, battalion chief, and chief of the first brigade

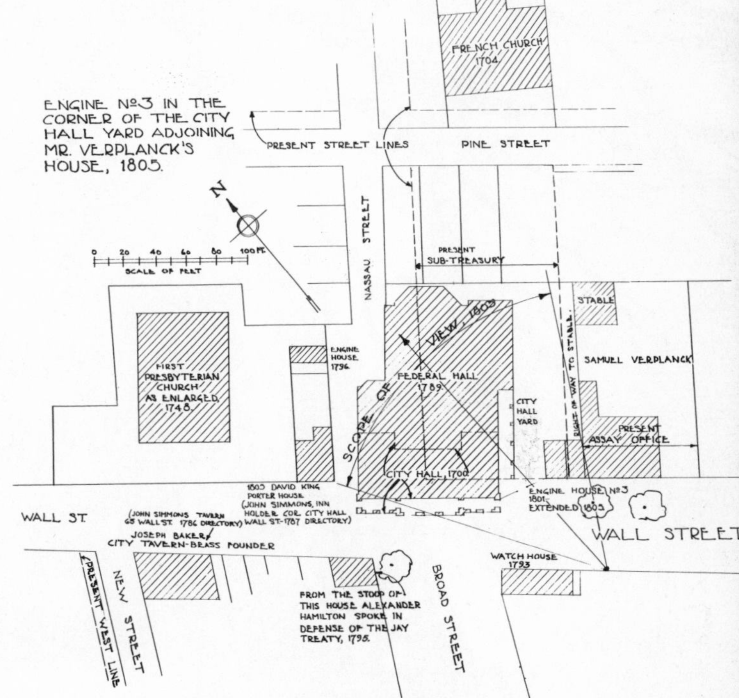

of the Paid Department. Later, as "surveyor" for the Home Insurance Company, he was responsible for the reorganization of the Boston and Chicago Fire Departments. He reported on the fire organization of more than 20 leading cities for the National Board of Fire Underwriters.

The most famous member of this company was Harry Howard, an orphan, who later became Chief Engineer of the department, and was one of the best-loved firemen in the city. Manhattan's Harry Howard Square, at the intersection of Canal, Walker, Baxter and Mulberry Streets, is named in his honor.

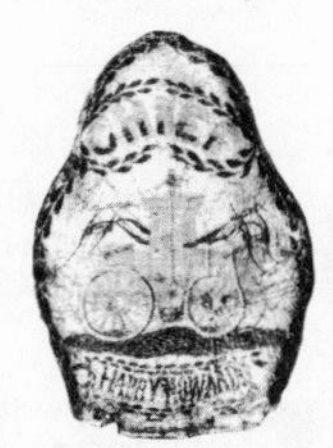

Harry Howard, Chief Engineer of the Volunteer Fire Department from 1857 to 1860. His hat front, shown at the left, is on display at the H. V. Smith Museum of The Home Insurance Company.

3

The Porterhouse Pets

Model of the "Old Brass Backs" engine.

In 1743 the first successful working fire engine ever built in this country was made by Thomas Lote, a cooper and boat builder of lower Maiden Lane. Called "Old Brass Backs" because of the lavish use of brass on the box of the machine, it became the envied possession of Engine Company No. 3. The success of this engine, along with the growing dislike for British imports, inspired local builders and inventors to produce machines of their own.

New York Engine Company No. 3 was organized in the middle of the 18th century, and its first recorded location was in the Presbyterian churchyard on the west side of Nassau Street, opposite the City Hall, where it remained until 1805.

A few steps away from the firehouse, on the northeast corner of Nassau Street, fronting Wall, was a favorite rendezvous of the members of No. 3, Simmon's Tavern or "Porterhouse." This was the tavern in which the first post-Revolution mayor of New York City, Mayor James Duane, was installed in office in 1784. Because of their fealty to David King, the proprietor, the men of Engine No. 3 became known as the "Porterhouse Pets."

Wall Street, then the center of New York's political life, was still the most fashionable place in the city, not yet invaded by the bootmakers, harnessmakers and ciderhalls of a later era. One of the neighborhood characters was a man called "Potpie" Palmer, who could never live down the fact that during the Revolution he was said to have entered a kitchen and run off with a potpie. His name became a byword and the boys used to harry him from a safe distance with the taunt:

"Potpie Palmer is a jolly old soul,
With a three-cornered hat and the pie he stole."

Another and more spectacular character was a mulatto known as "Dandy" Cox, described as a coat-scourer. "Dandy" drove a smart team of horses hooked up to a stylish two-wheeled business rig, and sometimes to a Stanhope. He was always well dressed and at the elaborate evening parties given by him and his wife often startled the guests by his bizarre costumes. His wife was in the habit of retiring several times during the evening, each time reappearing in a complete change of dress.

The view, which looks southwest through the burial ground of the Presbyterian Church to Wall Street, shows Nassau Street as it appeared in 1805.

If it is of any interest, many New Yorkers were startled by the exhibition of three "camel-leopards," the first of that ilk ever seen in America, in the livery stable of John C. Stevens in the rear of No. 4 Wall Street, in 1787. If of further interest, these were giraffes.

N. W. Stuyvesant kept a crack team of trotters in the old Bayard stable, a few feet east of Stevens' barn. These horses were so spirited they had to be led out into the roadway before taking off for a fast go on the Bowery Lane to Petersfield, or on to the fork in the road for a short visit to the Buck Horn Tavern.

Joseph Baker's City Tavern is visible at the southeast corner of New Street and Wall. At the southwest corner is the house where Washington Irving practiced law with his brother, John T. Irving.

The first Presbyterian Church in New York was built here in 1719, on the site of Peter Stoutenburgh's garden, and the land was conveyed to the general assembly of the Church of Scotland in 1730. The trustees received much needed financial assistance from a number of New England Sessions, particularly in Connecticut. Like the Dutch Reformed Churches, the first Presbyterian was used as a prison by the British while they occupied the city during the Revolutionary War. Its condition deteriorated to such an extent that it was finally torn down and replaced by a new structure, which opened for worship in 1811. This new church, designed by J. F. Mangin, burned in 1834 but was rebuilt in the following year. When its removal became necessary it was carefully dismantled and was re-erected in Jersey City, New Jersey, where it was demolished in 1888.

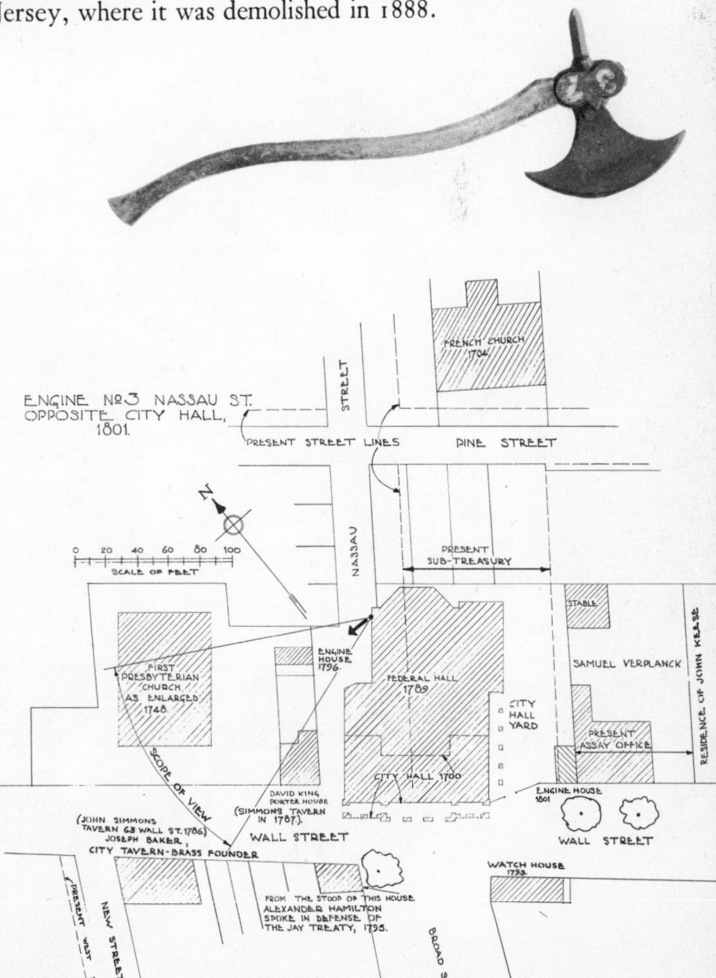

Old Slip

PRIOR to the fire of 1835, Old Slip, with its picturesque buildings, its humming market, colorful hucksters and grand view of the harbor, was one of the most interesting spots in the city to spend a few leisure hours. Along Pearl and Water Streets, "facing the ocean," the last remnants of New Amsterdam stood until obliterated by time, progress and the flames of the great fires. Many of these houses were of ancient Dutch construction with pediment walls marching up to the roof ends and presenting their squared ruffled gables to the street.

Nearby, at the head of Coenties Slip (71-73 Pearl Street), once stood the most famous Dutch building in the city, the ancient Stadt Huys, New Amsterdam's first City Hall which was originally built, in 1642, as the Stadt Herbergh or "city tavern." It was used as a City Hall in 1653. Here the burgomasters, schepens and magistrates presided over their various offices "as in the old country" and, believe it or not, in that long ago day and age, the place was over-run with commies*.

Oyster barges in 1878.

Engine Company No. 11, which moved to Old Slip in 1813, was involved in one of the most ironic situations in the city's history when that neighborhood was destroyed in 1835. As alarms rang out all over the city and engines rushed to the scene, the men of No. 11, closer to the fire than any other engine company, worked furiously over their engine which had frozen up during the night. Although their efforts to thaw it were unsuccessful and both the house and engine perished in the flames, her men went to the aid of other companies and helped to battle the fire.

**Dutch for government clerks.*

Franklin Market and Old Slip.

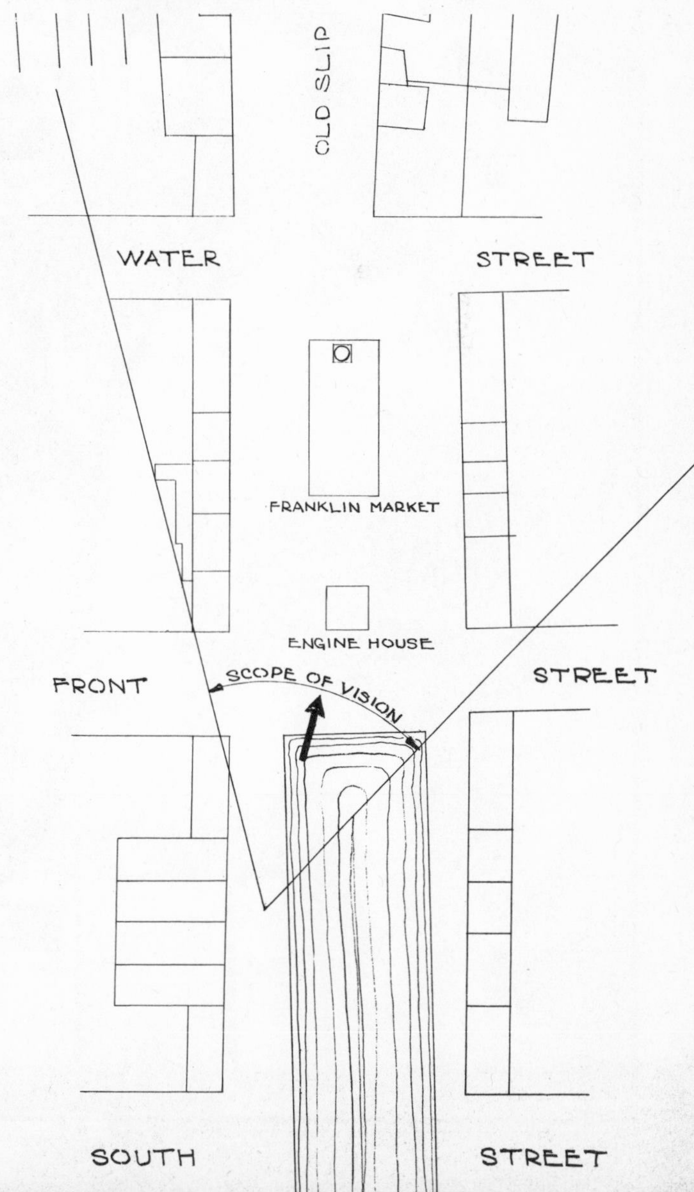

View looks up Old Slip to William Street, showing buildings in Hanover Square in 1823. The building with the cupola just behind the engine house is Franklin Market, "a small, neat, brick building with eight stalls." The Fire of 1835 wiped out all the structures in this view. Old Slip was later filled in and a new Franklin Market was built in 1836 on the site between South and Front Streets.

THE FIREMAN.

The Nobleman of our Republic

THE FIREMAN

The Fireman in Four Plates—Plate No. 3

The Nobleman of Our Republic

PUBLISHED BY T. W. STRONG CIRCA 1850 COURTESY OF HAROLD V. SMITH, ESQ.

"One, two, three, four!
The panting Foreman's trumpet bellows
'Pull her along and jump her, fellows!'
Down through the cobbled avenues roar."

The view shows what is perhaps an exact demonstration of the pumping of an old goose-neck hand-engine after the introduction of suctions. With the oral assistance of their foreman, who is shouting encouragement and directions through his speaking trumpet, the men are working with the brakes to pump a stream of water while their pipeman, on the ladder, is playing on the flames. Hand engines of this type were normally operated at 60 strokes a minute but were sometimes speeded up to double that or more at a bad fire or if the machine was in danger of being "washed" by a rival working in line behind her.*

Engine 21 reported in its minutes of March 10, 1828, a fire which destroyed a blacksmithy, two livery stables and a dwelling in New Street near Beaver, and noted that "during the fire a Gentleman attach'd to one of our torches a piece of Ribbon as a mark of distinction for having play'd 128 strokes in a Minute."

On one occasion a fire company was said to have reached 170 strokes a minute, an amazing feat and a tremendous strain on the firemen. A stroke consisted of one up and down motion of the brakes. At a normal pace, the men could last at the brakes for about ten minutes but when rushed, one, two or three minutes were all a man could stand. Firemen frequently suffered torn fingers and broken arms while jumping in to give relief when the engine was working at top speed.

* So-called from the shape of the emission pipe on the top of the air chamber.

Lords of Hanover Square

TODAY a center of the marine industry, including the offices of shipping companies and marine insurance underwriters, Hanover Square was, until about 1845, the principal business center in New York City. It extended from Old Slip to Wall Street and the names of the occupants of buildings in and around it were among the most prominent in the New York financial and commercial worlds.

Where the Cotton Exchange Building now stands, William Bradford set up the first printing press in New York, in 1693, at the "Sign of the Bible." Bradford was appointed printer of the Acts of Assembly and Public Papers and in 1725 he published the first issue of New York's pioneer newspaper, *The New York Gazette*.* Bradford was over ninety when he died and was buried in the grounds of Trinity Church.

The sign of a somewhat less reputable publisher, Hugh Gaine, hung for forty years at the "Bible and Crown" in Hanover Square.** Gaine published the first American edition of "Robinson Crusoe" and founded a newspaper, *The New York Mercury*, in 1752, which shortly before the Revolution became *The New York Gazette and New York Mercury*. He wanted to put his money on the winning horse and could never make up his mind whether to be a Whig or a Tory. He finally compromised by keeping his New York paper neutral but printing another paper in Newark, New Jersey, in which he represented himself as a true patriot. The greater the British successes became the more his patriotism decreased. He was allowed to remain in New York after the British were driven out but his hypocrisy had earned him the hearty dislike of the population and his papers soon died a natural death.

In early days, Burger Jorissen, one of the first Dutch blacksmiths, had his forge in Hanover Square. Until the last he urged that Peter Stuyvesant fight the British if and when they attempted to take possession of the town but he was outvoted. He was never reconciled to this and left Manhattan immediately after the British arrived. During the height of his career he was a highly skilled worker and did much work for the city. Many a shackle was turned over the horn of his anvil. The lower end of William Street, a deep depression sloping down to the water's edge, was called the Burger Jorissen Path.

Two customs houses stood near the square, one, a confined place on the north side of South William Street, the other, of more respectable size, on the northwest corner of Moore and Front Streets. The latter, which today is the site of a grocery store, was once the celebrated "Stuyvesant Huys."

The earliest Jewish synagogue in New York was also located in South William Street. Some years later this site, which contained a fresh spring where ritual cleansings and ablutions had been performed, became a mill seat. The bark mill of Ten Eyck is believed to have stood there.

The residence of many naval officers of distinction arriving in New York was Beekman's house, on the northeast corner of Sloat Lane. Here, under the guardianship of Admiral Digby, lived Prince William Henry, the Duke of Clarence—later, King William III of England. He used to be seen about New York in the common garb of a midshipman's "roundabout," and liked to go skating with the boys on the Kolck Pond. He was only a knock-kneed, naive, natural lad at this time, unmindful of his station, and on board his ship in the harbor he once offered to lay aside his star and settle a controversy with a fellow midshipman by a fist fight.

Somewhat back from the square stood the house of Govert Loockermans which from 1691 to 1696 was occupied by the famous Captain William Kidd. When he first moved into Loockermans' house he was a privateersman, for there is a record showing that in 1691 he paid his fees to the governor and the king. But four years later, when he returned from England with the King's Commission, he was charged with piracy. In 1699, he was decoyed to Boston, arrested, sent to London and there ended his career at Execution Dock. Col. Robert Livingston had recommended Kidd to

**James Rivington, one of New York's outstanding printers and publishers, opened business in Hanover Square in 1760 as "the only London bookseller in America." In 1773 he began to publish "The New York Gazeteer," a paper so good that only John Peter Zenger's paper could compare with it in later years. The merchant Peter Goelet conducted his business at the Sign of the Golden Key. On the corner of Sloat Lane was the house built by Henry Remsen, whose son, also named Henry, became cashier and later president of the Bank of Manhattan. About 1798, Isaac Gouverneur built a grand house on the west portion of the same lot, and from 1809 to 1813 it was occupied by General Moreau. The famous India House stands between Pearl and Stone Streets, near what was in Dutch days called "The Burgers' Path."*

***One hundred and fifty years ago New York was already distinguished for the number and magnificence of its signs. Probably more signs appeared around Hanover Square than in any other part of the city. Every device was resorted to, and no expense was spared to make them attractive. They were crowded on every story and even on the tops and ends of some of the houses. A small house in Beekman Street had twelve signs advertising lawyers, and at 155 Pearl Street the names of Tilldon and Roberts were painted on the stone steps! "In truth," Watson wrote in 1829, "it struck me as defeating its own purpose, for the glare of them was so uniform as to lose the power of discrimination. It is not unlike the perpetual din of their own carriage wheels unnoticed by themselves, though astounding to others." For later advertising style see appendix.*

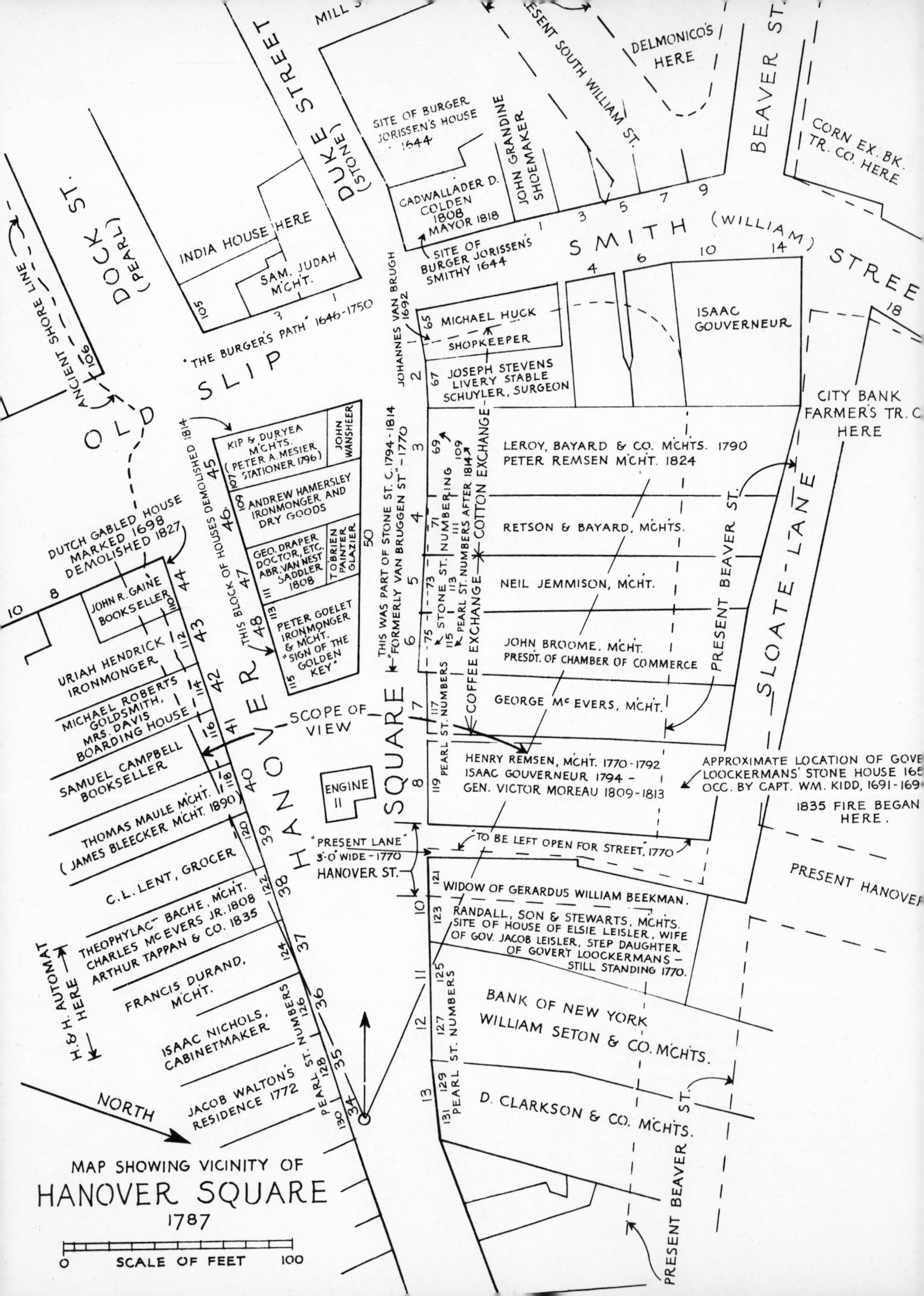
MILL
BEAVER ST.
DELMONICO'S HERE
PRESENT SOUTH WILLIAM ST.
CORN EX. BK. TR. CO. HERE
DUKE STREET (STONE)
SITE OF BURGER JORISSEN'S HOUSE 1644
JOHN GRANDINE SHOEMAKER
CADWALLADER D. COLDEN 1808 MAYOR 1818
SITE OF BURGER JORISSEN'S SMITHY 1644
SMITH (WILLIAM) STREE
DOCK ST. (PEARL)
INDIA HOUSE HERE
SAM. JUDAH M'CHT.
ANCIENT SHORE LINE
"THE BURGER'S PATH" 1646-1750
JOHANNES VAN BRUGH 1692
MICHAEL HUCK SHOPKEEPER
JOSEPH STEVENS LIVERY STABLE SCHUYLER, SURGEON
ISAAC GOUVERNEUR
OLD SLIP
CITY BANK FARMER'S TR. C HERE
KIP & DURYEA M'CHTS. (PETER A. MESIER STATIONER 1796)
JOHN WANSHEER
ANDREW HAMERSLEY IRONMONGER AND DRY GOODS
GEO. DRAPER DOCTOR, ETC. ABR. VAN NEST SADDLER 1808
T. O'BRIEN PAINTER GLAZIER
PETER GOELET IRONMONGER & M'CHT. "SIGN OF THE GOLDEN KEY"
THIS BLOCK OF HOUSES DEMOLISHED 1814
THIS WAS PART OF STONE ST. C. 1794-1814 "FORMERLY VAN BRUGGEN ST" -1770
STONE ST. NUMBERING
PEARL ST. NUMBERS AFTER 1814
COTTON EXCHANGE
COFFEE EXCHANGE
LEROY, BAYARD & CO. M'CHTS. 1790
PETER REMSEN M'CHT. 1824
RETSON & BAYARD, M'CHTS.
NEIL JEMMISON, M'CHT.
JOHN BROOME, M'CHT. PRESDT. OF CHAMBER OF COMMERCE
GEORGE Mc EVERS, M'CHT.
PRESENT BEAVER ST.
SLOATE LANE
DUTCH GABLED HOUSE MARKED 1698 DEMOLISHED 1827
JOHN R. GAINE BOOKSELLER
URIAH HENDRICK IRONMONGER
MICHAEL ROBERTS GOLDSMITH, MRS. DAVIS BOARDING HOUSE
SAMUEL CAMPBELL BOOKSELLER
THOMAS MAULE M'CHT. (JAMES BLEECKER M'CHT. 1890)
C. L. LENT, GROCER
THEOPHYLACT BACHE, M'CHT. CHARLES Mc EVERS JR. 1808 ARTHUR TAPPAN & CO. 1835
FRANCIS DURAND, M'CHT.
ISAAC NICHOLS, CABINETMAKER
JACOB WALTON'S RESIDENCE 1772
H. & H. AUTOMAT HERE
SCOPE OF VIEW
HANOVER SQUARE
ENGINE 11
PEARL ST. NUMBERS
HENRY REMSEN, M'CHT. 1770-1792 ISAAC GOUVERNEUR 1794 - GEN. VICTOR MOREAU 1809-1813
APPROXIMATE LOCATION OF GOV LOOCKERMANS' STONE HOUSE 165 OCC. BY CAPT. WM. KIDD, 1691-169
1835 FIRE BEGAN HERE.
"PRESENT LANE" 3'-0" WIDE - 1770
HANOVER ST.
TO BE LEFT OPEN FOR STREET, 1770
PRESENT HANOVER
WIDOW OF GERARDUS WILLIAM BEEKMAN.
RANDALL, SON & STEWARTS, M'CHTS. SITE OF HOUSE OF ELSIE LEISLER, WIFE OF GOV. JACOB LEISLER, STEP DAUGHTER OF GOVERT LOOCKERMANS - STILL STANDING 1770.
BANK OF NEW YORK
WILLIAM SETON & CO. M'CHTS.
D. CLARKSON & CO. M'CHTS.
PRESENT BEAVER ST.
NORTH
MAP SHOWING VICINITY OF
HANOVER SQUARE
1787
0 SCALE OF FEET 100

Old Molly pulling 11's goose-neck through the snow drifts

the Crown officers as "a bold and honest man," and during the trial many suspected him of being in Kidd's confidence.

The gazettes of pre-Revolutionary days were filled with the exploits of Captain Kidd and other pirates. In 1723 a pirate ship appeared off the coast of Long Island, commanded by a Bostonian named Lowe. A year later, startled New Yorkers awoke to a Merry Christmas in the middle of July. The same Lowe had brought his ship, accompanied by a consort, right into the harbor. There she lay, a brigantine of 300 tons, flying a black flag, with her name, "Merry Christmas," painted boldly on her bow and £150,000 in silver and gold buried in her holds. The pirates were engaged by the *Greyhound* of His Majesty's Navy, commanded by Captain Solgard who captured the smaller vessel and took 43 prisoners. They were tried and executed in Rhode Island. Lowe himself escaped and revenged himself by the vicious savageries which he afterwards inflicted on all Englishmen whom he captured.

In 1724, William Bradford published a general history of pirates, including the stories of two *women* pirates, Mary Reed and Anne Bonny.

As far back as 1750, a committee was named to get an adequate house built for a fire engine on the square, but it was not until December 28, 1780, that Oceanus Company No. 11, was finally established there. This was a jolly and sociable company and enjoyed many a "supper and liquors" at the taverns nearby. But their high spirits were matched by their sense of discipline. At a meeting on November 9, 1784, at Doughty's Tavern, Hasking and Moore were struck off the list for non-attendance at washings. The company appointed a moderator for their meetings and each man signed his name to the following pledge: "That upon first call of silence by the Moderator the person refusing to keep silence shall pay one shilling, and upon second call of silence by the Moderator the person refusing to keep silence shall be fined two shillings, or be expelled from the company if the majority shall so determine."

One of the famous "volunteers" of the earlier days was an old Negro woman named Molly, a slave of John Aymar, one of the last of the old Knickerbockers of New York. Aymar was himself a famous figure in New York in those days. A trim old gentleman, he continued to wear the style of dress common among the wealthier old aristocrats—a long-tailed coat, knee-breeches, silver shoe-buckles and the inevitable queue.

The boys of the company used to call Molly "Volunteer No. 11." She considered herself a very important member and often was seen running at the sound of an alarm in her calico dress and checkered apron, a clean bandanna handkerchief neatly folded over her breast, and another wound about her head. Once during a blinding snowstorm there was a fire in William Street, and it was hard work to draw the engine; but the first to take hold of the drag-rope that day was Molly, pulling away for dear life. This may have been the only time that she took hold of the rope, but afterward, when asked what engine she belonged to, she always replied, "I belongs to ole 'Leven; I allers runs wid dat old bull-gine." Later, one of 11's boys wrote, "You could not look at Molly without being impressed by her really honest face—it was a beaming lighthouse of good-nature."

All of Hanover Square, with the exception of one stone warehouse, toppled in ruins during the Great Fire of 1835. The section never regained its former preeminence.

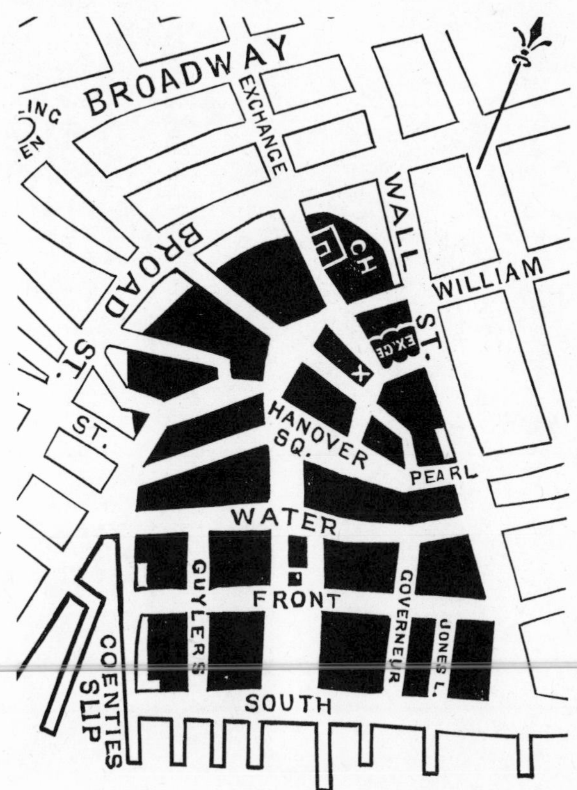

Between Broad and Broadway

JOHN WATSON reported in 1828 that Dr. B. Franklin once said, you could always tell a New Yorker by his awkward gait in walking on Philadelphia's smooth paving, "like a parrot upon a mahogany table." In Manhattan, Stone Street (Brouwers Straet) was the first to have been paved at all (1652) and then it was cobbled roughly as were others to follow. When "Stony Street," as it was often called, was to be widened, improved and repaved in 1784, folks bitterly exclaimed against exposing to public view the "dark recesses" of that street and of the "hallowed mysteries" of Petticoat Lane, now Marketfield Street. The general area below Wall Street, east of Broadway and containing Broad Street, included the earliest part of the settlement of Manhattan. The upper portion was originally the Schapen Weytie or Sheep Pasture.

Verlettenbergh (hill) rose to the east of Broadway at the present Exchange Place, and was named for Nicholas Verlett, husband of Anna Bayard, nee Stuyvesant. Verlett was commissioner of imports and keeper of public stores in 1657. In English times Verlettenbergh Street between Broadway and Broad Street became "Flatten Barrack" Street. This, together with Garden Street between Broad and William Streets, is now all Exchange Place.

In old Dutch days, Broad Street was a narrow canal, Der Heere Graft, and it is reported that oyster, vegetable and wood boats came up there to sell their wares. Two footbridges crossed the canal. The land was too valuable, however, and the region so busy, that the canal was filled in at an early date. On Broad near Water Street, in 1752, stood the Royal Exchange, or New Exchange, an arched brick structure.

...bombeler ...st Dutch ...oolmaster.

From 1693 until the fire of 1835 the old South Dutch Reformed Church stood, east of Broad Street, on the north side of that part of Exchange Place, first known as De Warmoes Straet (Street of Vegetables), then Dutch Church Street. The "Dutch Free School," now the Collegiate School* of The Dutch Reformed Church, was across the way from the church from 1748 to 1828. This educational institution, probably the oldest in America, has been in continuous operation (interrupted only by the Revolutionary War) for more than three centuries having had its primitive start under the West India Company, with Adam Roelantsen, schoolmaster, in Governor Van Twiller's time in 1633. After occupying many locations during its career it now carries on at 241 West 77th Street, Manhattan.

**January 16, 1854, Washington Irving wrote to the Headmaster, H. W. Dunshee, as follows: "There is one historical fact of which you make no mention, and possibly know nothing. A war once raged between the Dutch school and the school to which I belonged (kept by Mr. Benjamin Romaine, on Partition, now Fulton Street, below St. Paul's Church), and more than one doughty battle was fought, in which, on the whole, I rather think we of Partition Street came off the worse. However, these were feuds of the last century, and have long since passed away. I have no longer any pugnacious feelings towards your school, etc."*

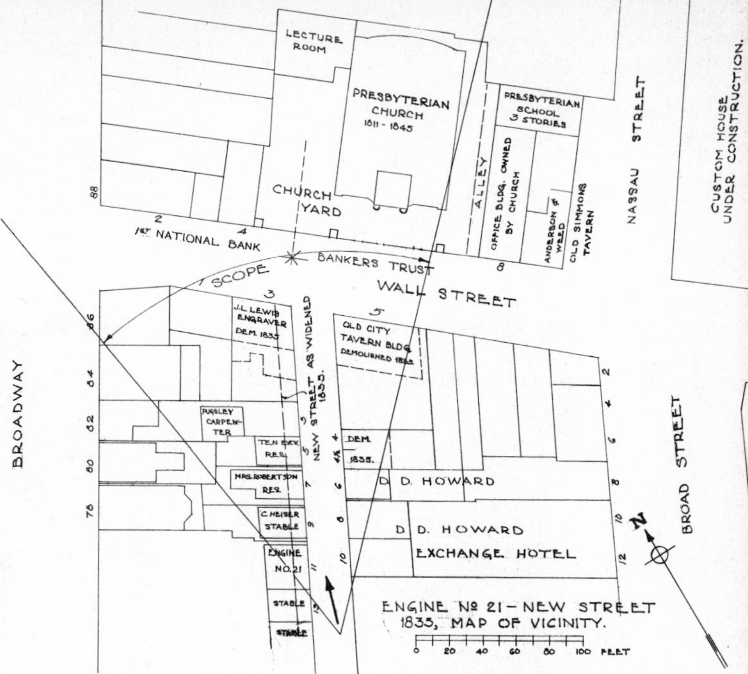

Tablets commemorating New Amsterdam's first school and the first French Huguenot Church in Petticoat Lane (1688) can be seen in the court of the present Produce Exchange.

Fraunces Tavern is now probably the oldest existing structure in Manhattan. It was built in 1719 "upon the range of the Coffee House," on the northeast corner of Broad and old Great Dock Streets, as the residence of Stephen Delancey. Later it was occupied by Col. Joseph Robinson until his death in 1759. This old house has had both glorious and ignominious moments; few buildings have survived so many physical vicissitudes. Its walls rang with the voices of our nation's first great leaders. On December 4, 1783 Washington bade a sad, eloquent farewell to his faithful officers at war's end; in later years here was a German tenement house with a noisy lager beer saloon on the first floor. It has been a storehouse, and has housed the War Department of the United States Government under the Hon. Henry Knox in 1789. During the Revolution a round shot from His Majesty's Ship of War, *Asia*, striking the inn, was

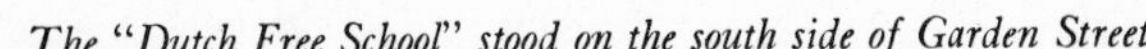

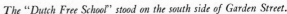

The "Dutch Free School" stood on the south side of Garden Street.

Washington's farewell to his officers at the Fraunces Tavern, December 4, 1783.

implanted in its walls; the roof burned off in 1832; two stories were added twenty years later, and it was disfigured by alterations again in 1890.

Samuel Fraunces, who purchased the property in 1762 was destined to become a famous host. By the sign of the "Queen's Head," the hospitable doors of the Queen Charlotte Tavern were thrown open. In that house the witty and devoted daughter of Fraunces exposed a venomous plot to assassinate the "Father of His Country." Fraunces sold the place in 1785 to George Powers of Brooklyn, and retired to a farm in Monmouth County, New Jersey. He returned, however, to become Steward to President Washington's household during the latter's residence in Franklin Square in 1789. Samuel Fraunces died three years before Thomas Gardner bought the old inn from Powers in 1801. The historic building remained in the possession of Gardner's descendants until after 1901. The Sons of the American Revolution bought the place in 1904. It was restored to its original condition from plans of William H. Mersereau and opened to the public as a museum in 1907. Today it remains as one of New York's most interesting museums and eating places.

New Street, between Broad and Broadway, in 1835, as shown on page 54, was lined with small houses and stables, and was only 25 feet wide between its building lines. In 1835-6 it was increased to its present width of 35 feet by demolishing the buildings along its west side. Engine 21, "Old Slippers," moved from Tryon Row to No. 11 New Street in 1831.

Engine 21 had many distinguished members, among whom were James Gulick, Chief Engineer of the Volunteers and Joseph Curtis, a friend of Robert

Fraunces Tavern before its restoration. The tavern then offered "first class regular dinner for 25¢, for gentlemen only."

The Fraunces Tavern today, completely restored and operated as a club, museum and restaurant.

The original Fraunces Tavern as it appeared in early days.

Fulton. The latter, who was a secretary of the company, was in the hardware business in Maiden Lane in 1804. He is credited with having invented the trap for waste pipes and for perfecting and carrying the first oil fire torch. His face was frozen while driving a sleigh to Albany in 1817 to secure passage of the Manumission Act to free the slaves in New York.

Many changes were taking place in Wall Street, in 1835. The two houses at the corner of New Street had been demolished to make way for new buildings, thus providing a clear view of the corner of the Presbyterian Church. The old step-gabled Ten Eyck house at No. 5 New Street was the subject of an anonymous fairy tale published in 1833. It seems likely now that Washington Irving was the author, as he would have been in a position to know about the place, since his brother, Judge John T. Irving, owned the corner house and the two of them had their law offices there from 1807 until 1810. It was in New Street that the fire of 1845 originated, as described in the diary of George Templeton Strong.

Behind the two three-story houses shown in the Chrystie view was the famous Exchange Hotel in Broad Street. When the celebrated Indian Chief Black Hawk and his companions visited the city in 1833 they took residence there, a few days after a visit by President Andrew Jackson. This was the site where, in 1864, the Stock Exchange erected the first of their three buildings.

41-43 Broad Street, 1831. Horatio N. Ferris, a cooper, occupied an old Dutch structure at 41; Joseph Meeks, a cabinet and chairmaker, was at No. 43.

The residence of Gen'l Horatio Gates, 1768, later of Gen'l Alexander (Lord Stirling), 69 Broad Street.

Broad Street to this day continues to play a foremost part in the commercial and financial life of the city.

The present Broad Exchange Building* at 25 Broad Street is situated about where the old canal ended, and near the site of early Curb Exchange activities.

**Here today are the office of The City Investing Company of which Robert Dowling is president. Mr. Dowling, who is also a director of The Home Insurance Company, is one of the great civic leaders of New York and serves as Chairman of the Greater New York Civic Center Association, and director of the Federal Hall Memorial Associates.*

The southeast corner of Broad and Pearl Streets, showing the Fraunces Tavern (at right) after alterations in 1852 had been made. It was known at this time as Beaumeyer's Broad Street House.

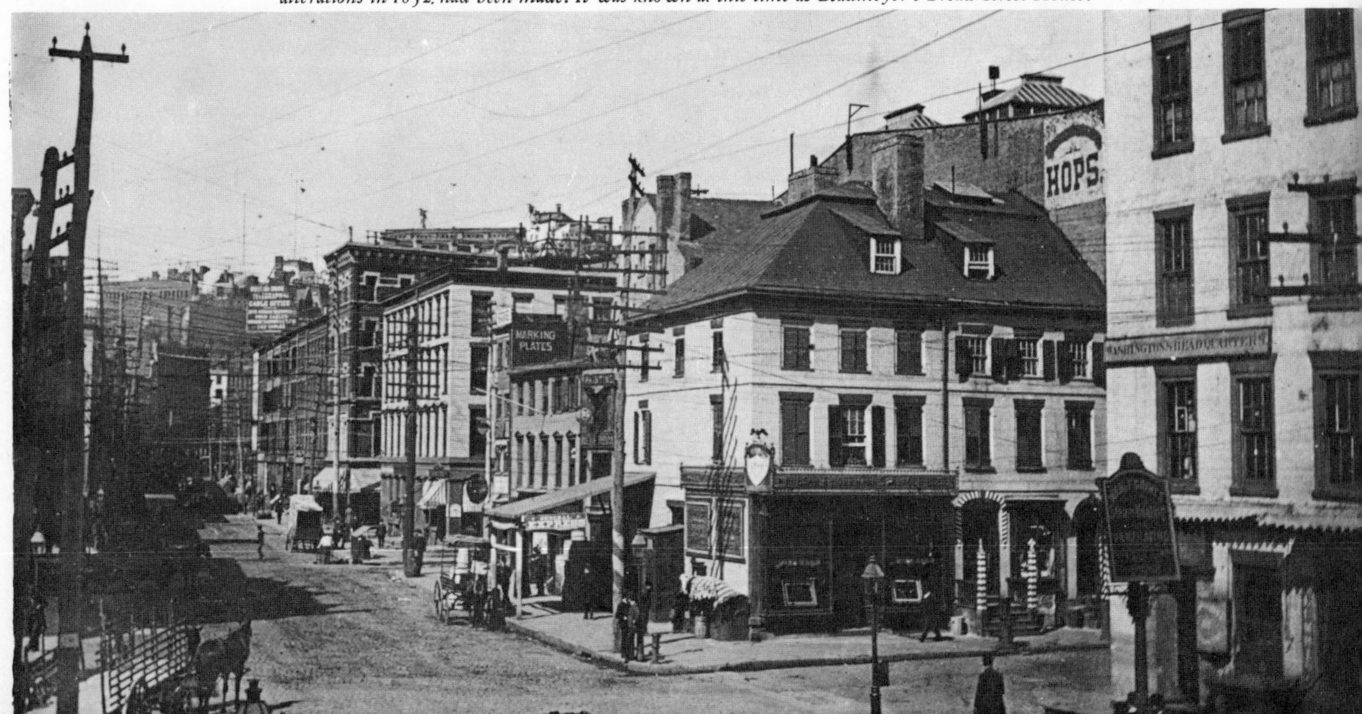

The "*Famous Beau of Old Manhattan*"

The Peabody view of their establishment at 233 Broadway. The home of Philip Hone is at the right of the view.

Philip Hone

On Broadway opposite City Hall Park stood the home of one of New York's most colorful mayors, Philip Hone. Born in Dutch Street in 1780, Hone developed into a striking figure of a man, tall and spare, distinguished in bearing and courtly in manner. Descendant of an old New York family, he was a master of the social graces and an intimate of the great men and women of his day. Within the walls of the cheerful mansion, Hone entertained such illustrious men as Daniel Webster, Henry Clay and Charles Dickens, as well as a host of other celebrities of the time.

Hone and his gracious wife, Catherine, who was much admired for her beauty and charm, lived at 235 Broadway until 1836 when Hone sold the mansion to Elijah Boardman for $60,000. It was to be "converted into shops below and the upper part to be an extension of the American Hotel kept by Edward Milford." Hone then moved to the "Colonnade Houses" at 714-716 Broadway between West 4th Street and East Washington Place. In 1850, his beloved Catherine died there and the following year, Hone passed away after reaching the age of seventy-one.

Philip Hone is known today chiefly through his diary which has served as a boon to generations of historians as well as providing a fascinating glimpse into the mores and morals of the young metropolis during the first half of the nineteenth century.

Some idea of Hone's many interests and activities can be gleaned from the fact that in addition to his service as Mayor of New York—to which he was elected in 1826—he was an important insurance and bank director, a governor of New York Hospital, trustee of Columbia College, vice-president of the New York Historical Society, the American Seaman's Fund Society, founder and governor of the Union Club and a member of the Vestry of Trinity Church.

Philip Hone was an important man. He was also an intelligent and an observing man, and fortunately he kept a diary. Perhaps he was writing for posterity. At any rate, through his painstaking efforts we learn much about the man and more about the city in which he lived.

During his time occurred the worst conflagration that had yet struck New York or any other American city. Through his eyes, we are privileged to witness the Great Fire of 1835. Here is his account, as he wrote it along with added observations (in italics) by other eye-witnesses:

The "Colonnade Houses," Nos. 714-716 on the east side of Broadway, between West 4th Street and East Washington Place. Courtesy of Museum of the City of New York.

Primitive fire escape

"FIRE!"

"December 17.—How shall I record the events of last night, or how attempt to describe the most awful calamity which has ever visited these United States? The greatest loss by fire that has ever been known, with the exception, perhaps, of the conflagration of Moscow, and that was an incidental concomitant of war. I am fatigued in body, disturbed in mind, and my fancy filled with images of horror which my pen is inadequate to describe. Nearly one-half of the first ward is in ashes, five hundred to seven hundred stores, which, with their contents, are valued at $20,000,000 to $40,000,000, are now lying in an indistinguishable mass of ruins. There is not, perhaps, in the world the same space of ground covered by so great an amount of real and personal property as the scene of this dreadful conflagration.

"The fire broke out at nine o'clock last evening. I was writing in the library when the alarm was given, and went immediately down. The night was intensely cold, which was one cause of the unprecedented progress of the flames, for the water froze in the hydrants, and the engines and their hose could not be worked without great difficulty. The firemen, too, had been on duty all last night, and were almost incapable of performing their usual services. The fire originated in the store of Comstock & Adams, in Merchant Street—a narrow, crooked street, filled with high stores lately erected and occupied by drygoods and hardware merchants, which led from Hanover to Pearl Street."

William H. Macy (later president, Seaman's Bank for Savings), a member of the Supply Engine Company, with Chief Engineer Gulick, entered the second building to catch fire. He held a nozzle on the Tontine steps and, more than any other fireman, was credited with preventing the flames from crossing the street at Wall and Water Streets. His greatcoat was burned through and his leather hat roasted to a crisp. Engine 13, at Wall and Pearl, was ordered to play a stream on him to keep him alive (the temperature stood at 17 degrees below zero).

"When I arrived at the spot the scene exceeded all description; the progress of the flames, like flashes of lightning, communicated in every direction, and a few minutes sufficed to level the lofty edifices on every side. It crossed the block to Pearl Street. I perceived that the store of my son was in danger, and made the best of my way, by Front Street around the old Slip, to the spot. We succeeded in getting out the stock of valuable drygoods, but they were put in the square, and in the course of the night our labours were rendered unavailing, for the fire reached and destroyed them, with a great part of all which were saved from the neighbouring stores; this part of Pearl Street consisted of dry-goods stores, with stocks of immense value, of which little or nothing was saved."

A view of the Great Fire of 1835 taken from the roof of the Bank of America, corner of Wall and William Streets. In this aquatint by N. Calyo, the Merchant's Exchange is at the left center of the print, completely destroyed.

Sorely needed reinforcements were sent for. Engine companies from Philadelphia and New Jersey reached or attempted to reach the scene to relieve the exhausted men of the New York companies.

"At this period the flames were unmanageable, and the crowd, including the firemen, appeared to look on with the apathy of despair, and the destructon continued until it reached Coenties Slip, in that direction, and Wall Street down to the river, including all South Street and Water Street; while to the west, Exchange Street, including all Post's stores, Lord's beautiful row, William Street, Beaver and Stone Streets, were destroyed."

Charles King, editor of the American, heroically crossed the ice-filled East River to the Navy Yard in an open boat to procure gunpowder. Loading a brig with kegs of gunpowder, sailors and marines made a risky passage to the flaming Manhattan side. Wrapping the kegs in blankets to protect them from flying sparks, the gallant sailors and marines, joined by firemen, coolly proceeded through the rain of fire to place demolition charges in Bailey's and several other large buildings and established a fire-break to the northeast.

"The splendid edifice erected a few years ago by the liberality of the merchants, known as the Merchants' Exchange, and one of the ornaments of the city, took fire in the rear, and is now a heap of ruins. The facade and magnificent marble columns fronting on Wall Street are all that remain of this noble building, and resemble the ruins of an ancient temple rather than the new and beautiful resort of the merchants. When the dome of this edifice fell in, the sight was awfully grand; in its fall it demolished the statue of Hamilton, executed by Ball Hughes, which was erected in the Rotunda only eight months ago, by the public spirit of the merchants."

South Dutch Church was consumed and the old bell imported from Holland destroyed. The adjoining school went too.

The Journal of Commerce building, in the rear of the old South Reformed Church in Garden Street (Exchange Place), was saved by the timely use of vinegar, a number of open hogsheads of which were stored in the adjoining yard.

"It would be an idle task to attempt an enumeration of the sufferers; in the number are most of my nearest friends and of my family; my son John, my son-in-law Schermerhorn, and my nephew Isaac S. Hone, and Samuel S. Howland were all burnt out.

"The buildings covered an area of a quarter of a mile square, closely built up with fine stores of four and five stories in height, filled with merchandise, all of which lie in a mass of burning, smoking ruins, rendering the streets indistinguishable."

Engine 33 was run on the deck of a brig at the foot of Wall Street and was supply engine pumping from the river into Engine 26, which passed it on to Engine 41. A pipe of brandy was put in the machine to keep it from freezing.

Panel presented by the New York Fi Department to the Franklin Fire Company of Philadelphia, which rus to New York with their machines du the fire of 1835.

"The Mayor, who has exerted himself greatly in this fearful emergency, called the Common Council together this afternoon for the purpose of establishing private patrols for the protection of the city; for if another fire should break out before the firemen have recovered from the fatigues of the last two nights, and the engines and hose be repaired from the effects of the frost, it would be impossible to arrest its progress. Several companies of uniformed militia and a company of United States marines are under arms, to protect the property scattered over the lower part of the city.

"I have been alarmed by some of the signs of the

The monument erected on 90 Pearl Street to commemorate the Fire of 183 When the building on which it was erected was recently demolishe the statue was damaged. It is now in the hands of the H. V. Smith Museun (through the courtesy of the Grace Line) where it is being restorea

times which this calamity has brought forth; the miserable wretches who prowled about the ruins and became beastly drunk on the champagne and other wines and liquors with which the streets and wharves were lined, seemed to exult in the misfortune, and such expressions were heard as, 'Ah! they'll make no more five per cent dividends,' and 'This will make the aristocracy haul in their horns.' Poor, deluded wretches!—little do they know that their own horns 'live, and move, and have their being' in these very horns of the aristocracy, as their instigators teach them to call it. This cant is the very text from which their leaders teach their deluded followers. It forms part of the warfare of the poor against the rich—a warfare which is destined, I fear, to break the hearts of some of the politicians of Tammany Hall, who have used these men to answer a temporary purpose, and find now that the dogs they have taught to bark will bite them as soon as their political opponents.

"These remarks are not so much the result of what I have heard of the conduct and conversations of the rabble at the fire as of what I witnessed this afternoon at the Bank for Savings. There was an immediate run upon the bank by a gang of low Irishmen, who demanded their money in a peremptory and threatening manner.

"December 19.—I went yesterday and today to see the ruins. It is an awful sight. The whole area from Wall Street to Coenties Slip, bounded by Broad Street to the river, with the exception of Broad Street, the Wall Street front between William and Broad, and the blocks bounded by Broad Street, Pearl Street, the south side of Coenties Slip and South Street, are

The Journal of Commerce at 19 Beaver Street was saved by the use of vinegar taken from hogsheads stored in an adjoining yard.

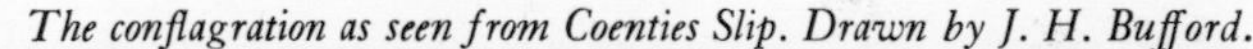

The conflagration as seen from Coenties Slip. Drawn by J. H. Bufford.

One of the rarest and most interesting views of the Great Fire, drawn by Alfred Hoffy.

1. Chester Huntingdon
   Police Officer
2. John Jacob Schoonmaker
   Keeper of the Battery
3. Nathaniel Finch
   Member of Fire Co. No. 9
4. Matthew Bird
   Member of Fire Co. No. 13
5. James S. Leggett
   Assistant Foreman, No. 13
6. Zophar Mills
   Foreman of Engine No. 13
7. Wm. H. Bogardus, Esq.
   Counsellor at Law
8. Col. James Watson Webb
   Editor of Courier and Enq
9. A. M. C. Smith
   Police Officer
10. James Gulick
    Chief Engineer
11. John Hillyer, Esq.
    Sheriff of City & County of
12. Oliver M. Lownds, Esq.
    Police Justice
13. Charles King*, Esq.
    Editor of the American
14. Hon. C. W. Lawrence
    Mayor of the City
15. James M. Lownds, Esq.
    Under-Sheriff
16. James Hopson, Esq.
    Police Justice
17. Edward Windust
    Of Shakespeare, Park Row
18. Thomas Downing
    Of Nos. 3, 5 and 7 Broad S
19. Jacob Hays, Esq.
    High Constable
20. H. W. Merritt
    Police Officer
21. Peter McIntyre
    Of Montgomery House, Ba Street, formerly of Washi Hall

A Key to the print of the Great Fire of the City of New York, published by the proprietor, H. R. Robinson, 48 Cortlandt Street, embracing original likenesses, taken from life, of all the parties herein named, and who rendered themselves conspicuous through their exertions in quelling the awful conflagration.

now a mass of smoking ruins.

"It is gratifying to witness the spirit and firmness with which the merchants met this calamity. There is no despondency; every man is determined to go to work to redeem his loss, and all are ready to assist their more unfortunate neighbours."

Though there have been many authentic and, in some cases, dramatic accounts of the most destructive fire ever to visit the city—James Gordon Bennett's report, a detailed summary of the after-effects by Augustine Costello, and several eye-witness accounts by contemporary merchants—the most penetrating and exciting analysis is undoubtedly the foregoing account handed down by the "Beau of Old Manhattan."

N.B.—The gentleman running up the Exchange steps is Mr. Patterson, of the firm of Patterson & Gustin, who wished, if possible, to preserve the statue of Alexander Hamilton, which was totally destroyed a few minutes afterwards.

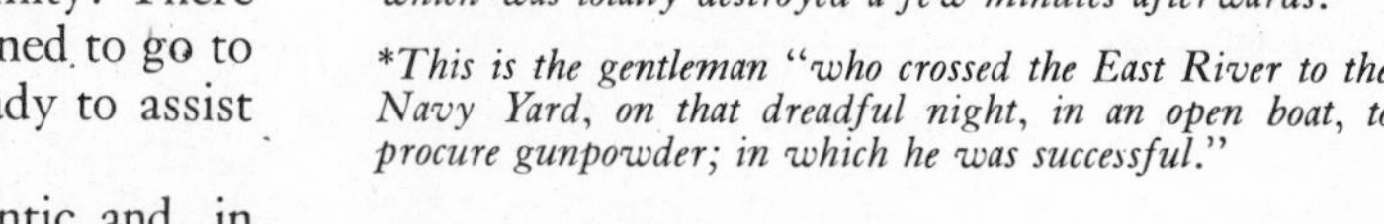

**This is the gentleman "who crossed the East River to the Navy Yard, on that dreadful night, in an open boat, to procure gunpowder; in which he was successful."*

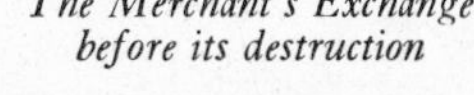

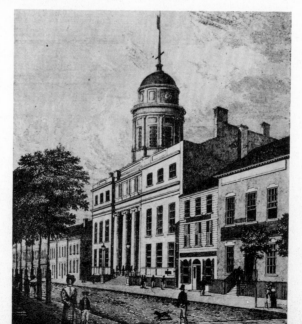

The Merchant's Exchange before its destruction

James Gulick, Chief Engineer of the Volunteer Department from 1831 to 1836, who displayed great heroism during the Great Fire.

Trinity

THE site selected for New York's first Trinity Church was in "the ground lying on the southwest side of the burial place of the Citty" which lay along the west side of Der Heere Wegh (The Broad Way). This land was included in the "Company" farm in Dutch days and became the "King's Farm and Garden" under British rule. The original site was leased to Trinity Vestry in 1697. After accession of Queen Ann, these lands, then the "Queen's," were conveyed outright to the Vestry in 1705 by Lord Viscount Cornbury,* her majesty's governor, appointed in 1701.

The first Trinity Church, commenced in 1696, was completed in March 1698. Like St. Paul's Chapel today, its front with tower and steeple faced the beautiful Hudson. The Vestry records that during construction in 1696, the "partner" of that old "Coote," the Earl of Bellomont, and Robert Livingston, the unfortunate Captain William Kidd, "has lent a Runner and Tackle for the hoisting up stones as long as he stays here and Resolv'd that Capt. Clarke doe take Care to gett the Same" —a strange quality indeed for a "pirate."

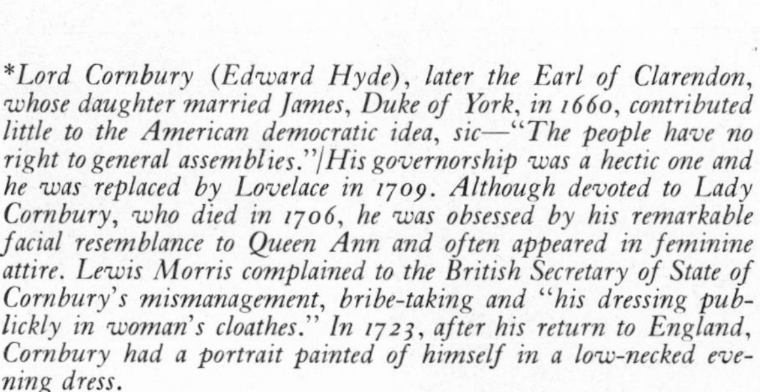
esent Trinity was completed in 1846

The fire of 1776 completely burned out the roof and interior of the first Trinity Church, and the sombre ruins remained an eyesore on the scene until the walls were razed in 1788. The second structure was erected between 1788 and 1790.

When the present Trinity edifice was being constructed in 1841, George Templeton Strong, who could keep an eye on proceedings from his Greenwich Street house, wrote—

"March 29, Monday—Moonlight and a clear sky tonight—that's one comfort. I went over (figuratively speaking) the rising glories of Trinity Church on Saturday. It's going to be a glorious affair provided they don't deform it with galleries. If they'd leave them out and the *pews* likewise, put the altar at the East End and have cathedral service there per die, I'd be quite satisfied."

**Lord Cornbury (Edward Hyde), later the Earl of Clarendon, whose daughter married James, Duke of York, in 1660, contributed little to the American democratic idea, sic—"The people have no right to general assemblies."/His governorship was a hectic one and he was replaced by Lovelace in 1709. Although devoted to Lady Cornbury, who died in 1706, he was obsessed by his remarkable facial resemblance to Queen Ann and often appeared in feminine attire. Lewis Morris complained to the British Secretary of State of Cornbury's mismanagement, bribe-taking and "his dressing publickly in woman's cloathes." In 1723, after his return to England, Cornbury had a portrait painted of himself in a low-necked evening dress.*

The locust grove on the hill (left) was a famous trysting place

Near the Broadway fence, north of the church, unread by most passersby, stands the Monument of the Empire Engine Company No. 42. It was erected to the memory of the gallant members of this company who gave their lives in line of duty in both war and peace.

The remains of many distinguished people lie sleeping in the Trinity graveyard. William Bradford, Marinus Willett, Albert Gallatin, Robert Fulton, Alexander Hamilton and his son, Philip, both killed in duels by the same set of pistols, Hamilton's wife Eliza, who lived on fully half a century after her husband's death, Captain Lawrence of the Chesapeake and many others too numerous to mention here.

Let us hope that the present surviving edifice of the trinity of churches built on this ground will long remain a familiar landmark.

The first church was enlarged twice

Hunter's Hotel

City Hotel replaced the old City Tavern.

1655 – *The farm of Henry Van Dyck * looked peaceful enough. Too often, however, the delicious fruit of the heavily-laden peach trees was an irresistible lure to the occasional Indian passing by.*

"Devil kids, dirty squaws," stealing peaches from his orchard; he'd show them! Hidden, the ex-schout watched a graceful young form silently move through the dusk. She stopped. Then the lurking Van Dyck shot the squaw as she reached into the tree. Ga-nek-yehs, like falling snow . . . she gently slumped to the earth. . . .

In the deep sigh of utter silence that nature takes just at the brink of dawn, the graceful prows of the leading canoes slid along the high-banked shore of the Hudson. Softly, as willow leaves, they came out of the mists. Sixty-four craft, sitting deep in the water with armed and painted warriors, strung out down the current for a distance of four bow shots. The noiseless, swift movement of nearly five hundred men seemed to intensify the hush.

Below De Waal, the column swung to shore. The men silently fanned out through the town. Homes were searched, a few blows were struck, but there was a strange lack of bloodshed. Then they left... to come back after guard-mounting. There he was, the hated Van Dyck. A pogamoggin whirled and the guard was knocked aside. With the cry, "Nihil'li oc ka'yu nihillowet" ("Death to the woman-killer"), a warrior loosed an arrow which struck deep into the breast of their quarry.

The cry "Murder!" rang through the town. The Dutchmen charged to the strand. The red men left three of their dead on the beach below the peach trees.... in three days the entire countryside was in flames, and there were a hundred dead Dutchmen—all for a few peaches. —Dave De Camp

1794—Near the site of an old peach orchard, Hunter's Hotel at 69 Broadway (the house with lamps in front) was in 1794 a favorite meeting place for the boys of Engine 13. This building was probably the same one, repaired or rebuilt, which was damaged in the 1776 fire. Up to that time, it had been the Lutheran Church House, parsonage and meeting rooms.

Robert Hunter** lived at 69 Broadway until 1798 and in the following year he became keeper of the State Prison. John Lovett then took over at 69 Broadway until he became proprietor of the famous City Hotel, which was on the west side of Broadway, between Thames and Cedar Streets.

To the right of Hunter's Hotel was the old Lutheran Church building, which was erected in 1729 and also gutted in the fire of 1776. Although never again used as a church, it was leased in 1792 to the church treasurer, David Grim (the reminiscer), a merchant who repaired the building sufficiently to use it as a storehouse. In 1806, both it and Hunter's were removed to make way for the building of Grace Church.

Adjoining Hunter's Hotel on the south was the new mansion of John R. Livingston, the younger brother of Chancellor Robert R. Livingston, who also resided on Broadway at No. 5.

The scene at the left also shows the second Trinity Church, which was completed in 1790. When it was demolished in 1839 to make way for the present church, the bones, coffin and coffin plate of Lady Cornbury were found under the tower.

During the Revolution, the area along Broadway in front of the "English" Church Grounds, was used as a military parade ground, which the lobsterbacks called "The Mall." Military bands played there to spectators assembled on the opposite side. It was here that the first umbrellas, carried by British officers, were seen in New York, a fact which, in those days, did not tend to better relations with the local boys.

The City Hotel, also referred to as the Tontine City Hotel, was the first building in the city to have a slate roof. It remained one of the principal hotels of the city until torn down and replaced by stores in 1849. It was the meeting place of many distinguished groups and its ballroom was the scene of the city's foremost social functions. A number of the early semi-annual concerts of the first Philharmonic Society were heard here by the music lovers of the city. Gilbert Stuart's portrait of Washington was first exhibited here on February 1st, 1798.

In 1801, the property was sold to Ezra Weeks, who was a brother of the unfortunate Levi Weeks.† Although partly destroyed by fire in 1833, the hotel was repaired and continued to hold its prominent place in the affairs of the city until the end.

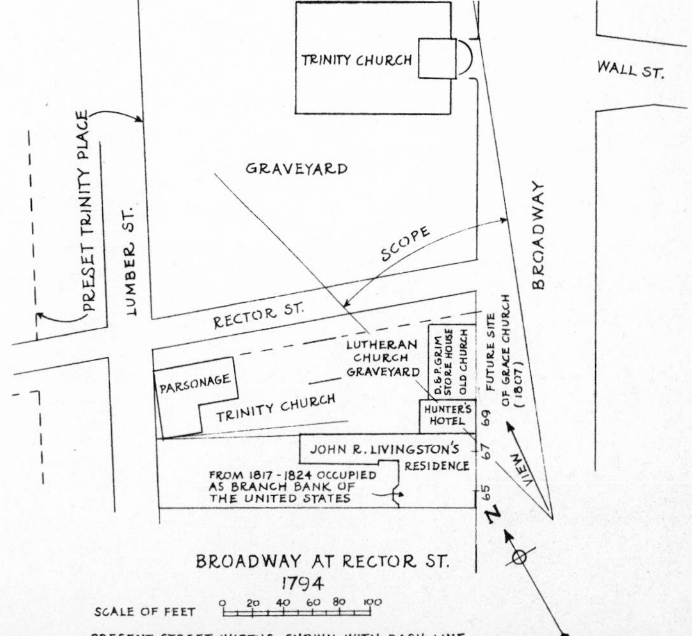

**The farm lay to the west of the Broad Way and extended below the present Exchange Alley.*

***In 1816, a Robert Hunter owned property in William Street, one lot north of Maiden Lane, now included in the land site of The Home Insurance Company.*

†*Levi Weeks, the suspected murderer of his fiancée, Alma Sands,* *see page 204.*

CORNER OF GREENWICH STREET, 1810

Greenwich and Dey Streets, 1810

WATERCOLOR BY BARONESS HYDE DE NEUVILLE

STOKES COLLECTION–NEW YORK PUBLIC LIBRARY

The view shows the corner of Greenwich and Dey Streets in 1810, with the latter street in the foreground. The three-story residence at the right was built circa 1786 by Isaac Stoutenburgh, son of Jacobus Stoutenburgh, a gunsmith who was one of New York's original firemen—appointed in 1738. Jacobus succeeded Jacob Turk, also a gunsmith, as Overseer or Chief Engineer in 1762 and held that office until 1776. When the Revolutionary War broke out, the entire Fire Department, consisting of one hundred and sixteen men was detailed as a home guard under General Washington, with Stoutenburgh in command at fires. Few of the original firemen survived the war. This site was later occupied by the North River Bank which acquired the property in 1829.

The two adjoining houses on Greenwich Street were owned by Robert Campbell and Leonard De Klyn, respectively. An unmistakable indication of an oval firemark, probably one of New York's first marks, may be seen below the middle, upper-story window of Campbell's home.

The large two-story structure in the center foreground, at the northwest corner, belonged to John Dey, who sold it a few months after this view was painted to John Wood, a tin-plate worker who also owned the adjoining house. The house at the extreme left, only a section of which appears in the view, belonged to a merchant named Robert Hyslop, who had purchased it in 1785. The site of the Dey and Wood houses became an open space when College Place, now West Broadway, was extended south from Barclay Street forming the present plaza at the intersection of Greenwich and Dey.

In accordance with the propriety demanded of a lady in those early days, the Baroness de Neuville kept her artistry as inconspicuous as possible. She painted the scene from the window of her rooms in a boarding house at 61 Dey Street, one door from the corner, in January, 1810. The Baroness' husband, G. Hyde de Neuville, along with Victor Moreau, the celebrated French general who lived at 119 Pearl Street in 1811, established a French school for children at which each gave lectures. Moreau later returned to France and was killed at the battle of Dresden. Baron de Neuville became Minister from France to the United States after Louis XVIII became king.

In The Governor's Yard

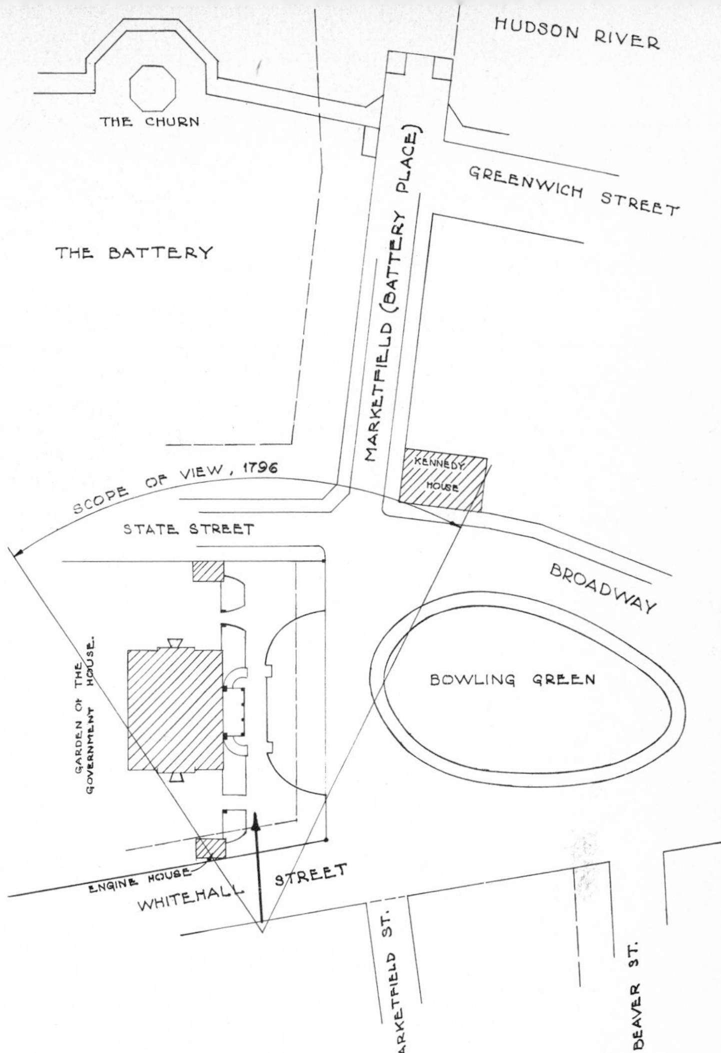

In 1658 Peter Stuyvesant began the construction of his "Great-house" on land now covered by the intersection of Moore and South Streets. This home, conveyed to T. Delavall in 1677 by Judith Stuyvesant was acquired by Governor Thomas Dongan in 1697 and named "Whitehall" shortly afterwards. The Governor's official manor for several years, the building was destroyed by fire, October 25, 1715.

Looking west from this spot during the British occupation, one could see out over the Bowling Green and the park of the Lieutenant-Governor to the Battery and the Old Dutch Fort, which stood on the site of the present Custom House. Nearby, at No. 1 Broadway, was the Kennedy House, which served as a town residence for General Howe, General H. Clinton and Sir Guy Carlton, and in the third house on the west side of Broadway, lived Benedict Arnold after his perfidy.

Opposite the Fort was the Bowling Green. Originally called The Parade, it was leased to John Chamber, Peter Bayard and Peter Jay at the rent of one peppercorn annually for the purpose of enclosing and making a bowling green with "Walks therein for the Beauty & Ornament of the Said Street."

In 1788 the old fort was demolished and levelled to prepare a site for George Washington's presidential mansion. During the excavations the doorstone of the old Church in the Fort, named for St. Nicholas, the patron saint of New York, was discovered in the ruins of the ancient Dutch chapel. The historic relic was removed to the South Dutch Church in Garden Street, where it was destroyed in the fire of 1835. President Washington never occupied this house, and the imposing structure became instead the official home of Governors DeWitt Clinton and John Jay. Bowling Green was for a time used as the governor's private garden.

Bolivar Engine No. 9, the first additional fire company to be organized after the British evacuated New York, was originally located at the Beekman Swamp in Leisler Street. From 1793 to 1796 it was located in the Governor's Yard, in Whitehall Street. Five new companies were established when the city was free and the entire department was reorganized, forming a fire-fighting force of thirteen engine and two hook and ladder companies.

The Bowling Green, circa 1890

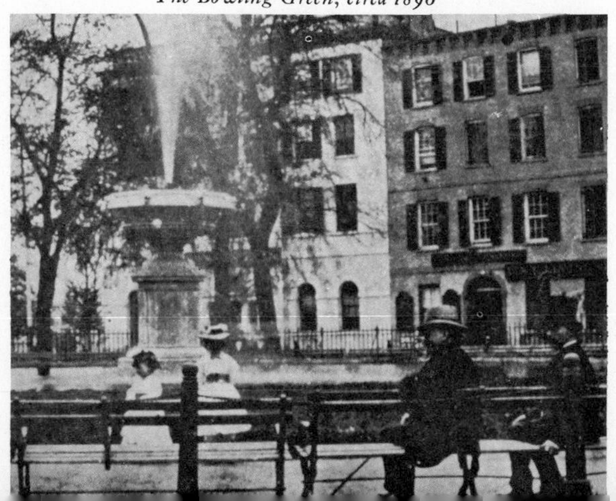

For many years Bowling Green Park was embellished with a wooden statue of George Washington, painted to imitate the colors of life. In 1843 the city's art critics prevailed upon the city authorities to remove it and it passed into the hands of a private collector who kept it until his death, forty years later. Sold for three hundred dollars, it was placed on top of the wooden arch at Washington Square that preceeded the marble structure. It was removed in 1889 and sold to the owner of a cigar store, where it stood in place of the usual wooden Indian.

A few days after a group of colonists disguised as Indians had dumped history's most famous cargo of tea into Boston Harbor, New York had a quieter but equally effective tea party of its own. Led by the Sons of Liberty, a company of citizens met the ships *Nancy* and *London* at the docks and, at the risk of their lives, forced them to turn back with all their tea still on board. Not a leaf of it was landed on the city's shores.

Burns' Tavern or Coffee House,* which was the scene of many patriotic meetings in the stirring days preceeding the War of Independence, was at 115 Broadway. Here, at the height of the Stamp Act agitation in 1775, a group of prominent merchants

**One of the oldest societies in the country, The St. Andrew's Society, was founded November 19, 1756, in New York City. Their first annual meeting was held at "Scotch Johnny's" tavern which was located at Whitehall Slip.*

Statue of George III in the Bowling Green, dragged from its pedestal after the Declaration of Independence

drew up and signed the first agreement not to import English goods so long as the Act remained in force.

When, at the last moment, this hated measure was repealed, the grateful citizens petitioned the assembly to erect a statue to honor their defender, William Pitt, and requested also a statue of King George III for the Bowling Green. Little did they foresee the results of their enthusiasm. A few leaden fragments of a horse's tail in the Museum of the New York Historical Society are all that remain today of the statue of the proud king and the statue of William Pitt has fared little better.

Erected with great public acclaim the equestrian statue of King George III was presented to the colonies on August 21, 1770. Five days after the signing of the Declaration of Independence, when the patriotic fury of the rebelling citizens was at fever pitch, horse and rider were dragged from their pedestal and broken to bits. Most of the pieces were sent to Oliver Walcott, Connecticut's patriot governor. Melted down and, it is said, molded into 42,000 bullets by his wife and daughter, the lead was distributed to American forces throughout New England and it was estimated that

French version of the Great Fire of 1776

Fragments of the statue now in the New York Historical Society

no less than 400 British soldiers were killed by bullets molded from a statue of their own king.

In retaliation for the outrage against the King's statue, British soldiers struck off the head and arms of the marble statue of William Pitt, which stood at the corner of Wall Street and William, and was the work of the same sculptor. They left the headless trunk standing until after their evacuation in 1783. After removal to the Bridewell Yard and then to the repair yard near the Collect it was acquired by a Mr. Riley, who displayed it for many years in front of the Fifth Ward Hotel, along with a section of the saddle cloth of the King's statue, which Governor Walcott's family were supposed to have hidden in a marsh near their home. The remains were finally placed in the New York Historical Society Museum.

A house in Whitehall Slip is believed to have been the starting point of the catastrophic fire which broke out about midnight on September 21, 1776, during the British occupation of New York. Spreading with intense rapidity, it destroyed almost a third of the town before it was extinguished. The regular firemen had left with the American military forces. Most of the church and other alarm bells had been hidden or carried off by the retreating American army. Fire engines and pumps were not in good order. An increasingly strong wind, veering from south to southeast, sent the flames raging through the city.

It was a night of holocaust and terror. Enraged British soldiers, in the belief that the Americans had deliberately fired the city, killed a number of citizens. An amiable loyalist carpenter who had had a drop too much was hanged from a sign post on suspicion of trying to spread the fire. The sailors who were called to extinguish the blaze pillaged all the houses they could reach.

The British claimed that the fire had been planned with the knowledge of General Washington but Washington and the Continental Congress denied the

Equestrian statue of George III

Watchman spreading the news of Cornwallis' surrender

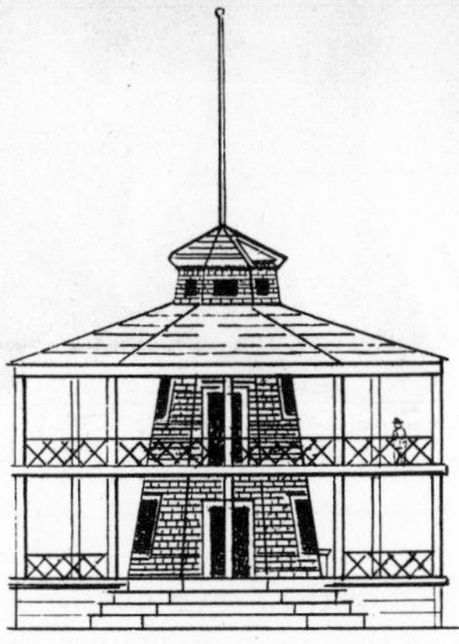

existence of such a plot. The rational and lucid account of the disaster given by David Grim and others was probably much nearer to revealing its true origin than the British accusations. Grim's manuscript notes that the fire began "in a small house (a low dive), near the Whitehall Slip." The Pennsylvania Journal stated similarly that Lord Howe's seamen who

Flag staff and "churn" at the Battery

Foot of Whitehall Street, 1830

had gone ashore for a frolic had carelessly set fire to a house on Whitehall Slip. It is possible that once the fire had broken out, there was a spontaneous attempt to spread it, but one thing is certain—that the British used the occasion as a pretext to inflict great cruelties on the population and that throughout the terrifying night they made no great attempt to distinguish between innocent and guilty.

One of the great William Pitt's closest friends, the Earl of Shelburne was the only man who knew the answer to a Revolutionary War mystery which baffled historians for more than 150 years. During the years 1769-1771 New York was stirred by a series of letters reprinted from a London newspaper and signed "Junius," which burningly opposed British policies, engendered the spirit of revolt, and were used by the revolutionary firebrand Tom Paine as a model for his own writings. But the identity of the author was never discovered. Dr. Francesco Cordasco, an Assistant Professor of Long Island University believed that the answer lay in the private papers of the Earl, who had promised to reveal the secret before his death. Last June his search for the Earl's library, lost since 1805, came to an end in a little house in Glasgow, Scotland. Among the Earl's papers, completely unnoticed by their owners, was a note in the Earl's own handwriting which revealed that the unknown "Junius" was the Earl's own private secretary, a Scotsman named Laughlin MacLeane, who had been a regimental surgeon with the British Army in the French and Indian Wars. And so, a few days before the 174th anniversary of the Declaration of Independence, the world was once again reminded of MacLeane's words of warning to a tyrant—"Whilte he plumes himself on the security of his title to the crown, he should remember that as it was acquired by one revolution it may be lost by another."

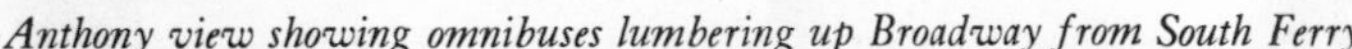

Anthony view showing omnibuses lumbering up Broadway from South Ferry

St. Brigid's Crypt at 7 State Street, former home of General Jacob Morton. Still standing

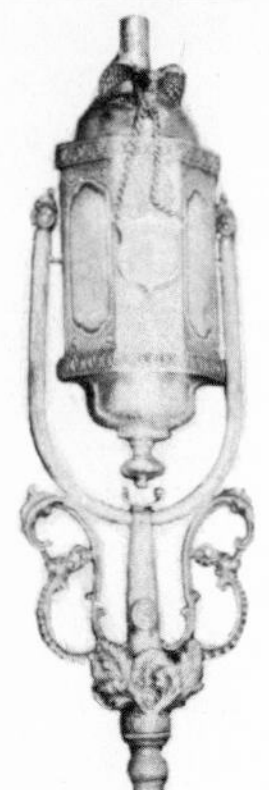

Sweet Jenny Lind

QUIET little Jenny Lind, the Swedish singer, must have been a bit startled when the steamer *Atlantic* was warped up to its berth at the foot of Canal Street on September 1, 1850. The curious, milling populace, stirred by months of Barnum's brassiest brand of bombastic blasting and blaring ballyhoo, just about mobbed her. But they liked her on first sight. Her modest and gentle mien appealed immediately in a way that folks in later years were to feel so deeply about their own "Sweetheart," little Mary Pickford.

While it was hoped that her debut would take place in the new Tripler Hall then being built in Mercer Street, that structure was not quite completed in time. Castle Garden was chosen as a substitute and the date set for September 11, 1850. The clamor for tickets would have been a scalper's delight. As it was, choice seats were auctioned at previously unheard-of prices. So great was the desire to see her that some people who failed to obtain tickets formed parties in rented rowboats and resting on their oars, listened from the river outside of the Garden during the performance.

Painting of the beautiful Jenny Lind

Her opener was a soul-stirring sensation which exceeded even Barnum's most elaborate claims. Though her voice was of no remarkable power or beauty, she was an artist to the fingertips, and her

Jenny as she appeared at the time of her Boston concert. From an old daguerreotype. Courtesy Metropolitan Museum of Art.

P. T. Barnum—He never forgot her either . . .

"Mr." Jenny Lind was a young German pianist named Otto Goldschmidt, whom Jenny married in Boston, where the photograph above was taken by Southworth and Hawes. Courtesy of The Metropolitan Museum of Art.

performance approached perfection in refinement and finish.

Little Catherine Havens wrote: "We are a musical family, all except my father; but he went with my sister to hear Jenny Lind in Castle Garden, and when she sang 'I Know That My Redeemer Liveth,' the tears ran down his face. My sister took me too, and I heard her sing 'Comin' Through the Rye,' and 'Jon Anderson, My Jo,' and a bird song and she is called 'The Swedish Nightingale,' because she can sing just like one."

Mlle. Lind's first concert brought her twelve thousand six hundred dollars but, still feeling the warmth of her reception, she donated the entire sum to New York charities, beginning with three thousand dollars for the Fire Department Fund for Orphans and Widows.

Such conduct heightened public affection throughout the land. In New York crowds followed her wherever she went, so that, in order to secure some degree of privacy, she was obliged once to abandon her hotel, the Irving House, to find refuge in more restful quarters.

On the 16th of December, 1850, John S. Giles, the treasurer of the Fire Department, received Jenny's contribution. "We are instructed by Miss Lind," wrote her lawyers, Messrs. Jay & Field, "to request your acceptance of the enclosed donation of three thousand dollars to the Firemen's Fund, and to express to you at the same time the warm interest with which she regards this charity, belonging to a city where she has been received so kindly, and affecting a body of men whose services she highly appreciates." In accepting this gift, which was "without a parallel in the annals of the Department," representatives assembled in Firemen's Hall and adopted a resolution containing the following:

"We, . . . deem it proper to express our heartfelt, sincere and unfeigned thanks to Mlle. Lind for the magnificent donation from her to the Fire Department Fund . . . which we feel that no language which we can express is capable of rendering more exalted the reputation of one whose brilliant virtue, whose deeds of Charity and benevolence of heart as well as simplicity of character have always secured to her, wherever her name has been heard, a fame whose value has no limit. In testifying our profound gratitude to Mlle. Lind for her kindness we beg to assure her that when her voice shall no longer charm the ear, her memory will be affectionately cherished . . ."

It was decided to present the resolutions to the singer in an elaborately engraved box manufactured of pure California gold, and also to present the "Nightingale" with an added gift—an original "elephant" edition of Audubon's "Birds of America"—a rather neat touch.

A committee of presentation headed by the officers of the meeting, Zophar Mills, President; George T. Hope, Vice-president; and Charles McDougall, Secretary, was appointed. Everything being ready, the delegation waited upon their "Sweet Jenny Lind" at her hotel.

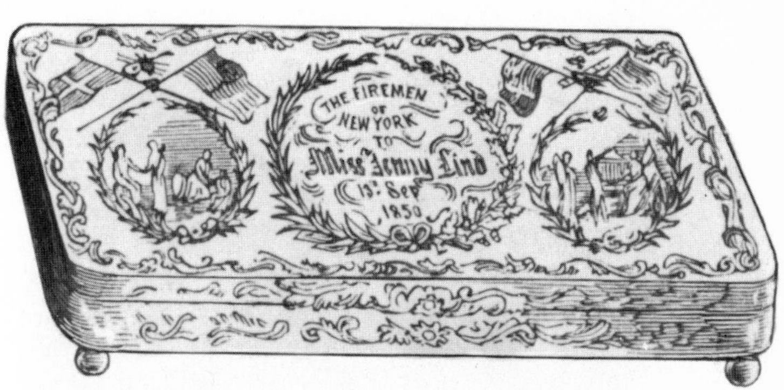

Gold box presented to Jenny Lind in September, 1850, by the firemen of New York. Courtesy of Antiques Magazine.

The generous woman, deeply touched by this courtesy and the devotion of her brave admirers wrote a few days later:

"New York, 6th June, 1851

"To Zophar Mills, Esq., President,
Fire Department, N. Y.

"Gentlemen,—I thank you most heartily for the kind wishes and warm regard evinced in the copy of resolutions that was duly presented to me by your committee. I consider and shall esteem it as one of the highest evidences of sympathy and good feeling I have ever received.

First appearance of Jenny Lind in America took place at Castle Garden, September 11, 1850.

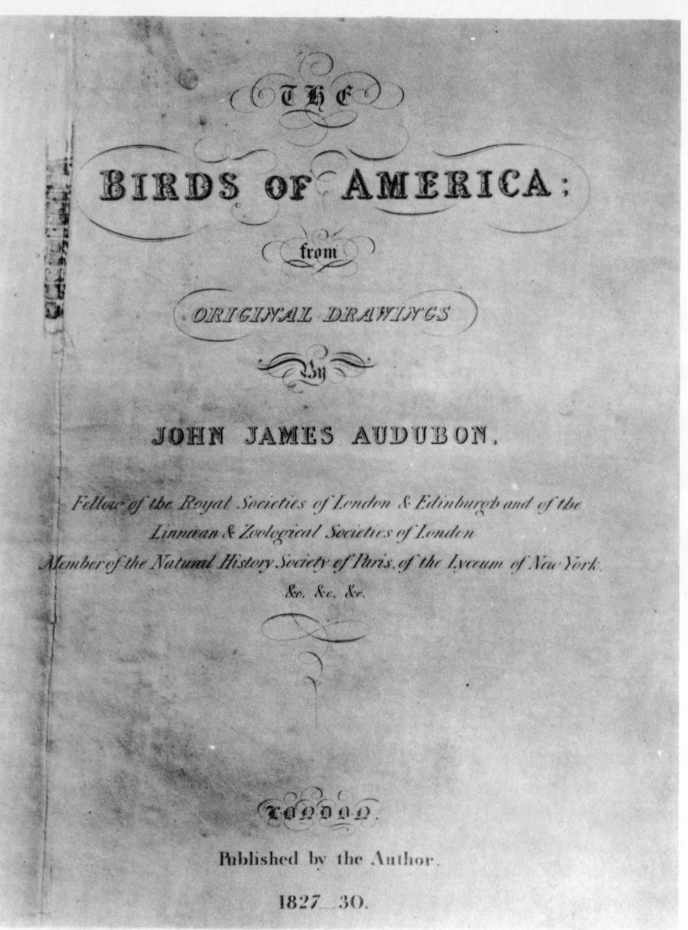

THE

BIRDS OF AMERICA;

from

ORIGINAL DRAWINGS

By

JOHN JAMES AUDUBON,

Fellow of the Royal Societies of London & Edinburgh and of the Linnaean & Zoological Societies of London
Member of the Natural History Society of Paris, of the Lyceum of New York,
&c. &c. &c.

LONDON.

Published by the Author.

1827 30.

Title page of Audubon's Birds of America

"The splendid edition of Audubon's 'Birds of America,' with which it was accompanied, I shall always look on as my most beautiful souvenir of America.

"In coming years I hope that I shall derive much pleasure from a perusal of the work, and that a better acquaintance with the 'Birds of America' will but deepen the grateful remembrance I shall always retain of the welcome I received from her people.

"I have always regarded with admiration the brave and useful body of men to which you belong—The Firemen of America—and whether in the execution of your very arduous duties, or in the general affairs of life, you always carry my warmest wishes for your success and happiness.

"I am, Gentlemen, with esteem and gratitude, yours,
Jenny Lind"

Miss Lind gave her farewell concert at the Garden on May 24, 1852.

Adieu, Sweet Jenny!

Castle Garden, where Jenny Lind made her American debut, was opened as a "place of resort" on July 3, 1824. Enlarged and improved as a concert hall, it was the scene of many important events in New York for thirty years. In 1855 it became a depot where

"The delegation waited upon their sweet Jenny Lind."

immigrants were received, sheltered, instructed and sent on to keep their destiny with the new world. A fire gutted the structure in 1876 but it was rebuilt. The Federal Government turned the property over to the city in 1891. In 1896 the Aquarium was opened to the public. Fort Clinton, the walls of which were used in the construction of Castle Garden, has only recently been declared a national monument and will be restored, as nearly as possible to its original condition to the delight of all history-minded New Yorkers.

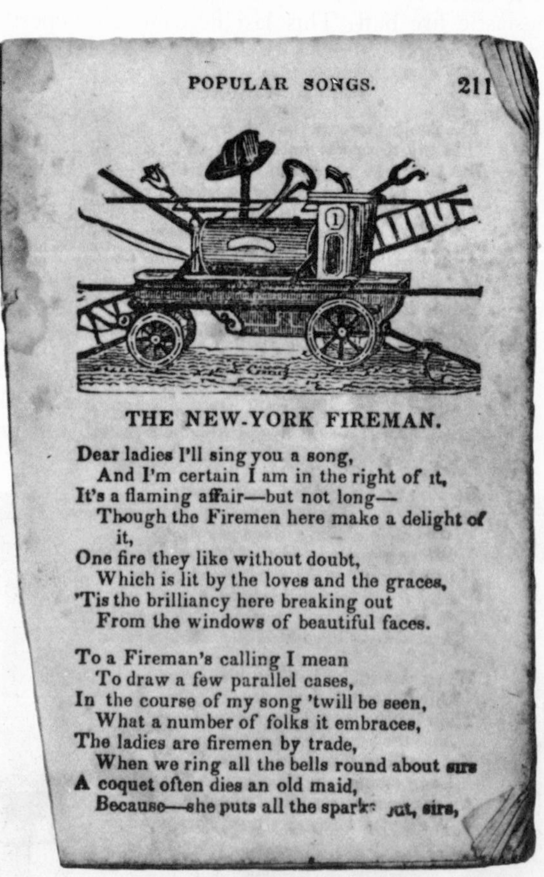

POPULAR SONGS. 211

THE NEW-YORK FIREMAN.

Dear ladies I'll sing you a song,
And I'm certain I am in the right of it,
It's a flaming affair—but not long—
Though the Firemen here make a delight of it,
One fire they like without doubt,
Which is lit by the loves and the graces,
'Tis the brilliancy here breaking out
From the windows of beautiful faces.

To a Fireman's calling I mean
To draw a few parallel cases,
In the course of my song 'twill be seen,
What a number of folks it embraces,
The ladies are firemen by trade,
When we ring all the bells round about sirs
A coquet often dies an old maid,
Because—she puts all the spark[illegible]ut, sirs,

Page from a fireman's song book

George Templeton Strong

Some men write in the hope of immortality, but George Templeton Strong was writing only for himself, and his diary* is the better for it. It is a warm, human document written in a flowing style years ahead of its time. Through it we are enabled to see the city as he saw it—the little things along with the big, the people and the way they lived. Through it, too, we get a clear picture of the man himself, a kindly, introspective sort of fellow with his regular attacks of "headaches and the blue devils," which sound like the depressing migraine headaches of today; his prescription of cayenne pepper and cream of tartar which he took when he was "moping like a sick cat," his reliance on catnip tea when he was suffering from a cold. He was an avid collector of literature and often stood on the Battery anxiously awaiting the safe arrival of a shipment of books from England.

George Templeton Strong

A graduate of Columbia University, Strong became a brilliant literary critic, a lover of music, a lawyer, a keen observer of the world about him—and an enthusiastic fire buff. This last he came by honestly, since he was the nephew of one of New York's most distinguished volunteer firemen, Benjamin Strong, who was foreman of Engine Company 13, Assistant Engineer of the Department, and one of the founders of the department's Charitable Fund, which he served as treasurer for eighteen years.

The New York in which George Templeton Strong walked was much different from the metropolis of today—but yet in some ways the same. He complains bitterly of the condition of the streets, the mud, the weather. On June 18, 1841, he says, "*June 18. Friday.* Very nice weather to be sure—for June—raining all day long like a collection of inverted Jets d'eau—the atmosphere penetrated by columns of descending water half an inch in diameter. Of course I had a grand pedestrian tour to make, and a very pleasant aquatic excursion it was. Ferry Street and Maiden Lane were positively sublime—the two gutters had coalesced . . ."

Old pump on Greenwich Street. Until its demise it was one of the few left in Manhattan

Later that year the weather changed, and we find him writing in July, "Let this day be infamous to after ages as the hottest a New Yorker ever perspired under . . . But Oh ye Gnomes and Salamanders from Archivarins Lindhorst in his flamecolored nightgown, down to the lowest of your 'lower orders' who may be revelling in the flame of my astral lamp or enjoying some other vulgar source of heat around me—impart to me some shadow of your elemental nature that I may endure the roasting night this is going to be, or of a truth there will be nothing left of me by tomorrow morning."

Croton water, soon to flow to New York, was a prime item of conversation in 1841 and we find Strong referring to it in several ways—mostly uncomplimentary. He says, ". . . the Croton water is slowly flowing towards the city, which at last will stand a chance of being cleaned—if water *can* clean it." And, later, "One consolation is that the Croton's fairly introduced into city society at last. I hope they'll keep the streets a *very* little cleaner, now."

Then, day of days, July 5, 1842: ". . . There's nothing new in town, except the Croton Water which is all full of tadpoles and animalculae, and which moreover flows through an aqueduct which I hear was used as a necessary by all the Hibernian vagabonds who worked upon it. I shall drink no Croton for some time to come. Post drank some of it and is in dreadful apprehensions of breeding bullfrogs inwardly."

The famous fountains of Croton water intrigued Strong. ". . . Took a walk uptown. Looked into Union Square. I find the fountain there is to be, as well as one can judge from its present appearance, just like that in the Park, viz. a circular basin with a squirt in the middle, and nothing more. A squirt of three or four inches in diameter and rising fifty or sixty feet will be a pretty thing—but we ought to have one or two fountains, at least, like those we commonly see in pictures where the water is carried up nearly to its greatest height in a pipe and then falls from two or three basins successively."

Later, however, he thought the "squirts" to be quite beautiful.

War was much in mind, then, too. In 1841, Strong's diary carries this dismal notation, "We're fated to

**The quotations from this remarkable diary are made with the gracious permission of Columbia University. It was presented to the university by the late Henry Taft of the legal firm of Cadwalader, Wickersham and Taft, successors to the Strongs.*

Engines drawing water from the fountain at the Bowling Green during the Fire of 1845

The old city's "white wings"

have a war, I do believe, and if we do, and if a British fleet bombards this city, I fear my library will stand a bad chance. We ought to move uptown, if for no other reasons but because No. 108 would be demolished in a twinkling by the very first broadside."

In 1845 occurred one of the worst catastrophes ever to visit New York—the Great Fire of that year. The fire started, modestly enough, in a packing box factory* on New Street. It spread, however, to an adjoining warehouse filled with saltpeter. At 3:45 in the morning of July 19, the warehouse exploded** with such violence that windows were shattered as much as a mile away. Tragically, some of the engines fighting the fire were close by and were destroyed, just at the time they were most needed. The fire then raged unchecked. Previously unpublished, Strong's diary provides us with a moving eye-witness account of the holocaust.

"Rather a notable day. Was waked at half past three this morning by a couple of explosions in quick succession that shook the house like an earthquake and must have blown me out of bed I suppose, for I was at the window before the roar had fairly died away. And the last remnants of sleep were pretty well knocked out of me by the aspect of things out of doors. The moon was shining full and bright; the dawn just beginning to show itself—and to the southeast there rose into the air a broad column of intense red flame, that made the moon look pale, and covered everything with a glare that passed every effect of artificial light which I'd ever witnessed. Didn't stop to analyze the phenomenon but hurried my clothes on, gave the alarm downstairs, and rushed out of the house as fast as possible. Hadn't far to go—the fire was in New Street and Exchange Place, and burning most fiercely. Everything in New Street as I looked down from Wall seemed withering away and melting down in absolute white heat. Only a few people assembled, and no engines visible. The explosion that had just taken place had taught people to keep at a deferential distance—shown them that fire was not to be played with. 'Bad fire,' said I to a fellow who was looking on with admiration from behind a lamppost, 'there must be a dozen stores burning.' 'O Lor a Golly,' said he, 'it's down to Broad Street already—this here a'nt nothin'.' Whereupon I went home to report the position of affairs, as I'd been requested to do.

"Came back to the scene of action, and seeing that all Broad Street on both sides from about No. 20 down was one grand solid substantial flame, most glorious and terrible to look at, and that the two or three fire companies on hand and hydrants open were likely to make about as much impression on it as His Honor the Mayor would have made by a singlehanded attack on the conflagration a la Capt. Gulliver, I pelted home very expeditiously to tell my father that the days of '35 had returned and that he'd better turn out and see the sport. When I got back the fire was crawling down Beaver Street and Exchange Place towards William quite unchecked, and then I came into Broadway just in time to see the iron shutters on the stores three or four doors south of Exchange Place beginning to grow hot, and the Waverley House beginning to disgorge occupants, furniture, and smoke from every convenient outlet—they said 'twas already on fire in the rear, but I didn't believe it, and thought it could have been saved had anybody made much exertion to save it. One of St. John's people whom I knew met me and told me awful stories of the Croton giving out and the fire having it all its own way. Soon a few snaky little curls of flame made their appearance in the stores on Broadway—the Waverley, abandoned to its fate, was burning slowly down, story after story, beginning at the top—fire appeared in great activity lower down

Stokes says the fire started in Van Doreon's sperm oil establishment at 34 New Street.

The jet in City Hall Park

Primitive view of the Croton Water Celebration in 1842. The City Hall jet is in the background and a goose-neck engine, drawn by six white horses, is in the foreground

Union Square and its elaborate jet, circa

Broadway and the engines were playing on the west side of Broadway where the house fronts were hissing hot already, and I began to consider where the fire was likely to stop, in a very serious kind of way, as a matter wherein I should soon have some personal interest. And I believe I went home and reported my apprehensions.

"Indeed it was shocking to watch the fire at this point—building after building taking fire, not in regular order, but as they caught in the rear, from New Street, where it was raging among stores and carpenter shops. Saw Cram and walked about with him, and at last went on top of one of the new Wall Street stores where his office is, and where we had as good a view of the burned and burning district as the smoke would allow. This was at about *seven*, and I then came to the conclusion that the fire was beginning to find its match, and that the worst was over. There was a fine show of the ropy thick black smoke that indicates a fresh fire, to the east, with beautiful flashes of red flame playing up through its columns and lighting them up with tints of purple and dark crimson, but this we took for one or two stores 'going' in Beaver Street.

"On the west we could make out some fire, but it seemed to be the Waverley and buildings to the south of it. We could look down New Street and Broad, and the fire had clearly burned itself out, there. Everything between Broadway, and Water Street, where active combustion was still going on, was covered with a uniform cloud of dim smoke which prevented our discovering that there was still terrible mischief in progress to the south, and enabled me therefore to go home and eat a comfortable breakfast. Came out directly afterwards and found out my mistake: the fire had reached the Bowling Green—all the east side of Broadway from Exchange Place to Whitehall Street was burned or burning. The Adelphi Hotel was a magnificent sight, blazing from roof to cellar.

"At the south end of Broad Street the sight was grand. Everything was going down before the fire and the wind rising. One side of South William Street burning fiercely and every prospect of its crossing the street. Everybody in Water, Front, Pearl, and the other streets about those parts moving out in frantic haste—the fire will go from river to river, a sure thing in all men's mouths. Stone Street and the narrow streets to the north of it, absolutely impenetrable, arched with fire, and the throngs of people in the streets all working for their lives, carrying off goods, rushing about with account books and papers and hurrying back for fresh loads—the indications of desperate terror and haste in every store one passed: seen in the boxes and barrels that were tumbling out of doors in such utter recklessness—the universal consternation that prevailed made even the streets that were yet untouched by the fire most exciting and rather perilous places.

"Went once or twice to the office, where my father was mounting guard, and returning by a circuit on the east side of the fire. Saw with dismay on reaching the Bowling Green that it had *crossed Broadway* near Morris Street. Couldn't get through the line of police to see exactly the extent of the mischief, but went

A view of the Fire of 1845 from the corner of Broad and Stone Streets

home and to Wall Street two or three times before I could get all parties possessed with the very disagreeable fact. Very disagreeable it was, for the wind was getting round to the south—the fire would probably soon cross Morris Street and the houses on the west side of Broadway, thoroughly heated by the fire opposite them, would soon communicate it to one of the stables on Lumber Street in the centre of the block running parallel with Greenwich and Broadway. Nothing of course could be done to stop them from

****Francis Hart, Jr., who was on the roof of a building blown out from under him, sailed on the rooftop to the street four stories below and escaped without injury. Mr. Hart's experience, as related by himself under oath, was certainly remarkable:*

"I was at the fire on the 19th instant; and was with the pipe of No. 22 on the rear of the fourth story of the chair factory in Broad Street, when that building took fire. An alarm being given, preparations were made to take down the pipe. I remained to light down the hose, and when I undertook to go down the flame and smoke were so great as to prevent my descending, and I went on the roof of the chair factory. I went along from that building to the corner of Broad and Exchange Streets, breaking each skylight as I proceeded over the roofs, but found no stairs leading from such skylights. Finding myself thus on the third building from the chair factory, without any means of getting down, I sat in the scuttle. I did not then consider myself in any danger. I had been there about five minutes when I heard the first explosion—a species of rumbling sound—followed by a succession of others of the same kind. The gable of the house next to the corner shook with the first and each successive explosion, so that I had prepared myself, if it threatened to fall, to jump through the scuttle of the corner house. After the small explosions the great explosion took place, the noise of which seemed to be principally below me. I perceived the flames shooting across the street. I felt the building falling under me, and the roof moved around so that a corner of it caught in the opposite side of Exchange Street, and was thrown off into Exchange Street, but without any serious injury to my person. As far as I could judge, the whole roof that I was on moved in one piece, and the walls under it crumbled down beneath it. I think there were some fifteen or eighteen small explosions. I could see our engine from the roof I was on, and know that the explosions occurred in 38 Broad Street. None of the explosions, before the great one, came through, or disturbed any of the roofs of houses in Broad Street."

Fire Bell

going, and they would make short work with the east side of Greenwich Street up to Rector, which would bring the fire into most unpleasant closeness to No. 108 Greenwich. Got thro' the line of police by special favor of one of them and went down Broadway. Five houses, beginning at the south corner of Morris Street were burning, among them Ray's and Brevoort's, old acquaintances of mine. Never shall forget the aspect of things there—the street wholly deserted, save by two or three firemen; all one side of it a mere chaos of ruin and smoke and flame still flickering over the wreck and here and there a single front wall still standing with the background of smoke visible through its naked windows—and on the other side these buildings on fire and quite past saving. The heat was too great to stay there very long, so off I came. Everybody in Greenwich and Washington Streets as far up as Rector moving out in hot haste, the crowd confusion and panic worse if possible than on the east side of the City.

"At about half past ten I thought the crisis had come, for the southeast corner of Greenwich and Morris '*took*' on the roof. But by dint of cutting off burning fragments and sending a stream into the building it was stopped, and about the same time the five houses on Broadway having burned themselves out without communicating the fire on either side to any great extent, the danger was pretty much over—to my great satisfaction, for the last hour had been one of more excitement than was altogether pleasant.

"By half past eleven the fire was stopped at all points—having burned near 300 buildings and done only not quite so much damage as the great fire of '35.

"*July 20. Sunday* . . . Not much thought of or talked of but this fire. Everyone still rigorously excluded from the burnt district by a cordon of the National Guards—very sensible arrangement. Much more so than Havemeyer's first move, in ordering out a troop of horse; a specially fit arm of the service is cavalry for checking a conflagration. Not knowing how to charge the burnt district, feeling that they ought to be doing something on so momentous an occasion, they merely made themselves ridiculous by rampaging about on the sidewalks and putting themselves in everybody's way.

"The New Police, by the by, did themselves infinite credit yesterday—very active and resolute, and what's remarkable, very civil and forebearing at the same time.

"Why were no buildings blown up yesterday? A few kegs of gun-powder judiciously ignited at five or six o'clock in the morning would have saved millions, and at that hour it was most apparent that if no extraordinary means were used the loss of property was going to be tremendous. One store in Stone Street was blown up I believe by the owners. (It was not—21st.)

"As for the explosion that caused all this mischief by spreading the fire instantaneously all around the building where it took place, is attributed to saltpeter, which explanation I shan't believe till it's pretty well authenticated.* Nothing but gunpowder could have made such havoc, torn to pieces half a dozen buildings, shattered half the windows in the Exchange, the plateglass in the windows of Griswold's buildings, and even the window frames in Atwater's place at the head of Wall Street. It roused John Anthon's people out of their beds at Staten Island and was heard at Rockaway by Fox so distinctly that instead of spending the day there as he'd intended, he came to the city in all haste to find out what was going on. Saltpeter would doubtless burn with a great fizz and a sort of quasi explosion, and be a very nasty ingredient in a conflagration, but it would hardly do such damage in whatever quantity it might be stored, nor would it go off with a couple of sharp, well-defined reports as this material did. If it was gunpowder, the people who stored it ought to be *hanged*. The heavy loss of life (extent not yet exactly ascertained) caused by their recklessness surely brings them within one of the common law definitions of Murder, and even by our Revised Statutes I think an indictment for some 'degree' of manslaughter could be maintained against them.

"Walked around the ruins with my mother after church, and looked at them a little tonight. The spectacle is sufficiently deplorable; but things might well be much worse. Half an hour before the fire was finally brought under control I could see no very obvious reason why it shouldn't go clean through from river to river. There had been three fires and two false alarms in the course of the night, the firemen seemed absolutely worn out; at no one point, except on the north line, did any impression seem to be made on it—the only points indeed where its progress was contested by anything that looked like an adequate force or with much appearance of energy were on Beaver Street and Exchange Place where it was advancing on William Street. There the fire was furious—the large stores packed with drygoods were vomiting masses of flame, and though the battle was most gallantly fought, and the firemen were giving ground only inch by inch, things looked desperate in that quarter.

"Farther to the south, toward the lower end of Broad Street the prospect was less encouraging if possible—the firemen were far from being strong-handed and seemed sadly in want of some systematic plan of operations. Many were wasting water on buildings that were behind the advances of the conflagration, and already past all salvation, while they left the head of the column that was moving down street to (exert) a force that was too weak to check its progress a moment. The appearance of the fire at this point, among the irregularly built houses, many of them cabinet-makers' establishments, was very picturesque, more so than among the uniform lines of

**Strong apparently was wrong in this opinion. It was pretty well established that saltpeter was the guilty agent. (Crooker & Warren's storage house for saltpeter.)*

Speaking trumpet

View of the explosion at Broad Street during the Fire of 1845. *See footnote explanation at the bottom of page 77.*

stores in Exchange Place and elsewhere—it came steadily down the street in superb style, with a crackle like an irregular fire of rifles and musketry, and a dull, monotonous roar underneath that sounded very much in earnest, and on the whole, when I left that point at a little before ten, there was no indication of a successful defence there.

"On Broadway the signs of the times were as bad as bad could be—the fire had got into Whitehall Street and was bound for the south in a determined way—a little farther progress would bring it to bear on the stables in the rear of Whitney & Co. and then it would make wild work with State Street and everything else, and the forces here seemed paralyzed and inefficient. On the west side of the street there was not a quarter part of the force that was required, and I fully agreed with the man who told me that everything in Morris Street 'would be enveloped' in a few minutes.

"So 'twas with almost as much surprise as satisfaction that I found the fire brought to a stand so soon afterwards. The fact was that the extent of the line was such that sufficient strength could hardly be brought to bear on any one point, but as soon as it *was* mastered in a single spot, reinforcements came to the assistance of the rest of the line, and they gained on it very fast. By ten o'clock they'd probably got everything safe on the north side where the wind helped them, and could send very valuable assistance to the other points.

"*July 22. Tuesday.* Very heavy thundershower last night that cooled and freshened the air a little—rarely saw lightning so continuous and vivid. It has just put out the embers of the great fire, though, as I know by a personal inspection just made. Fine sight rather, especially looking east from Broadway with the light of the full moon contrasting with the glare of the fires that are still burning by the score.

"There's another alarm of fire! The toll of the Hall bell has quite a dolorous sound to my ears just now. I've seen enough of conflagrations to last me for some time.

"Shall I turn out in pursuit? I'll give two minutes for the chance of a false alarm and see if the toll ceases. No, it doesn't, and there goes the quick ringing of St. Paul's. I'm off . . ."

This was the Great Fire of 1845, told by one who was there, who saw it with his own eyes and reported it while the smoke was still in his nostrils and who, apparently, couldn't get enough.

George Templeton Strong's "Aunt Olivia" lived around the corner from the Strong residence at 2 Carlisle Street in a building similar to these on Carlisle Street shown above.

Old doorway at 6 Carlisle Street

A view of 106, 108 and 110 Greenwich Street in 1845. George Templeton Strong's residence was at 108. Greenwich Street at this time was lined with handsome private residences which were occupied by many of the city's foremost merchants.

Road To Greenwich

GREENWICH Street, which ran along the shoreline of the Hudson River, was the route most frequently used in early times by those travelling to Greenwich Village from the lower part of the City. One of the principal streets in Manhattan for many years, the old road extended from the Battery to the Gansevoort Market just south of West 12th Street and traversed the section occupied by many of the city's cabinet-makers and the German segment of the town.

One of the early landmarks of the city was Mesier's

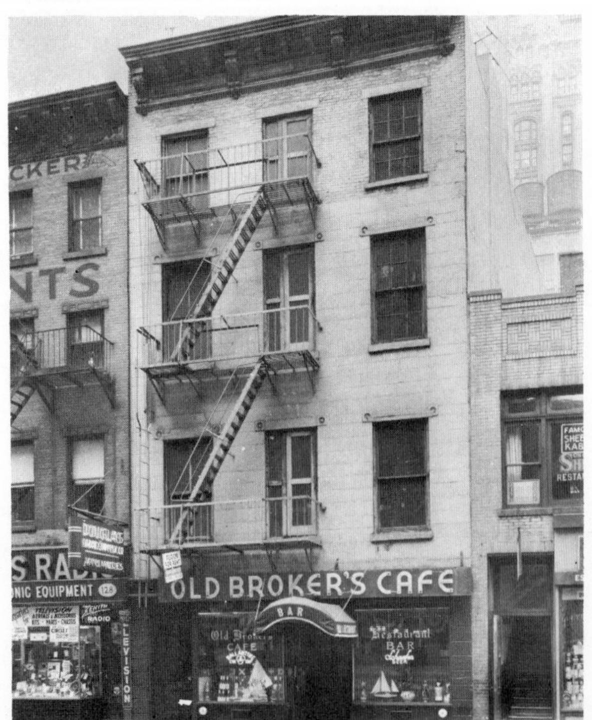

Edgar Allan Poe lived in this building (130 Greenwich Street) which was built in 1809

windmill which stood north of Greenwich Street near the present Liberty Street. Constructed in 1686, and demolished nearly a hundred years later, Mesier's mill can be seen in most early views of the city taken from the Hudson side and it was almost as well known to shipping and river craft as the lighthouse on Greenwich Street. The site of the latter, incidentally, was later occupied by the Lighthouse Tavern. In 1946, Joseph Costa, a nurseryman whose shop lies near the site of the old tavern on Greenwich Street, discovered several fine pieces of early delft ware, stone bottles and other Dutch relics.

George Washington Strong, father of diarist George Templeton Strong, lived at 108 Greenwich Street with his brilliant son. The elder Strong was an attorney and brother of the famous Benjamin, who has been described elsewhere as a noted fireman and foreman of Engine Company No. 13. During the years he was connected with the Fire Department, his sons and daughters always saw him off to a fire, running to find his fire cap at the sound of the alarm. In the last few years of his life he was always disappointed when a night alarm rang and he was not called from his bed.

The first elevated railway constructed in New York City was built on Greenwich Street in 1869-70 extending from Battery Place to 30th Street. An unofficial run was made in 1867 by Charles Harvey, originator of the plan on a half-mile line of experimental track which ran from the Battery to Dey Street. The line was built on a row of iron supports and consisted of a single track, the cars being operated to and from their destination by a cable. Locomotive power was added in 1871.

Mesier's windmill, one of old New York's most famous landmarks

Charles Harvey on his experimental run on the first elevated railway in New York, 1867

Northwest corner of Greenwich and Vesey Streets, circa 1900

The construction of the Curb Exchange in 1921. This was the former site of Planter's Hotel.

20
WASHINGTON
20

Down By The Hay Scales

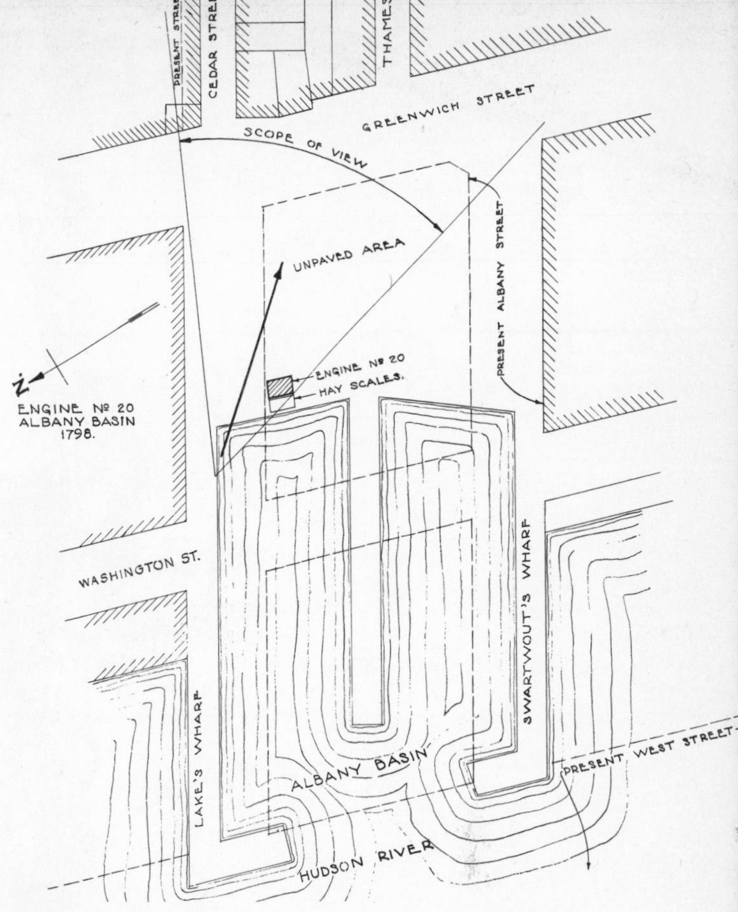

All of the present waterfront area of lower New York is formed by filled-in land. Indeed, the original shore line of the Hudson at several points extended beyond the present line of Greenwich Street.

In the middle of the 18th century, Ellison's Dock stood at the foot of Cedar Street and Comfort's Dock at the end of Thames Street. This slip served the river shipping which brought produce and hay from the farms of the Hudson River Valley. Most of the produce in these cargoes was used to supply the Oswego Market which stood up the hill in the middle of Broadway between Cedar and Liberty Streets. The hay was weighed by public scales which were out at the unloading point and distributed from there. The improved Albany Basin which replaced the old docks and slip in 1791 continued to serve the "Old Swago," then at the head of Maiden Lane, until the market was removed in 1811.

The hill from the river front to Broadway, up which present-day hurrying commuters huff and puff, was once even a steeper grade. Washington Engine Company No. 20, which was associated with this neighborhood throughout most of its career continually complained of the difficulty encountered in hauling its engine up the same hill.

The men of Engine 20 ran a number of engines at different times in their history—a goose-neck, a chain-box engine, a piano engine, a crane-neck engine and finally, a steamer which the Company had until the Paid Department took over in 1865. It was in the house of Engine 20 in Cedar Street that the invention of William Gleason, formerly of Engine 3, was first put to practical use in 1863. This was a preheating system for keeping the water hot in the boilers of fire engines so that they were constantly ready for quick firing to obtain steam in the event of an alarm.

The Company, which had its social side, was honored by having a Quadrille dedicated to it, the "Quadrille Fire-Set." This composition featured a fire-bell as part of the orchestration and by the use of Greek Fire in the ballroom created a rather uncomfortably realistic effect of fire, adventure, heroism—the high points in a fireman's life. After the disbandment of the Volunteer Department, the members of this company formed "The Washington Association" and continued to meet every three months at Old Tom's Restaurant, at the corner of Thames and Temple Streets.

In the view of 1798, from left to right, are seen the Middle Dutch Church cupola, the City Hotel, the Van Cortlandt residence and "Sugar House," the cupola of the City Hall on Wall Street, the top of the Wall Street Presbyterian Church, Trinity Church and the mansion of John R. Livingston, at the extreme right.

In 1836 the First Ward Hose Company No. 8 was organized at 74 Cedar Street and after 1859 was located at 39 Liberty Street. This company, known as "Old Cedar," ran a four-wheeled hose carriage which carried the "most beautiful lamps in the country surmounted by miniature solid silver cedar trees."

A century later, Cedar Street still retains some flavor of its past with several old brick buildings between Church and Washington Streets still standing. These include Ye Olde Chop House and the Firehouse Tavern at 126 Cedar Street which originally housed the machines of Engine No. 20 and is undoubtedly the oldest firehouse in Manhattan, having been built in 1819.

The boys of Engine 20 hauling their machine up Cedar Street.

In this view of 1829, 20's house (89 Cedar) is at right.

FERRY HOUSE

THE FOOT OF CORTLANDT STREET

OIL PAINTING — *Circa 1825* — ARTIST, UNKNOWN

REPRODUCED BY COURTESY OF THE NEW YORK HISTORICAL SOCIETY

This view, reproduced in color for the first time by permission of the New York Historical Society, shows the Cortlandt Street Ferry House and the Northern Hotel. It is of particular interest because of the care of the artist in depicting the vehicles, vessels, and costumes of the times. The ferry boat shown is of the type used in the earliest days of steam propulsion. The verdant shore of New Jersey on the other side of the Hudson River shows in the background. The Northern, a well-ordered establishment and an excellent eating place, was built by John Wilkins and opened May 20, 1809. It was one of the principal hotels in the city during the first half of the 19th century. Not only was it regarded as a rendezvous for owners and captains of both steam and sailing vessels during lay-overs, but it became very popular as a travellers' resort due to increasing traffic from the ferries as the years went on.

This vicinity is, perhaps, more familiar to generations of New Jersey ferry commuters than any other part of New York. There were regular ferry schedules since 1764, by rowed barges and, later, by horse treadmills (team boats), then steam. The first regular steam ferry to New Jersey, the "Columbian," was owned by John C. and Robert L. Stevens and took off from the Cortlandt Street slip to Paulus Hook (Jersey City) on July 25, 1812. The first regular Hoboken steam ferry was run in 1822 from the Corporation Wharf near the Washington Market. This service operated from the foot of Vesey Street, Manhattan to the foot of Vesey Street, Hoboken.

To many of the perennial commuters, who voyage most of their days to and from the shores of Manhattan, the airy, all-too-brief crossing at least provides a salty whiff of romance or adventure. The sight of graceful, soaring gulls, of incoming tramp steamers washed by the seven seas, the trim liners, the fishing schooners, both the sight and smell of strange cargoes, the sounds of romantic names and foreign tongues; and at other times, the beckoning thump of river ice, the eerie cloak of gray fog pierced only by the weary yet tense droning of horns and the short, sharp clapping of bell-buoys combine in various proportions to add a flush of adventure to lives too often bounded by monotony.

Once heard — who can ever forget the cheerful, chattering ring of pawl and ratchet proclaiming the safe crossing of a ferry boat and the taking up of the mooring lines to hold it snugly in its slip.

The Maiden Lane

'T MAAGDE Paatje, the Maiden's Path, was originally in a valley which Maiden Lane now traverses. The path followed a charming rivulet which ran through the little vale towards the East River. The stream, fed by a living spring, came tumbling down over the rocks, forming a series of pools. It descended on the north side of the valley as far as the spot now crossed by William Street. An early map shows it forming a fair-sized pond at that point, from which it made its way to the East River. The entire area was one of pastoral beauty and was one of the most picturesque spots in the city.

Maiden Lane undoubtedly got its name from the practice of the "goude Dutch vrouws" and their daughters who came through the fields from the old town to wash their family linen there. The fine run of spring water and the flat, smooth stones were excellent for this purpose and the adjacent grassy slopes were ideal for bleaching and drying.

The supply of water provided by the pond at William Street was probably the reason for the establishment of the Maiden Lane tan pits. These extended on the north side of the Lane between Smit (William) Street and the Rutgers Hill (Gold Street). For decades after the removal of the pits and the filling in of the pond, the intersection at Maiden Lane was a wet morass crossed by a "vlonder," or footbridge.

Although the families of some of the city's leading merchants and officials settled here rather early, this did not immediately result in bringing about public improvements. As late as 1820, the residents in the neighborhood asked the Common Council for a new bridge of flat stones to be laid across Maiden Lane from the westerly sidewalk of William Street to replace the old footbridge.

After the Revolutionary War, Maiden Lane became one of the leading business streets of the city and its dwellings furnished lodging and homes for a number of cabinet members and other government officials. Here also, the United States Gazette, the leading news vehicle for the newly formed government, was published, and in the block between William Street and Nassau, Thomas Jefferson made his home, directly across the street from a senator and two representatives from Virginia.

Prominent among the silversmiths, who in early times occupied Maiden Lane (many silversmiths and dealers still do) and whose marks are now eagerly sought by collectors, was Joel Sayre, who, in 1816, was located at 59 Maiden Lane, the present site of The Home Insurance Company building.

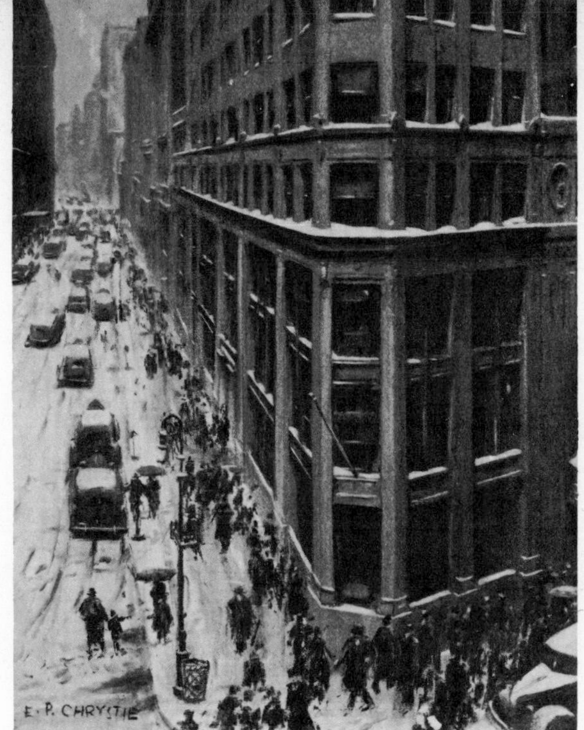

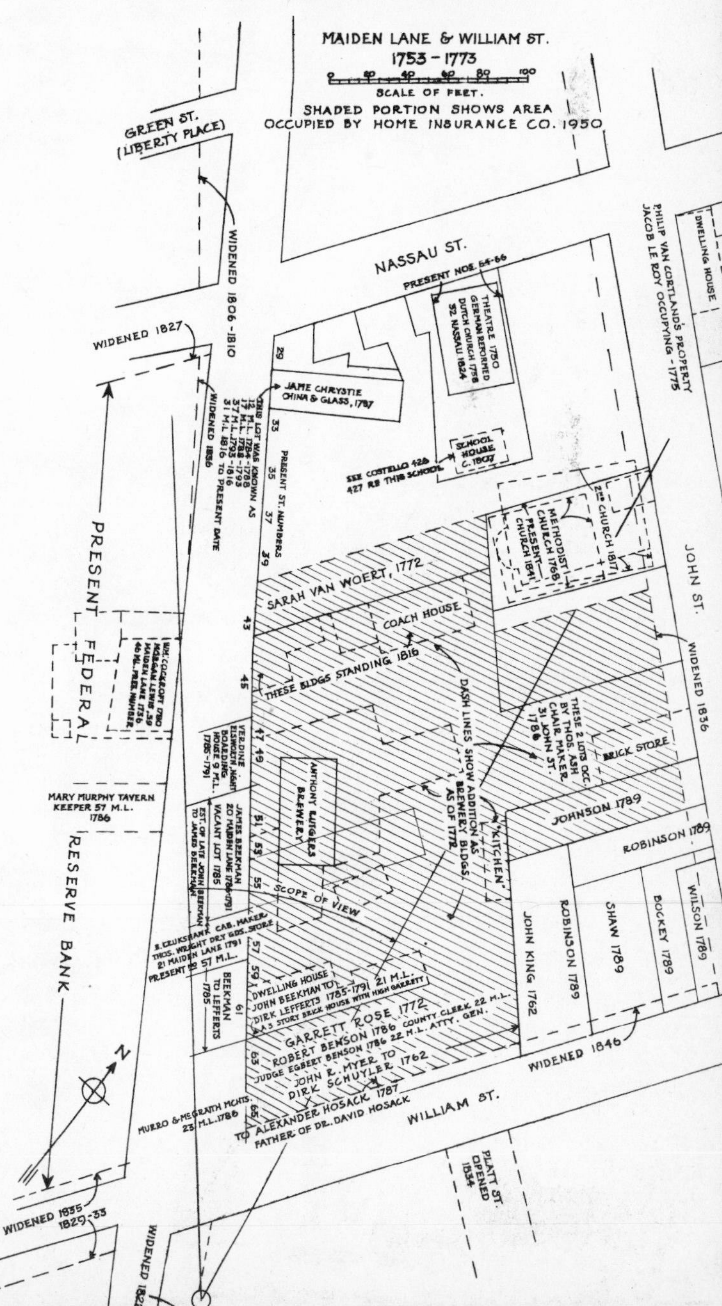

In the view at the left, the large building setting back from the Maiden Lane building line and near the westerly end of the present Home Insurance Company plot was Rutgers' brew house. This building, originally built by Anthony Rutgers, was used by the British as a storehouse during the occupation of the city in the Revolutionary War. At the right is a modern view of Maiden Lane and William Street showing the building of The Home Insurance Company.

EAGLE 13 ENGINE
THOMAS STEVENSON.
HORSE SHOEING

Stephenson's street view of Maiden Lane between Nassau and William Streets.

Lucky Thirteen

THERE was a time when the strand on the East River was so near to Maiden Lane that one of its earliest residents, who lived at the corner of Maiden Lane and Pearl Street, used to keep a boat tied to his stoop. This was probably the house (on the northwest corner) built in 1641 by Captain Lourens Cornelissen Vanderwel, who described himself in documents by the imposing title of "Skipper under God of the ship the Angel Gabriel." Through failure to improve his land, he forfeited a portion of it to one of the most romantic figures in the old colony, an Indian trader called Sander Leendertsen. The owner of this good Dutch name was really a tempestuous Highlander, Alexander (Sandy) Lindesay of the Glen, a Scotsman born and come a-venturing to the new world from Inverness. He was descended from Sir

Interior of "Pig's Cheek," coffee house at foot of Maiden Lane, 1867

David Lindesay, a 16th century Scottish poet who bore the office of the heraldic King-at-arms under James IV, and who is described in Sir Walter Scott's *Marmion*.

This Scot retained for many years the fine stone house which he had built on what was once the Skipper's garden. The remains of his well, which lay 50 feet northeast of the house, could probably still be found under the building which now stands at 215 Pearl Street.

In 1665, Sandy, or Sander, left New Amsterdam to become one of the pioneers of a new settlement—Schenectady—where he died five years before the massacre of this community by the French and Indians in 1690.

In the view, looking up Gold Street from Maiden Lane, showing the neighborhood as it appeared in 1790, the large building with the gambrel roof was the old brewery built by Harmanus Rutgers. This part of the property which he bought in 1720-22 thereafter became known as Rutgers Hill. On the east side of Gold Street appears the house of Daniel Bloom Coen and the stable of Thomas Pearsall. In the distance, above the blacksmith shop, is the tower of the North Dutch Church at Fulton and William, which remained steepleless until 1820.

Few have heard of the Screeching Woman of Maiden Lane. Long before the days of street lighting, when this deeply shaded street was still in a primitive neighborhood, the Lane, at night, was a rather forbidding thoroughfare. Certain mysterious goings-on had the whole town "gonzen"* over guarded whispers of an evil terror that strode abroad in the darkest pit of night to terrorize the residents of the area. The horrible shriek of this night-strolling female pariah of the shadows was enough to freeze the very blood

Good Dutch for "buzzing."

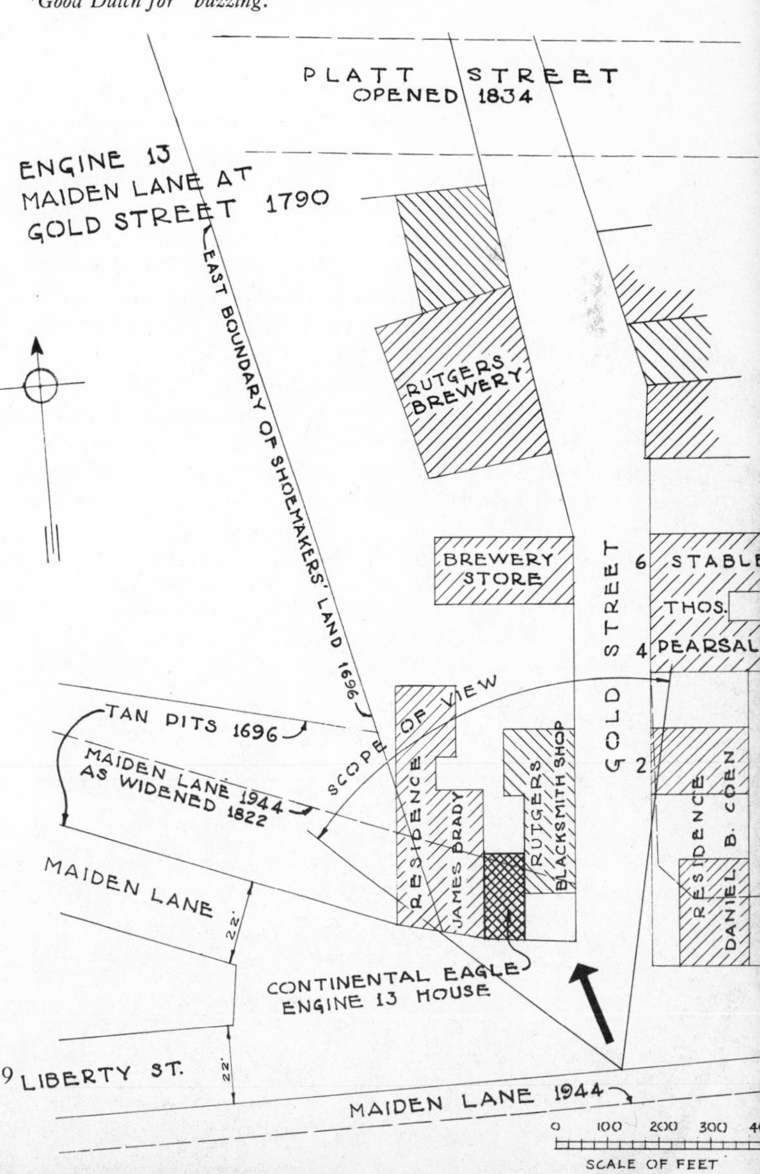

57-59 Maiden Lane, circa 1844, showing the shops of John Bucklass, tailor, and Gustavus Meyer, toy manufacturer.

of those who heard it. Watson, in reporting the vivid recollections of an elderly citizen in 1828, wrote:

"... she was a very tall figure of masculine dimensions, who used to appear in flowing mantle of pure white at midnight, and stroll down Maiden Lane. She excited great consternation among many. A Mr. Kimball, an honest praying man, thought he had no occasion to fear, and as he had to pass that way home one night, he concluded he would go forward as fearlessly as he could; he saw nothing in his walk before him, but hearing steps fast approaching him behind, he felt the force of terror before he turned to look; but when he had looked, he saw what put all his resolutions to flight—a tremendous white spectre! It was too much!—he ran, or flew, with all his might till he reached his own house by Peck's Slip and Pearl Street, and then, not to lose time, he burst open his door, and fell down for a time, as dead! He, however, survived and always deemed it something preternatural... The case stood thus when one Capt. Willet Taylor paced Maiden Lane alone at midnight, wrapped like Hamlet in his 'inky cloak,' with oaken staff beneath. Bye and bye, he heard the sprite* full-tilt behind him, intending to pass him, but being prepared, he dealt out such a passing blow as made 'the bones and nerves to feel,' and thus exposed a crafty man."

**The "apparition" was created by a little man with a white cloth draped over a tall wooden frame.*

Corner of Fulton and William Streets, circa 1870. Note early street lamp, book stalls and "bier" wagon.

The "screeching ghost of Maiden Lane"

Brass
gnal torch,
lated 1823

On the grounds of the old Tan Pits, next to Tom Stevenson's blacksmith shop, at the present northwest corner of Maiden Lane and Gold Street, stood the firehouse of Continental Eagle Engine No. 13, in 1790. To the Minute Book of this company, the earliest on record in Manhattan, we are indebted for a great deal of information on the customs and manners of the old fire laddies, as well as early fires and fire-fighting methods.

The first use of a real suction-pipe seems to have been made by Engine No. 13 at a fire in a ship at the foot of Pine Street. The minutes of the company for May 29, 1806, make special and proud mention of the fact that the engine "played by the means of the suction."

One of the famous members of Engine 13 was Benjamin Strong, who was the uncle of diarist George Templeton Strong and an associate of many of the city's great. He was president of the New York Sugar Refining Company for twenty-two years, president of the Dry Dock Company, one of the founders of the first Bank for Savings in New York City and president of the Seamen's Savings Bank for sixteen years.

Zophar Mills, a foreman of Lucky Thirteen, was thought by many of his contemporaries to be the most daring, able and brilliant fireman in old New York. Many of his exploits were remembered for decades after the Volunteer Department had been disbanded. New York's first modern fireboat, built in 1883, and then the most powerful in the world, was named the Zophar Mills in honor of this hero. Assigned as Engine 51, this remarkable vessel served in the Marine Division of the Fire Department until 1934.

Mills' energy and endurance were phenomenal. In a single night he would attend a meeting, leave for the ballroom and show his less adept buddies how to dance, run with his engine to a conflagration and then return and dance until morning.

Benjamin Strong

At a fire in Pearl Street in 1834, which Lucky Thirteen had cause to remember for many a day, Zophar* was buried beneath a falling wall, but as he himself explained later, "crawled out with only the loss of my cap." Two of the younger members of the company, F. A. Ward and Eugene Underhill, were not so fortunate and were crushed to death by the same wall. The handsome monument dedicated to them still stands in James J. Walker Park (the former Trinity Parish Cemetery at Hudson and Clarkson Streets) in Greenwich Village.

Hat fronts worn by Danny Donovan and James Brice, members of "Lucky Thirteen"

Memorial still standing in James J. Walker Park, dedicated to heroic members of Engine 13

**Adam Pentz, for several years President of the Fire Department, described Zophar Mills as follows: "Where the smoke was the thickest and the fire hottest, there he was. I don't believe there has been a fire in forty-five years that he has not been to. Even now he is like the old war-horse—as soon as he smells battle, he is off to it. He is the fire-king. He is the cap-sheaf. As the boys say, he 'takes the rag off the bush.' He is a wonderful man, and the truest man that ever breathed the breath of life."*

Mills was one of the founders of the Exempt Firemen's Company and Assistant Engineer of the Volunteer Fire Department from 1838 to 1842.

Lamp used by Engine 13. Now in H. V. Smith Museum

The Eagle on the Fly

THE first home of Eagle Engine Company No. 13 was close by the Fish Market, near the ferry stairs. This market, located just below the Fly Market at the foot of Maiden Lane, was among the foremost in the city and many a Long Island oysterman operated his own stall there.

On January 18, 1770, the Fly Market and the vicinity of Golden Hill were the scene of a serious skirmish between British soldiers and American patriots. Provoked by the 4th Liberty Pole which the Sons of Liberty had erected in the Common near Montagne's Tavern which stood on Broadway just above Murray Street, members of the 16th Regiment attempted to blow it up with gunpowder on January 13th. Several days later, the soldiers cut down the pole, sawed and split it in pieces and piled it in a heap before Montagne's door.

The next morning, 3000 infuriated citizens solemnly resolved to treat all armed soldiers found in the streets after roll call as enemies. The soldiers' answer was to print a handbill in which they called the colonists "robbers," "traitors," and "rioters," and dared them to carry out their threats. Isaac Sears and Walter Quakenbos, two patriots, attempted to stop soldiers posting the bills. One of the soldiers drew his bayonet and was knocked unconscious by Sears, wielding a wicked-looking ram's horn.

Other soldiers, however, ran for reinforcements and before long scattered fights were raging in the vicinity of Golden Hill and the Fly Market. At last the soldiers were driven back with clubs and they were finally ordered back to barracks by the mayor and a party of their own officers, but many patriots had already been wounded and a sailor run through the body and killed.

Skirmish between British soldiers and American patriots at the Fly Market, January 18, 1770.

The fighting did not subside until the next day, after a party of soldiers fell upon the mayor, but in a fierce struggle were driven off by the Liberty Boys. Thus, two months before the Boston Massacre, was the first American blood shed in the struggle for liberty.

The men of Eagle Company were extremely proud of their fine engine and honored her for her performance at fires, praising the machine rather than their own courage. "No. 13 did her part this night," is an entry found again and again in the early records of the Minute Book, begun in 1791.

A section of the Fly Market, corner of Front Street and Maiden Lane, 1816.

The tablet commemorating the fire of 1835 (now in H. V. Smith Museum) was erected on Pearl Street House as may be seen in early billhead.

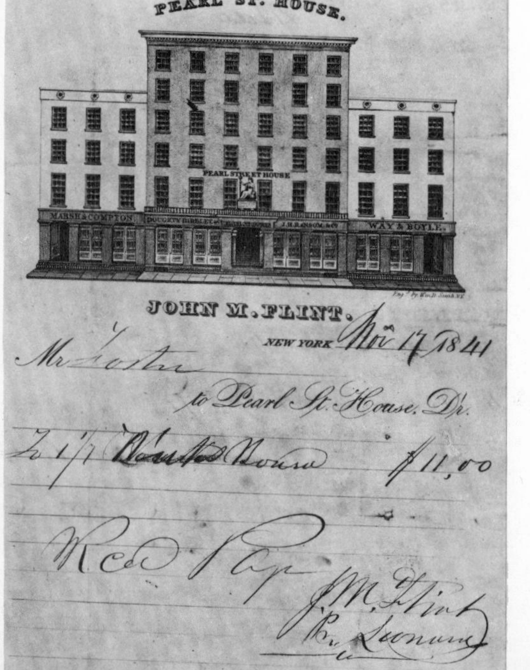

PEARL ST. HOUSE.

PEARL STREET HOUSE

MARSH & COMPTON

WAY & BOYLE

JOHN M. FLINT.

NEW YORK Nov 17 1841

Mr Foster

to Pearl St. House, Dr.

To 1/7 Board House $11,00

Rec'd Pay

J. M. Flint
pr. Leonard

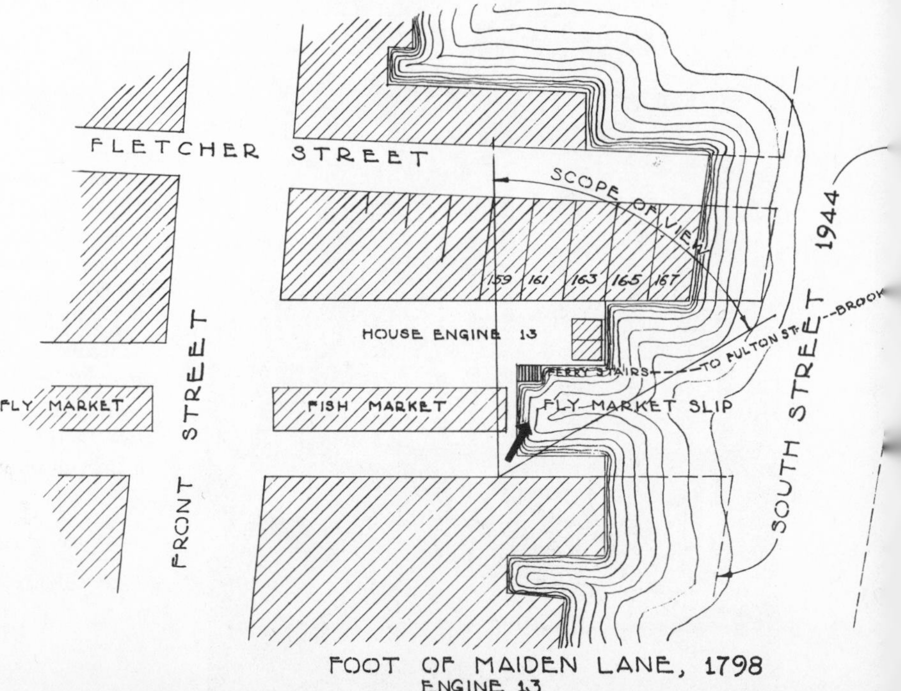

The foot of Maiden Lane in 1798 showing the house of Eagle Engine 13 and a section of the Fish Market in the foreground

Southeast corner of Maiden Lane and William Street, circa 1878

The heavy fines imposed for missing a fire led to some very amusing excuses for non-attendance being recorded in the Minute Book. On January 15, 1807, for instance, the minutes noted that "Harris Sage's excuse is received. He says at the time of the fire he was locked in someone's arms and could not hear the alarm."

On page 270 the book describes "the greatest display of elegant fire engines ever witnessed in the United States." The occasion was the 34th anniversary of the incorporation of the Fire Department, 1832. The lovely, black, gold-striped Eagle Engine was mounted on a stage drawn by four milk-white horses, splendidly caparisoned, and led by four Negroes in Moorish costume. The engines passed in review before the Mayor and the Corporation and then presented a novel and beautiful exhibition at the Battery, first playing in rotation from the river, by suction, and then playing in a line, in unison.

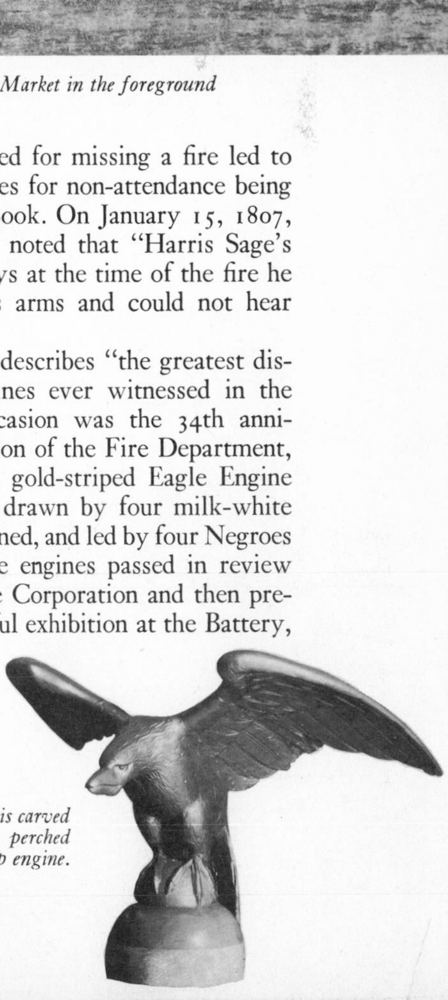

Symbol of Engine 13, this carved wooden eagle formerly perched atop an early hand-pump engine.

OSWEGO MARKET

Swinging Past Oswego

UPON the sounding of the alarms, the boys of Jackson Engine Company No. 24 liked to swing up Maiden Lane and with a flourish, turn to race up or down Broadway, according to the dictates of the signal. In the daylight hours they could always count on an appreciative audience of marketgoers. The machine they ran in 1811 was built by Hardenbrook in 1802 and was a very efficient piece of apparatus.

John Declew, the confectioner in Nassau Street, was a member of this company as well as Ahasuerus Turk, a grocer at the corner of Church and Duane Streets. The Turk family was connected with the fire department for several generations. Other members of this company included John Daddy, sailmaker; John Vandenberghs, a carpenter; Stephen Bonner, brass founder and the father of Robert Bonner, the owner of famous trotters; and John C. Hegeman, attorney, of Beekman Street.

The Jackson's house stood in Maiden Lane at the rear of the Oswego Market, "Old Swago" as it was sometimes called. This market was originally located in the middle of Broadway at the head of Liberty Street, until it was declared an obstruction to traffic. In 1773 it was moved into the Lane. Engine 24 moved there in 1808 and remained until the market was abolished in 1811 and replaced by the Washington Street Market. Engine 24 then took quarters in Tryon Row and Peter Schenck was authorized to build a limehouse on the old market and firehouse sites.

Maiden Lane was, for a short time, known as Green Lane. The present Liberty Place, which extends up a steep grade from opposite 21 Maiden Lane to Liberty Street, was Little Green Street. Near the northwest corner of this and Liberty Street, in early times, was the Quaker Meeting House and graveyard.

Oswego was in full bloom when Catharine Havens' mother moved to Maiden Lane with her parents, the Orange Webbs. Little Miss Havens recorded her mother's description of what was perhaps the earliest use of running water in a house in the city: "Everybody had a cistern for rain water for washing, in the back yard. And when she (my mother) lived in Maiden Lane, the servants had to go up to the corner of Broadway and get the drinking water from the pump there. It was a great bother, and so when my grandfather built his new house at 19 Maiden Lane he asked the aldermen if he might run a pipe* to the kitchen of his house from the pump at the corner of Broadway, and they said he could, and he had a faucet in the kitchen, and it was the first house in the city to have drinking (running) water in it. And after that several gentlemen called on my grandfather and asked to see his invention."

**This was a wooden pipe which led into a tank at the house. Each morning a servant would go up and pump the day's supply.*

During the water shortage in New York City in 1950, the need for a supply of water for air conditioning prompted Barthman and Company, the well-known jewelers on the northeast corner of Maiden Lane and Broadway, to seek water where the old

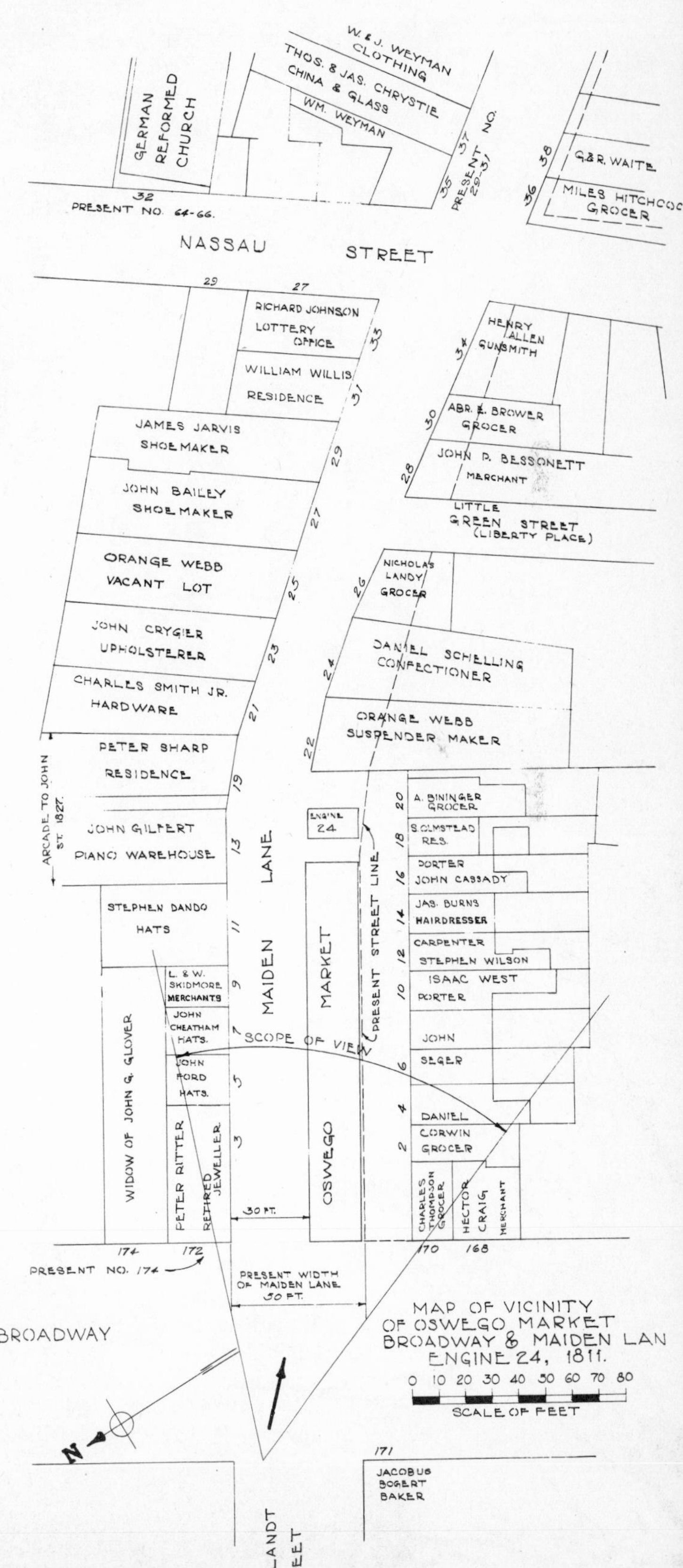

Maiden Lane east to William Street, 1822. Maiden Lane at this time was made up of small residences and shops. The merchants on the Lane usually established residence over their shops.

pump described by Miss Havens once stood. After several weeks their workmen were successful in locating a good supply of water 22 feet below the floor of their subcellar. This water* lay in a deep bed of brownish sand which contained a great deal of mica, and has been identified as a type of sand native to this vicinity. The Ritter house on this corner in 1811 was the home of a jeweler.

In 1799 Orange Webb failed in business, a circumstance which led to a surprising end. Probably few persons who have contentedly snapped a suspender brace ever suspected that nutmegs had anything to do with that pleasurable act. But let Orange Webb's daughter tell it,—

"The firm named was 'Webb and Lamb, Shipping Merchants,' corner Pearl Street and Burling Slip. One of their vessels loaded with nutmegs from Surinam was lost, and Mr. Lamb lost in it. This disaster caused their failure. In those days the laws were very rigid. My father had to go on what was called 'the limits,' until he could pay his debts. In 1799 my brother, Augustus Van Horn, was born, and a nurse, Mrs. Page, taking care of my mother, seeing my father was a very ingenious man, advised him to go into the suspender business, and showed him a pair which she had made herself. This was something

**The flow of water, while bounteous at first, finally levelled off to a little more than one gallon per minute, which, while steady, was insufficient for the purpose in mind. This was probably due to lack of natural surface cover except for St. Paul's Churchyard.*

The same site today showing The Home Insurance Company building left, and the Federal Reserve at the right of the view.

A firemen's parade up Maiden Lane (circa 1865) which featured a horse-drawn steamer. In the background may be seen a horse-drawn hose reel and a hook and ladder truck.

entirely new, and, there being no business of this kind in the city, my father made several improvements in the article, until he brought out something very handsome. I remember how he shut himself in his room, not admitting any of his family, until he had completed his invention. During T. Jefferson's administration, he went to Washington and took out a patent under name of 'Webb's patent suspenders.' His store was in front of his house, and his living room in the rear. His factory was in the basement. This was in our own house, No. 19 Maiden Lane."

Thus was born the modern, improved version of what was to become, in some ways, man's best friend.

Suspenders, or "gallowses," as they were called, really caught on, and the ones ordered by the firemen were as bright as everything else about them. The boys of No. 24 were special admirers of the new style of adornment. Equipped in this fashion, the lads figured they could get a mite more speed and could race a lot safer. These 19th century "life-savers" came in all colors of the rainbow and were fastened in the back by a leather clasp made in the shape of a shield, a star, an eagle's head or some other design. On this clasp or on the front braces was often displayed the number of the machine. Many a runner had his eye blackened by another who had lifted his jacket from behind and discovered on the clasp of his "gallowses" the number of a rival engine. The last pair of fire-braces made by Gratacap, the famous fire cap manufacturer, was an order from Mr. Brokaw, the clothier. It was a fine specimen, and cost seventeen dollars.

In earlier times Maiden Lane was the city's most important silver center. A few silversmiths and a number of dealers have remained, preferring to cling to the old ground rather than follow the commercial trend uptown. Now most of these enterprises dispense products that are conceived and executed elsewhere by present-day craftsmen. Although the shops which display these modern wares are not so picturesque as were those which occupied the same sites in 18th century New York, the silver itself in time probably will have acquired the same soft patina as the heirloom pieces once exhibited in the workshops of old Maiden Lane.

One of New York's most interesting commercial landmarks was the two-tiered arcade which, in 1828, stood about 120 feet from Broadway on the north side of Maiden Lane and occupied a 60-foot front. This arcade extended through to John Street and has been described as a beautiful edifice of white marble. Architecturally of Grecian influence, three graceful arches spaced the front and the semicircular vestibule was flagged with freestone.

Each passage through the building was 14 feet wide, having ten stores on each side, making a total of forty shops on the two floors. The building was devoted principally to retailing of fine clothing and fancy goods. The milliners were located on the second floor. It was one of the first buildings in the city to be lighted by gas. Its shops were among the best, perhaps the best, that milady of that day could find. Few passing through the present arcade in the Silversmiths Building at 15, 17, 19 Maiden Lane realize there was an earlier arcade next door, where the Jewelers' Building now stands.

Engine lantern

16

The Hounds and Their Kennel

AFTER the British occupied the city, prisoners of war and captured rebels were imprisoned in the Livingston Sugar House which stood on Liberty Street, next to the Middle Dutch Church. The conditions in this large structure, "a dark stone building, grey and rusty with age and of dungeon-like aspect," did more to arouse the ire and hatred of loyal citizens than any other acts executed by the British. As many as 800 Americans were crammed within its four walls at one time, suffering a tremendous amount of abuse and left with the choice of either starving or freezing to death. Conditions were so bad that many inmates carved messages and their names on the beams and walls. For years afterwards these "last wills" remained.

One contemporary tells of seeing the "Death Cart" come every morning to the prison to bear off six or eight of the dead, and another recalls the daring efforts of local citizens to smuggle food to the prisoners. It was here that the heroic Judge Thomas of Westchester County died in 1777 from treatment received at the hands of British soldiers; his body was later thrown into a ditch in Trinity Churchyard. Commodore Talbot, one of the first American naval commanders, was also imprisoned at the Sugar House, as were a group of young boys who borrowed a rowboat and a blunderbuss and fired a shot through the ports of a British frigate anchored off the Battery. The boys were caught and imprisoned, though it is said the young daredevils later escaped by tunneling under the street.

Old Post Office, formerly Middle Dutch Church.

As miserable and squalid as conditions were at the Sugar House, the American prisoners still managed to retain a sense of humor. Two young prisoners, Captain Lord and Lieutenant Drumgoogle, instituted a mock court, taking turns as judge and prosecuting attorney, and trying many of their comrades on charges of overeating and overdrinking. The ironic humor of the situation was naturally comic relief to the prisoners and their laughter was so loud that it pierced the prison walls, arousing the curiosity of a group of young English officers who were deeply touched by the sight of the starving men attempting to amuse themselves by ridiculing their own misfortune. The two American officers soon received an invitation to dine with their English visitors and, a short time later, were able to escape with the assistance of a colored slave. The Americans always believed that the sympathetic English officers were responsible for the fact that all the right doors were left open at the most opportune time.

The Sugar House was finally demolished in 1840. The Middle Dutch Church which stood a few yards away was also used as a prison at the beginning of the war, though it soon became a riding academy for British officers. New York's Liberty Bell, now hanging in the belfry of the Middle Dutch Church, Second Avenue and 7th Street, once hung in this church but was removed when the structure was demolished in 1882. Built in 1731, the church was restored after the war and continued in use until 1844 when it was altered for use by the Government as the city's main post office.

Engine Company No. 16, or "the Hounds" as they were called by their fellow firemen, was located on the church grounds from 1786 to 1827. The members of the company, made up of merchants in the vicinity of the engine house, were a distinguished group and meticulous in the care of their engine and devotion to duty. John W. Degrauw, an outstanding fireman, philanthropist and social leader, was foreman of this company and was instrumental in luring many of the city's most famous names into the roster of the department.

SURROUNDINGS OF MIDDLE DUTCH CHURCH, 1826. PRESENT STREET LINES SHOWN WITH BROKEN LINE.

South side of Liberty Street, just east of Nassau in 1826, on the grounds of the Middle Dutch Church. The large structure to the left is the Livingston Sugar House and the building next to it, the office of Seaman & Tobias, at this time proprietors of the Sugar House. The Middle Dutch Church, on the right, also called the New Dutch Church to distinguish it from the earlier church on Garden Street, was built in 1731 and demolished in 1882. In 1827-28 Liberty and Nassau Streets were widened and a handsome iron railing supplanted the old wooden picket fence. The widening necessitated the removal of the shed of Engine Company No. 16. The section of Nassau Street between John Street and Maiden Lane was once known as Kip Street and south of here as the "Pye Woman's Lane."

Early view of the church on Liberty Street.

The First Hydrant

IN 1808, vestiges of Manhattan's old Dutch-type architecture could still be seen on the west side of William Street, on the frontage extending from Liberty Street to Maiden Lane. In front of Mrs. Close's candy store on the northwest corner of Liberty Street, the city's first fire plug was installed in the same year.

Next to the confectionery, at 81 William Street, was James M'Kinley, hairdresser, and at 83, in the little house with its gable end to the street, Jeremiah Allee kept a shoe store. As late as 1823 this house was assessed for only $800.00. Allee had his residence around the corner on Liberty Street. The corner house, then known as 72 Maiden Lane, was occupied by Cary Dunn.

The buildings on this side stood until 1824 when they were replaced by three-story commercial buildings, which in turn disappeared eleven years later when William Street was widened on its west side from Wall Street to Maiden Lane. The building line on that side of the street has not been changed since.

Until 1829 the street below Maiden Lane was only 25 feet wide, and the first widening was made in that year at the expense of the buildings along the east side of the street. At that time the two houses occupying the block between Liberty Street and Maiden Lane, where the Wolfe Building stands today, were owned by the Wolfe estate, proprietors of a hardware business at 93 Maiden Lane. The character of William Street below Maiden Lane was of a miscellaneous nature. Above, where it was 10 feet wider, William Street was an important retail drygoods center.

On the northwest corner of Maiden Lane and William Street (the present Home Insurance Company corner) stood the old Dirck Schuyler-Alexander Hossack house—at this time known as 69 Maiden Lane and occupied by J. & N. Griffith, merchants.

While the first regular improved fire hydrant was placed in front of the dwelling of George B. Smith, a member of Engine 12, on Frankfort Street in 1817, the first experimental "working" hydrant in New York City was installed at the northwest corner of Liberty and William Streets in 1808.

Examined by the Common Council and described as of "very good benefits and use," the first hydrant was a welcome triumph for the volunteer firemen who had waged a long and loud crusade for adequate water sources. Until the city agreed to make this experiment, the engine companies were forced to get as close as possible to the docks along the North and East Rivers, or else draw their supply from wells and pumps which were often incapable of meeting their requirements. Quickest water was obtained by backing the machines as close as possible to the town pumps, and filling them by the aid of wooden vanes or troughs, one end of which was strapped to the pump spouts and the other fastened to the engine box. This once common piece of fire equipment is now all but forgotten. If the fire was at too great a distance from the supply, water had to be relayed by bucket lines to the engines. This primitive system prevailed until suctions were introduced and companies were furnished with copper riveted leather hose.

Hydrant companies were formed in 1831 to manage the hydrants and to protect the plugs from damage. After the hydrants were installed, the possession of those nearest the flames became the chief objective of the companies on the sounding of the alarms. A number of fights and many wild races took place among the eager fire lads. Costello tells the story of one of the most famous races, in which the Hydrant Company beat both the hitherto unbeatable "Old Maid's" boys of Engine Company 15 and the lads of "White Ghost," Engine 40.

The Hydrant Company, the story goes, struck Orange Street, bucked the curb and took to the walk. Each man firmly grasped his wrench, as with increased energy he pushed onward. A crowd assembled in front of Pete Williams' noted dance house, in Leonard Street, scattered like chaff before a gale as the Hydrant Company reached 40's tail screw. When Con Donahue's Democratic headquarters were reached, the "White Ghost" of Forty began to see the heels of the Hydrant Company. (Old-timers in the Fire Department long remembered Con, and also the war cry of his constituents—"Citizens of the Sixth Ward, turn out! turn out! puir Con Donahue lies a-bleeding on the pave forninst his own door.")

Song sheet commemorating Hydrant Company

As the Hydrant's boys turned Vultee's corner into Chatham Street they saw Engine 15's back* ahead and heard the shout from her trumpet ring clear in the night, "Come on, old Peterson! Now you've got 'em!"

The youngsters on 15's rope never made better time, but their best was not good enough. The Hydrant Company came up in a flurry, passing the "Old Maids" at the corner of Pearl Street.

In this "race" Lady Washington Engine No. 40 was hauling a crane-neck piano engine probably weighing more than a ton-and-a-half, and Engine Company 15 had on their ropes a Ludlam gooseneck machine weighing almost as much. The Hydrant Company's "apparatus" was one hydrant wrench per man.

**Her painted back panel.*

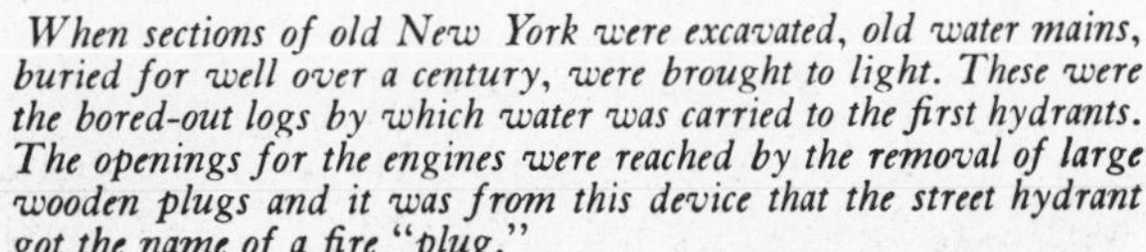

When sections of old New York were excavated, old water mains, buried for well over a century, were brought to light. These were the bored-out logs by which water was carried to the first hydrants. The openings for the engines were reached by the removal of large wooden plugs and it was from this device that the street hydrant got the name of a fire "plug."

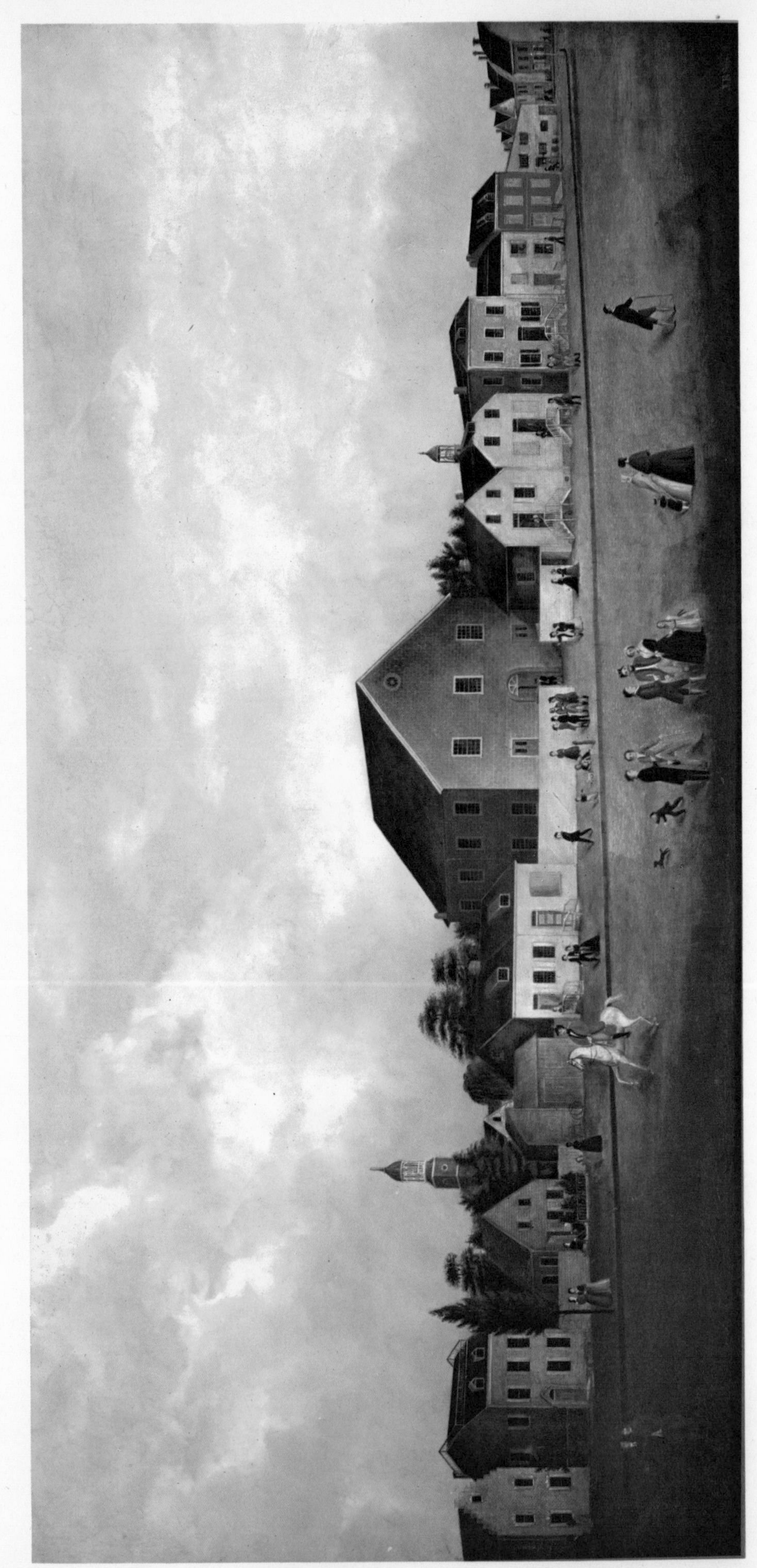

JOHN STREET, 1768

OIL PAINTING BY JOSEPH B. SMITH

MUSEUM OF THE CITY OF NEW YORK

OWNED BY MISS ETHEL HOWELL

A rare and beautiful view of the south side of John Street, from Nassau to William Street, in 1768. (From the original painting by Joseph B. Smith, in the Museum of the City of New York, the property of Miss Ethel M. Howell who graciously permitted this first color reproduction.) The view shows the original Wesley Chapel (John Street Meeting House) which was opened October 30, 1768. This was the first Methodist Church structure in America. The parsonage at the right and partly in front of the church had been built many years before. The shop of Thomas Ash, Windsor chairmaker, was in the east end of the double house adjoining the Chapel yard fence in 1789. The second church was built on this site in 1817-1818. The present building, the third, was erected in 1841.

The steeple to the left is that of the Middle Dutch Church in Nassau Street and the smaller steeple, to the right, is that of the First German Reformed Church which was built in 1765 on the front of the deep lot containing the old Nassau Street Theatre which was in the rear of Nos. 64-66. This church was sold in 1822 to the South Baptist Church.

John Street, during this period, contained a mixture of homes and small shops of wood-workers, painters, and professional men:

Blasius Moore, the clown of Engine No. 5, lived at No. 2; James Seaman, victualling house, No. 8; Elbert Kip, (of Kip's Bay) No. 11; Ruthven and Son, Ivory turners, at No. 14; P. Bailey, coach and heraldry painter, No. 16; Amos Root, grocer, No. 16; Widow Margaret Roosevelt, No. 19; John Hyslop, No. 27; Charles and William Rollinson, engravers, at No. 28; John Scott, bookbinder, No. 31; William Colgate, tallow chandler, (founder of the Colgate Soap dynasty), lived at No. 34; Francis Hall, painter, lived at No. 41; Alex Patterson, chairmaker, at No. 54; Samuel B. Harper, grocer, at No. 57; Charles Fraser, painter, at No. 69; Jesse Scofield (of Keeler & Scofield, and later Scofield, Phelps & Co., Merchant Tailors) home, No. 73; Philip I. Arcularius Jr., chairmaker, No. 75.

The street numbering of John Street prior to 1794 started on the north side, beginning at the house adjoining the Broadway corner, running consecutively to William Street, then back from William Street to Broadway on the south side of John.

99. JNO. N.
GENTLEMEN'S FURNISHING WAREHOUSE
WHOLESALE & RETAIL
STOCKS
SCARFS
CRAVATS
COLLARS
BOSOMS
GLOVES
HOSIERY
97
97
C.B. HATCH
J.M. WHITE
93
J.N. BAUR
99 CLOCKS

Smith's Valley

William Street, world famous as an insurance center, is a thoroughfare that is rich in history and tradition. Its origin goes back to the days of Dutch New York, but since then, parts, as well as the whole of it have been renamed many times.

The section of William Street north of Maiden Lane was named Horse and Cart Lane after the Horse and Cart Tavern, located between John and Fulton Streets, while the section extending to Wall Street from Maiden Lane was titled Smee Straat (Smith Street). It was later called Shoemaker's Street and finally adopted its present appellation in honor of William III.

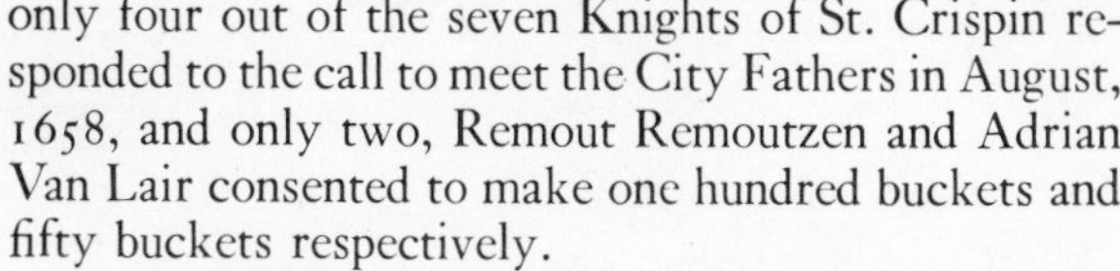

e print of old house at 178 m Street and adjoining shop

In early times William Street catered, among other things, to man's convivial proclivities. On it were such taverns as The Black Horse Inn, the Knight of St. George, The Three Pigeons, The Dog's Head-in-the-Porridge-pot, The Blue Boar and the Bunch of Grapes.

Toward the latter part of the year 1657 the need of regular leather fire buckets to fight fire was much felt. It was decided, therefore, to invoke the aid of the city shoemakers, but the shoemakers of those primitive days lacked confidence in their ability to perform the task assigned them and only four out of the seven Knights of St. Crispin responded to the call to meet the City Fathers in August, 1658, and only two, Remout Remoutzen and Adrian Van Lair consented to make one hundred buckets and fifty buckets respectively.

At 80 William Street lived a Frenchman named Francis Adonis, who displayed a sign reading, "Hairdresser from Paris," a decided curiosity for those times. Adonis, whose customers were principally French refugees, had earned for himself a notorious reputation because of his insistence on bearing his hat under one arm until the restoration of a Bourbon (Louis XVIII) to the throne of France. He claimed to have been the hairdresser of Louis XVI.

General James Robertson, once commandant of the city, lived in William Street, near John Street, as did Washington Irving who was born at 131 William Street. Soon after, however, Irving and his family moved across the street to No. 128, where he passed much of his boyhood. He had a reputation as a prankster and loved to climb to the ridgepoles of neighboring houses to drop large stones down the chimneys to the utter consternation of the occupants who believed their chimney was falling apart or that they were receiving the visitation of evil spirits.

Modern view of William Street showing Home building

It was reported that young Irving and some of his friends once lassoed a watchbox in which a "leatherhead" (policeman), was sleeping and rushing off, dragged the box with its clamoring occupant down the street.

Before the Revolution and for a short time afterwards, William Street from Maiden Lane up to Pearl Street was the great mart for drygoods sales and was thought of as "the proper Bond Street for the beaux and shopping belles."

"Here's your fine Rockaway Clams! Here they G-O!"

At the northeast corner of Frankfort and William Streets stood the Carleton House. Edgar Allen Poe is said to have lived there for a time. It was evidently a respectable hotel at its inception but, as time went on, its reputation became sordid and many grim tales were told about it. These tales received shocking confirmation in 1884, when workmen, cleaning out the subcellar, unearthed from the ashes and rubbish the mouldering skeleton of a woman, around whose neck was a strangling band of calico, and over whose face was a great stone. An Englishman named Benjamin Gray, who was found in the Trenton Prison under sentence for an attempt to murder another woman, was believed to be the murderer but the Carleton House case could not be proved against him. Later the house gained a humble decency by the patronage of retired newspaper men, who lived there in old-time relationship, discussing their younger days with so much vigor and color as to attract the attention of the city's literary men who went there occasionally to commune with the old boys. This curious old house was built on the site of the Lutheran Church of 1767, which had a graveyard where the remains of a number of the Hessian officers in the English army were interred.

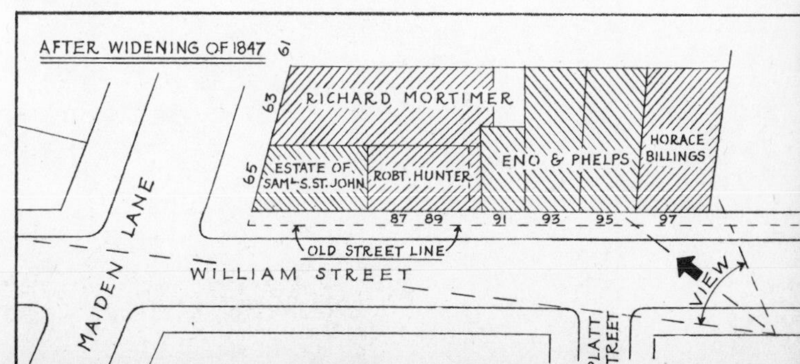

Old John Street

NESTLING quietly in the shadow of downtown skyscrapers on "Golden Hill," the John Street Church, oldest Methodist society in continuous existence in America, still serves the spiritual and meditative needs of the neighborhood. Built in 1841, it is the third Methodist church on this site. Its foundations and walls seem to indicate that parts of the first two structures are built into the present edifice.

Still retaining many of the original furnishings—the old clock; the pulpit desk built by its first pastor, the versatile Philip Embury; the altar rail installed shortly after its completion, and a valuable library which includes numerous prints, books and manuscripts—all in the present "Wesley Chapel" in the basement of the church, it serves as a reminder of the dignity and simplicity which were an innate characteristic of the city's early citizens and worshipers.

The first sermon, in 1766, and weekly ones to follow, were given by Philip Embury at his own home in old Barrack Street near City Hall Place, on the insistence of his cousin, Barbara Heck, who decried the noticeable lack of religious discipline in the city. Embury, who was said to be of a shy and reserved nature, but who later developed into a stirring preacher, delivered his sermon to an enraptured audience of five, one of whom was Betty, slave to Barbara Heck. This initial meeting, re-created in a painting by J. B. Whittaker, hangs today in the church, a reminder of the society's humble beginning.

Joined by a barrackmaster, Captain Thomas Webb, whose unorthodox preaching aroused enormous comment and attention, the society grew until eventually it was forced to move to larger quarters in a rigging loft at 120 William Street. Here Embury and Webb developed an enthusiastic congregation. Webb, who lost his right eye at the siege of Louisbourg in 1758, was a speaker with an amazing gift for directness. He would lay his sword across the pulpit to emphasize the militant nature of his terms with Jehovah in demanding the unconditional surrender of all sinners. His flaming red coat, huge black patch over one eye and more than double the amount of fire glaring out of the other, rather over-awed but definitely impressed a large following. "The old soldier," as he was fondly referred to by John Wesley, deserved much credit for helping to insure a popular support of the young church.

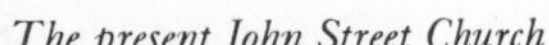
The present John Street Church

The first John Street Church built October 30, 1768

The first "Wesley Chapel" on John Street, dedicated on October 30, 1768, was built of rough stones covered with light blue plaster. The interior was furnished with backless benches and the high pulpit built by Embury along with the "Wesley" clock which is, incidentally, still keeping excellent time. A fireplace stood in one corner and ladders went up the rear to the gallery where slaves and "people of color" attended the services.

After the opening of the first church, Francis Asbury, a zealous young itinerant minister, appeared on the scene. He probably did more to shape the future of the society in America than any other man with the possible exception of Wesley himself. It was he who was called to Federal Hall four blocks away at Wall Street to invoke a blessing on George Washington and on the Congress then in session. Francis Asbury is hailed today by many as one of the world's great religious leaders and organizers.

Peter Williams, a former slave of Benjamin Aymar, tobacco merchant, was bought by the trustees of the church for forty pounds sterling and was installed as first sexton. Peter married the substantial and happy Molly (another Aymar ex-slave), whose heart was closely attached to "Moll's Boys" and that "Ole 'Leben engine" in Hanover Square. Peter earned his freedom after "keeping" the chapel with his wife, Molly,

Peter Williams

for many years. They were much respected citizens and attracted to Wesley Chapel a devoted group of their own people, who later formed the first colored Methodist congregation in New York. Emulating his ex-master, Peter Williams, already an expert cigar-maker, went into the tobacco business, and became an early "American success story." He used a considerable part of the fortune he amassed, as well as his time and energy, in making possible the building, in 1801, of the first Methodist church for Negroes in New York. The former slave laid the cornerstone of this, the Zion African Methodist Episcopal Church, located at Leonard and Church Streets, on July 30, 1800.

Union Engine Company No. 18, whose first engine house was in Water Street, near Fulton, in 1787, moved to quarters "on the Hill at John Street, near Pearl" in 1796. They were known as the "shad-bellies" in those early days but later preferred to be called "Drybones." Try as they might, they could not win the affections of Molly Williams away from "her" Engine 11, even after she moved to John Street.

When yellow fever struck New York in 1822, John Street was in the center of its most virulent devastation. People dropped on the streets like flies and everyone who could fled to Greenwich Village or anywhere else free from the contamination. Living in John Street at that time was an old colored woman named Chloe, who sold flowers and did odd jobs. She was a great favorite of the lawyers in the vicinity, whose offices she often cleaned. As the John Street folks prepared to leave, Chloe obstinately refused all who offered to take her with them. After the plague had run its course and the residents returned to their homes and shops, they found that Chloe had remained for a very definite purpose. There she was in her small quarters surrounded by all the dogs, cats, goats and birds which had been abandoned. She had faithfully tended and fed them.

Everyone was so touched and gratified that enough money was quickly collected to have Chloe's portrait painted, surrounded by the pets whose lives she saved. The artist chosen was no less than William Dunlap, historian, actor, artist and local celebrity, who wrote the "History of the American Theatre," as well as the tremendous work, "History of the Arts and Design in America." The finished painting must have delighted everyone because an engraving was made from it so that prints could be distributed to the folks who had contributed toward the painting. It is strange that the original painting, as well as all of the engravings, has disappeared but as long as the memory still lives, who knows but that some day Chloe and her pets will come to light again.

"Chloe" of old John Street

Note: *During the Revolutionary War, Golden Hill, named for the wheat fields there (John Street), and vicinity, down to the Fly Market at the foot of Maiden Lane, was the scene of a number of conflicts between the Liberty Boys and the British soldiers. The Battle of Golden Hill, in which American blood was shed four months before Lexington, was one of these.*

John Wesley

John Summerfield

Captain Thomas Webb

Philip Embury

Fire Patrol

WHILE the natural instinct of New York's earliest citizens was to assist in saving (often at risk of their lives) each other's property in case of fire, the first effort to form regular companies to carry on this work began in the middle of the 18th century. Both the "Heart to Heart" and the "Hand in Hand" Fire Companies were of this era. The oldest known record of the latter is in the form of a notice to Lord Stirling of a meeting at City Arms Tavern, March 3, 1762, signed by Isaac Roosevelt. On April 8, 1781, the Friendly Union Fire Company was formed.

These companies were to aid in the removing and securing of personal property at fires. They were equipped only with the usual hand salvaging tools plus stout linen bags in which removable property was to be protected and removed. Members of these groups were authorized to wear round leather hats with black brims and white crowns upon which the insignia of the various companies were painted.

These members also were to be exempt from the usual bucket lines or pumping the engines of the regular Fire Companies and they were also not to be chosen for sentry or garrison duty by the military.

In 1803, the Mutual Assistance Bag Company was organized. Among the distinguished names in this company were those of the Bleeckers, Beekmans, Cuttings, DePeysters, Irvings, Laights, Roosevelts, Stuyvesants, Swartwouts, and Ten Eycks. These volunteer "assistance" companies were predecessors of the present Fire Patrol.

When the Bowery Theatre caught fire on September 22, 1836, it was such a tinderbox that in spite of the heroic efforts of the firemen, especially the members of Engine Company No. 26, it was totally destroyed in less than thirty minutes. T. J. Parsons, employed by Messrs. Benedict & Benedict, then of Wall Street, was struck in the head by a beam and severely injured.

It was at this fire that the plan of "old Matt" Carey, (an appropriate enough name), to cover adjacent buildings and exposed property with wet mats, carpets, blankets, etc., was first put into use, and this resulted in the saving of the adjoining buildings. Matt's ingenious plan was the forerunner of using covers to protect property by the Fire Patrol.

A paid Fire Patrol company was organized in New York in the latter part of 1839, although four years earlier in May, 1835, the Association of Fire Insur-

Toy Fire Patrol wagon

Famous "Pie Wagon" in action at a fire along the waterfront

ance Companies, successor to the Salamander Society* of 1819, employed four men at $250 a year each "whose duty it was to attend all fires and protect the interests of the Fire Underwriters by preserving property exposed to fire and damage thereto by water."

In November, 1839, the Association employed 40 men as a Fire Police Force in the mercantile district on a night patrol. These men were volunteer firemen or ex-volunteers, since the Underwriters recognized the necessity of having men in the patrol who had been trained to fight fire. The members of this first patrol were called "Red Heads" because of their red leather fire caps.

The Patrol first went into action at the fire of the Smith Tea Warehouse in Water Street, between Roosevelt and Dover Streets in 1839.

In 1845, still searching for more modern and efficient methods to safeguard property, improved covers for the protection of merchandise were purchased, while in 1851 ten roof or skylight covers were secured. In 1852 Company No. 1 was formally installed in Dutch Street, near Fulton, and later moved to 41 Murray Street in the downtown section of Manhattan. A small hand-drawn wagon was kept in the Dutch Street headquarters. The wagon carried six covers, six buckets and four brooms and was called the "Pie Wagon" because it resembled the vehicles in use at that time by pie bakers. Later the term "Pie Plates" was applied to the large metal and leather badges the men wore. During the day the wagon was stored on the top floor but promptly at 7 o'clock each evening it was lowered to the street, where it remained until 5 o'clock in the morning when, like a tired little bird it was hoisted to its perch under the roof. In those days and for many years to follow, the Fire Patrol worked only at night. The Company was divided into two sections which went on duty alternate nights.

**The Salamander Society, the first association of the Fire Insurance Underwriters, was active between 1819 and 1826. There were eight insurance companies in the organization at the time of its origin. What was left of the Old Salamanders was merged with the Association in 1839.*

The Fire Patrol was equipped with horses even before Manhattan's Paid Fire Department

In 1855 Patrol Company No. 2 was organized and stationed near the Marion Street or Centre Market Bell Tower, but later moved to 175 Elm Street. Each of the two Patrol Companies had a specific area to cover.

John Cornwell, of Hook and Ladder Co. No. 4, joined the Fire Patrol as a private in 1852. Although he retained his membership with Truck No. 4, he was promoted to Captain of the Fire Patrol in 1856 and eventually succeeded Alfred W. Carson as Superintendent. He is credited with being the first to urge the use of horses to haul the heavier patrol wagons and for a two months' period supplied teams at his own expense in order to demonstrate the feasibility of his plan. The result was that the Fire Patrol was equipped with horses some years before the Paid Department.

The earliest Patrol houses were equipped with a special chute so that the drivers could slide down into the seats of the wagons from the bunk rooms above while the rest of the boys hit the brass poles. The house of No. 3 Company at 240 West 30th Street still has one of these chutes.

Underwriter's badge

Toward the end of the Civil War a steam pumping engine for drawing water from flooded cellars was secured by the Patrol to replace the metal hand pumps which had been previously used. A few years later the staff of officers and men was enlarged and the areas covered by each of the companies were extended.

Upon disbandment of the volunteer fire companies the city moved rapidly forward in securing better fire equipment. The Patrol added horse power to its equipment and increased the capacity of the Corps' trucks. Patrol No. 3 was organized and installed on West 29th Street in February, 1868.

The Fire Patrol as it exists today, received its formal Magna Charta and recognition of its rights and duties by an act of the New York State Legislature in 1867, which chartered the Board of Fire Underwriters and granted power "to provide a patrol of men, and a competent person to act as Superintendent, to discover and prevent fires, with suitable apparatus to save and preserve property or life at and after a fire; and the better to enable them so to act with promptness and efficiency, full power is given to such Superintendent and to such Patrol to enter any building on fire, or which may be exposed to or in danger of taking fire from other burning buildings, at once proceed to protect and endeavor to save the property therein, and to remove such property, or any part thereof, from the ruins after a fire."

By the same act of the Legislature, every fire insurance company doing business in the city, whether a member of the Board or not, was compelled to pay a per centum tax upon its premium income within the city limits for the support of the Fire Patrol, which formerly was supported by voluntary contributions.

Under the new charter, Alfred Carson, a former Chief Engineer of the New York Volunteer Fire Department was appointed Superintendent of the salvage corps. Carson, once a member of old Engine Co. No. 12, was a frank, honest and courageous public official.

He was widely known for his temperate habits, sarsaparilla being his favorite beverage, and because of that he was affectionately known by the fire laddies as "Old Saxaparill." He was also the inventor of "Carson's patent capstan," which was used—without too much enthusiasm because it was hard to pump—on some of New York's hand-worked fire engines.

After the double set of men and officers working on alternate nights was abolished in 1870, one company for day and night duty was established at each house. There was a special reserve for extra night duty.

Undaunted by the terrors of nature, in the bitter winter of 1874-1875, when telegraph wires were down all over the city, a mounted force of the Fire Patrol, "The Cowboys," covered the streets at night. Meanwhile, new companies were constantly added as the city grew. In 1876, the fourth patrol, for service above 59th Street, was organized, and in 1882 a regulation uniform for all men and officers was established. Company No. 5, for service in Harlem, was organized in 1891 while Company No. 6, for service in the Bronx, came into existence in 1901. Company No. 7 was added to Manhattan in 1906.

Hat front worn by member of Salvage Corps.

While the salvaging of property or its protection from damage or theft has little of the glamour of fire fighting the Fire Patrol has a memorable history that precedes the formation of the regular paid fire department by many years, and in spite of the regular routine nature of their duties, a number of members of the patrol have given their lives in the protection of life and property; and indeed most of the members have had their dramatic moments under the most exciting conditions.

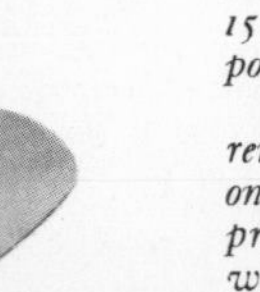

"Hand in Hand" leather fire hat

Note: *The Patrol today maintains a school of instruction at the house of Company No. 6 at 256 West 156th Street in the Bronx. The present equipment consists of 19 pieces of apparatus, 1 chief's car and portable gas line pumps.*

Although supported by property insurance companies the Fire Patrol, in responding to alarms, renders service regardless of whether or not property insurance is maintained. More covers are spread on uninsured property, particularly in private dwellings and apartment buildings than on insured property. There is no way of telling the countless millions of dollars in valuable goods and properties which have been saved by the Fire Patrol throughout the many years of its existence.

FIREMEN'S HALL

Firemen's Hall

THE first Firemen's Hall of which there is any record, was built in 1816 and located in a pie-shaped lot in the rear of the present 71 Fulton Street. The property had been occupied until 1815 by the residence of Rev. John Stanford, chaplain of the Alms House. Here Engine Companies Nos. 13, 18, 21 and 24 were stationed side by side. The Common Council erected a second story over the joint houses. The new quarters were furnished at a cost of one hundred and fifty dollars, each fireman having been invited to contribute for the purpose the sum of twenty-five cents.

Long before the city ordered the building of regular public cisterns, the fire companies at the Hall raised three hundred dollars among themselves and constructed a cistern large enough to hold a hundred hogsheads of water, under the Fire Alley or entranceway to the Firemen's Hall*.

Fair Street was the old name for Fulton east of Broadway. Then only twenty-five feet in breadth at this point, it was shortly afterwards widened and renamed Fulton Street. The companies moved out to other locations and the building was sold at auction in 1829 but it remained until 1852 behind another building which was erected on the front of the lot.

Venerable old Adam Perry Pentz of Hydrant Company No. 1 and Excelsior Engine No. 36, a Fire Warden and President of the Fire Department, told the following fire story to contemporary historians. He lived in nearby Beekman Street and a number of his family were active and renowned firemen. His father took part in the Fire Department Organization of 1796 and also was one of the founders of Mariners Temple on Roosevelt Street.

"One of the first fires I remember happened about the year 1820, in a collection of frame houses on the site now occupied by Fulton Market. The cold was intense, and the firemen were obliged to draw water from the river until all the buildings on the blocks bounded by South, Fulton, Front and Beekman Streets had been consumed. South Street then extended no farther than Peck Slip. While the fire was at its height, threatening destruction to all that portion of the town, and its appearance something almost frightful, a volunteer aid to one of the companies, by the name of Bill 'One-armed' Burke, became crazed with excitement to such an extent that, with shouts of 'The whole city is to be burnt! The world is coming to an end!' he endeavored to throw himself headforemost into the flames, and would certainly have done so, had he not been caught and held back by some firemen

**A section of flat curbing in front of 71 Fulton Street still shows where the alley was located.*

On the skyline of this scene are shown the towers, belfries or cupolas of the following historical buildings from left to right: The North Dutch Church before the cupola was added in 1820, St. Peter's Church at Barclay and Church Streets, the dome of Columbia College at Park Place and Church. The high steeple is that of the Brick Presbyterian Church on Park Row and next to the Chapel of St. George. The large roof of the Park Theatre lies below St. Peter's belfry.

Firemen's Hall on Fair Street was the scene of many a collation and soirée, the most gala being held on Christmas Eve.

near him. This Burke was a well-known character about town. He had but one arm, the other having been lost while he was on a privateer in the War of 1812. Being somewhat of a rough, he was in the habit of placing a stone in the end of his armless sleeve, and when in fighting difficulties, as was often the case, he used the same with prodigious effect. Having been nominated as a candidate for alderman, more by way of joke than in earnest, he polled so heavy a vote in the beginning of the canvass that the friends of his opponent, John Y. Cebra, had to use the greatest exertion to insure their candidate's election.

"'Bill,' said one of his comrades, 'you ain't fit for an Alderman.'

"'Why not?' he replied. 'I've got plenty of eddication. I can speak the English language scientifically and grammatically. Why, I know the man who made up the English language.' He referred to a speaking acquaintance with Mr. Lindley Murray."

In 1790 Duncan Phyfe came to New York from Albany at the age of twenty-two and opened his first shop on Broad Street. In 1795 he finally settled in Partition Street (Fulton), not far from the Common, where he resided and worked the rest of his life. Phyfe's home was at 32 Partition Street, now 193 Fulton, the present house of Hook and Ladder Company No. 10. He leased the property from Margaret Mackaness Ludlow, great-grandmother of the artist, Edward P. Chrystie.

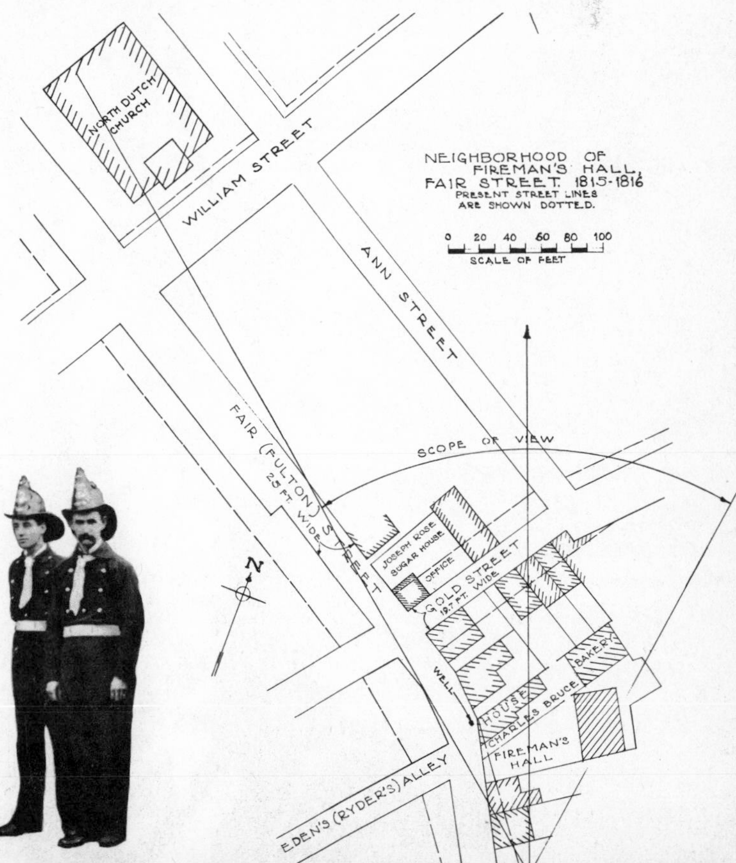

BUDD 105
PROTECTION
5
5

The Hive of the Honey Bee

Fulton Street in 1849 showing James Gordon Bennett's Herald Building and the first offices of the New York Sun at the left. The steeple of St. Paul's in the background.

Along with Eagle Engine 13 and Oceanus Engine 11, old "Honey Bee," Protection Engine Company No. 5, was one of the finest fire companies in the city. It was, moreover, one of the oldest. As early as 1762 No. 5 is mentioned as operating in Smith's Valley, which was then a fashionable place of residence in the city. "Honey Bee" probably served as a bucket company before acquiring their goose-neck engine from the Corporation. Like the "Eagle" on Maiden Lane, most of the company were brilliant firemen as well as outstanding citizens. William C. Conner, the type-founder of Ann Street, also Sheriff of New York County, and his partner and brother James, were typical representatives of the high calibre and integrity of its members. It was James Conner, treasurer of the company, who boasted that "Five's men used to patrol the streets all night and often beat the Insurance Patrol at preserving property . . . many a time I slept on a drygoods box out-doors all night . . . to watch for a fire." Besides the Conners were men like Chauncey M. Leonard, who later became Mayor of the city of Newburgh, New York; J. Murray Ditchett, Police Captain of the Fourth Ward; Joshua Abbe, Fire Commissioner; and a host of others who enjoyed local success.

Perhaps the most popular member of the company was Blasius "Blaze" Moore, once New York's most talked-of prankster whose antics provided the firemen with an almost inexhaustible store of laughing material. "Blaze" once led a big Holstein cow into an Ann Street tavern where his fellow firemen were dining, firmly securing it in the only entrance, a narrow hallway. When an alarm sounded the raging firemen had no choice but to climb over the plunging bovine or to drop through the windows in order to reach the street and their engine. Another time Moore hired two laborers to paint a coat of whitewash over the almost sacred brownstone of City Hall. The unsus-

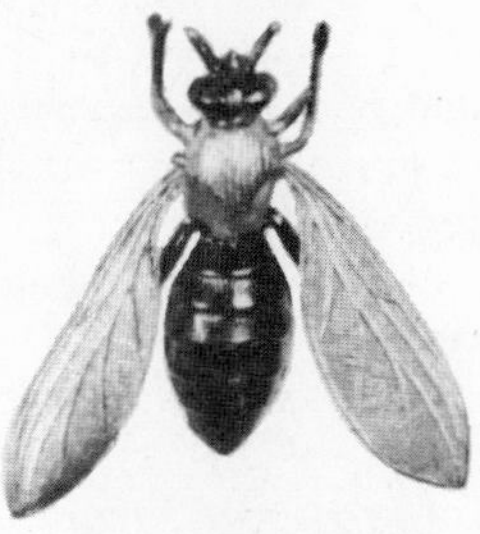

Pin of Honey Bee, emblem of Engine No. 5

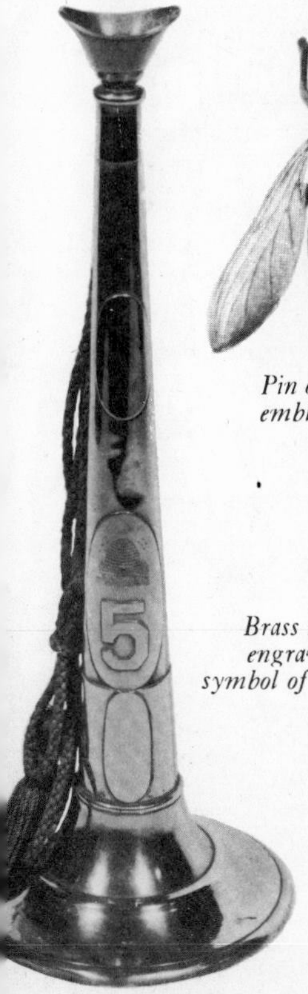

Brass trumpet with engraved bee hive, symbol of Protection No. 5

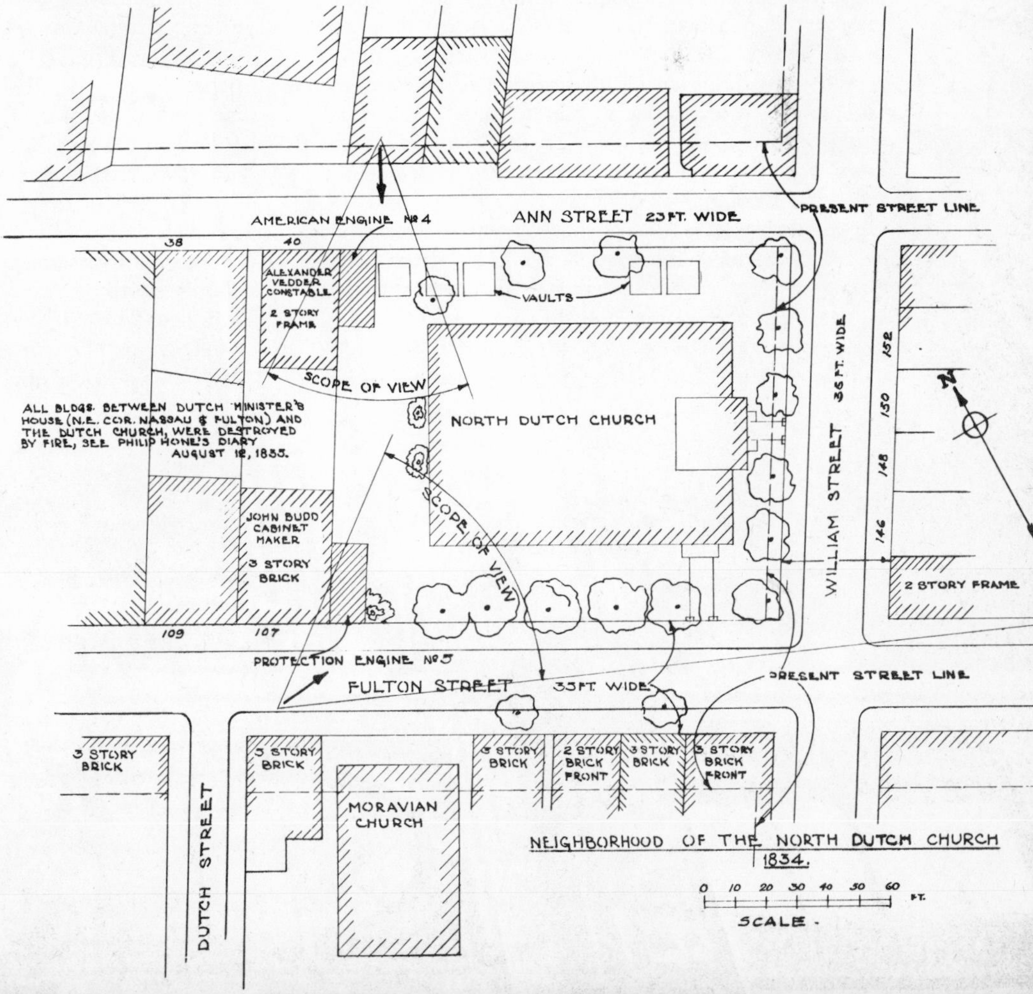

AMERICAN ENGINE No 4

ANN STREET 23 FT. WIDE

PRESENT STREET LINE

ALEXANDER VEDDER CONSTABLE 2 STORY FRAME

VAULTS

SCOPE OF VIEW

ALL BLDGS. BETWEEN DUTCH MINISTER'S HOUSE (N.E. COR. NASSAU & FULTON) AND THE DUTCH CHURCH, WERE DESTROYED BY FIRE, SEE PHILIP HONE'S DIARY AUGUST 12, 1835.

NORTH DUTCH CHURCH

JOHN BUDD CABINET MAKER 3 STORY BRICK

WILLIAM STREET 36 FT. WIDE

2 STORY FRAME

PROTECTION ENGINE No 5

FULTON STREET 33 FT. WIDE

3 STORY BRICK

MORAVIAN CHURCH

2 STORY BRICK FRONT

DUTCH STREET

NEIGHBORHOOD OF THE NORTH DUTCH CHURCH 1834.

0 10 20 30 40 50 60 FT. SCALE.

pecting victims proceeded with a will until interrupted by a group of apoplectic aldermen, who couldn't appreciate the humor of the situation. Moore was, by this time, safely hidden in a nearby grog shop relating his latest exploit to an appreciative group of firemen.

Another member of the company with an interesting history was David Scannel, who was a friend of Dave Broderick. He served as a Captain in the Mexican War, became a gold miner in the days of '49, and was appointed the first Chief of the San Francisco Fire Department. He was elected Sheriff of San Francisco County, in 1855.

Scannel was an inspiring leader and made many notable contributions in the development of adequate fire-fighting facilities in California. A brave and skillful fireman, known for his reckless courage, Scannel's arm was broken three times, his collar-bone fractured and two ribs broken while fighting fires.

The "hive" of the Honey Bee during its most active years, was at 105 Fulton Street, in the shadow of the beautiful Old North Dutch Reformed Church. People once said that on an alarm of fire the members could throw their boots out of the windows of the fire house and reach the street before the boots did. Though this was probably a slight exaggeration on the part of admiring friends, the company was known, nevertheless, for their amazing speed in getting to a fire. When the Thirty Dollar Prize for the first company at a fire was in vogue, the "Honey Bees" won twenty of the prizes in five years, a record second to no other company in the city.

Not among the lesser contributions of No. 5's boys was that they were the first in the city to wear red flannel shirts and this custom with various stylings spread throughout the entire department. In 1849 the famous Garibaldi, living in Staten Island, was fascinated by the dash and color of the fire-laddies and often was seen at the curb in Broadway watching the red-shirted companies parading or running to fires. He admired this feature of the uniform so much that he later adapted it for military use in his native land. The New York fireman's red flannel shirt became known in Europe as the "Garibaldi shirt."

Like many other engine companies in the old volunteer department, the Honey Bee was located near a prominent building, in this case, the Old North Dutch Church (see color Plate VI). This interesting church stood for 108 years at the corner of William and Fulton Streets. Like the Middle Dutch Church on Liberty Street, it was used as a prison during the Revolutionary War, confining a great many American prisoners. Living conditions were even more squalid here, however, because the prisoners had no fuel or bedding during two of the severest winters New York City had ever known and many perished from cold or starvation. (In 1779 the harbor froze so solidly that British artillerymen drove heavy guns and caissons across the ice to Staten Island.) The only pity and meager assistance they seemed to have received in this Tory neighborhood came from the hands of the compassionate naughty-pack of Ann Street, who smuggled food to them and what blankets they could muster, often at the risk of their own lives.

Home of Jacob Stoutenburgh on Fulton Street.

Close by, at the southwest corner of Nassau and Fulton Streets, in 1809, was the famous eating place, the Shakespeare Hotel. Here the literati of old New York gathered, along with actors from the Park Theatre, firemen, politicians and merchants, to discuss the issues and gossip of the day. In this house, probably the most popular one of its time, the National Guard of New York, was organized in 1824 by officers of the infantry battalion of the 11th Regiment.

American Engine No. 4, a back-to-back neighbor of the "Honey Bee," was on Ann Street, also in the rear yard of the North Dutch Church. No. 4's engine, which had a handsome panel on its back portraying the Indian, Red Jacket, was one of the most attractive in the city. Members of the company spent over ten thousand dollars on their engine and fire house, a fabulous sum for that period. The view of 4's house, looking south to Fulton Street, shows the Second Moravian Church, built in 1829, one door east of Dutch Street. The rear end of Protection Engine Company No. 5's house appears on the Fulton Street side.

The boys of No. 5 hauling their goose-neck engine past the fence of City Hall Park and up Chatham Street to a fire on Beekman Street. In the foreground is the goose-neck of Engine Company No. 11, her foreman preparing for a race.

4
AMERICAN
E. P. CHRYSTIE

FRANKLI
HOUSE

Along Broadway

For many years after 1800, the west side of Broadway, between Dey and Fulton Streets, was closely related to the drygoods trade and to the activities of various members of the Haight family, whose business enterprise had much to do with its development.

David L. Haight,* the builder of the Franklin House, had been engaged in the harness business at 169 Broadway as early as 1796, a few doors removed from his brother, Benjamin, who had opened a cap and saddle store the previous year.

The Franklin House, finished in 1817, was considered second in importance only to the City Hotel. It remained a popular resort until closed in 1850. The building was remodelled for commercial use in the following year and was demolished around 1873 to make way for the Western Union Building. By that time, all the small houses extending to Fulton Street had been replaced by commercial buildings. All of these eventually gave way to the present monumental building of the American Telephone and Telegraph Company—the first portion of which replaced the old Western Union Building.

In the 1860's, the intersection of Broadway and Fulton Street was considered the busiest and most dangerous in the city. The Common Council was persuaded by Philip Genin, a hatter then occupying the southwest corner, to erect a pedestrian bridge over the crossing. Built in 1867 and known as the Loew Bridge, it immediately drew the wrath of Knox, a rival hatter on the northeast corner, whose shop was cast in constant gloom by the massive structure. The feud between the mad hatters of Manhattan ended in a victory for Knox and the eyesore was removed in 1868.

This legalistic skirmish was mild, however, compared to the epic feuds of the engine companies in this district. Benjamin J. Evans of Engine Company No. 31 related the following particulars of one of their battles, the big fight of Sunday morning, July 26, 1846:

"We were called out from our quarters in West Broadway on a still alarm . . . our foreman singing out 'Come, pull away boys, for West Broadway's away!' As we were running home opposite the park, we met Engine No. 6, and she commenced to bark. Then along came Equitable, and she thought she would help No. 6, but found that she was in a mighty pretty fix herself.

"Number 5 just turned up near Ann Street and the 'Short Boys' cry greeting their ears, they said, 'Let us go and help old 31 and make the Short Boys feel sick.'

"Chief Engineer Anderson was standing, at the time, on the Astor House steps. Number 6, which then lay in Reade Street, began to bear at us and a fight resulted. In the midst of it, No. 1 Engine, which lay at the foot of Duane Street, appeared and sided with No. 6. Then Engine 23 of Leonard Street turned up and sailed in with us. No. 36 quickly followed, taking 6's side.

"Pipes, axes, and any weapon on hand was used in the fight. It was a terrific fight and lasted a long while. At last Anderson succeeded in putting a stop to it and made us go down Canal Street instead of through Chambers, so as to avoid our foes.

"Before the fight No. 5 sent their engine home in charge of a few men and this precaution saved them from being disbanded."

The neighborhood of the Franklin House was a busy one—the ferry to Paulus Hook was at the foot of Cortlandt Street and the North River steamboat landing was just above it.

A familiar character of this neighborhood was Little Billy the Fiddler (William Hofmeister), a dwarf about four feet six inches high, who announced in August, 1784 that, "being incapable of other employment, he would teach music of almost any kind, having taken a room at No. 101 Broadway, corner of Fair (Fulton) Street." Clad in a large cocked hat and a huge pair of boots, he is said to have presented a most ludicrous appearance; but his services were

Present day view of Broadway between Dey and Fulton Streets, showing American Telephone & Telegraph building, 195 Broadway.

**A son of David L. Haight, David H. Haight, was the builder of the famous St. Nicholas Hotel, partially demolished in 1833, but one section of which exists today.*

1831—At 235 Broadway (the house with the arched entry and high stoop) stood the mansion of Philip Hone, one of New York's most colorful mayors. Hone's residence was a showplace in the city until 1836 when he sold it to Elijah Boardman for $60,000. This site is now occupied by the Woolworth Building.

1845—Broadway looking south from Cedar Street. Site of Equitable Office Building

engaged for many parties. He claimed to have been a friend of Mozart and to have composed one of his sonatas.

In 1818 these streets were paved with round cobbles. Sub-surface street drainage had not come in as yet, and the water from Broadway and other streets on high ground ran into the gutters of side streets leading to the North and East Rivers. The old sidewalks were usually paved with brick, but as new buildings were erected, the brick was replaced by blue stone flagging. Along Broadway, in the vicinity of the Park, the oil lamps used to light the streets were of the special type shown in the view of 1845. Elsewhere they were generally similar in appearance to the gas lamps that were introduced in 1827, many of which remained until recent times.

The stage line for Albany—a two day journey—was operated by Thomas Whitfield from the old two-story gambrel-roofed house at the southwest corner of Cortlandt Street and Broadway. These stages operated between Bowling Green and Greenwich Village, beginning in 1816.

Water carts were familiar sights along Broadway at this time as they made their way up to the Tea Water Pump at Chatham (Park Row) and Roosevelt Streets for a fresh supply of New York's most dependable drinking water.

Among the residential properties torn down in the 1840's to make room for the city's mushrooming commercial establishments were three homes at 116-118-120 Broadway (view of 1845), which were removed to erect the Equitable Life Assurance Society's building, predecessor of the present 120 Broadway structure.

In the view of 1846 the street indicated by the curbing at the left is the present western extension of Exchange Place. The early Dutch dubbed this old thoroughfare Tuyn Paat, or Garden Path.

American ways in the misuse of Dutch words* reached a high in phonetical freedom when in later years the path became familiarly known as Tin Pot Alley. But the English knew it also as Pastie Lane or Oyster Pastie Lane as it led directly to a mount of the same name rising above Greenwich Street about opposite where Edgar Street is now.

An armed redoubt was built upon this hill and stood for many years. When the land to the west of Broadway extending from Rector Street as far north as Cortlandt Street was finally graded and levelled in many places (1794) the stone ruins of this old fort were excavated.

In 1842, when Charles Dickens visited New York, Trinity was still under construction. Rathbun's Hotel stood on the site of the present 165 Broadway Building. Rathbun's was one of a great number of hostelries that lined Broadway but was distinguished from them because south of Rathbun's was Putnam's bookshop and a few doors to the north was a frame building that housed Bogert's well patronized bakery, a bootmaker and a jeweler. This frame building was demolished about 1835 and replaced by the Benedict Building, said to be the first cast iron front in the city.

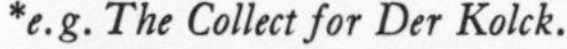
**e.g. The Collect for Der Kolck.*

1846—Broadway looking north from Exchange Alley: Site of the Adams Building

1848—Broadway looking south from Cortlandt Street: Site of 165 Broadway Building

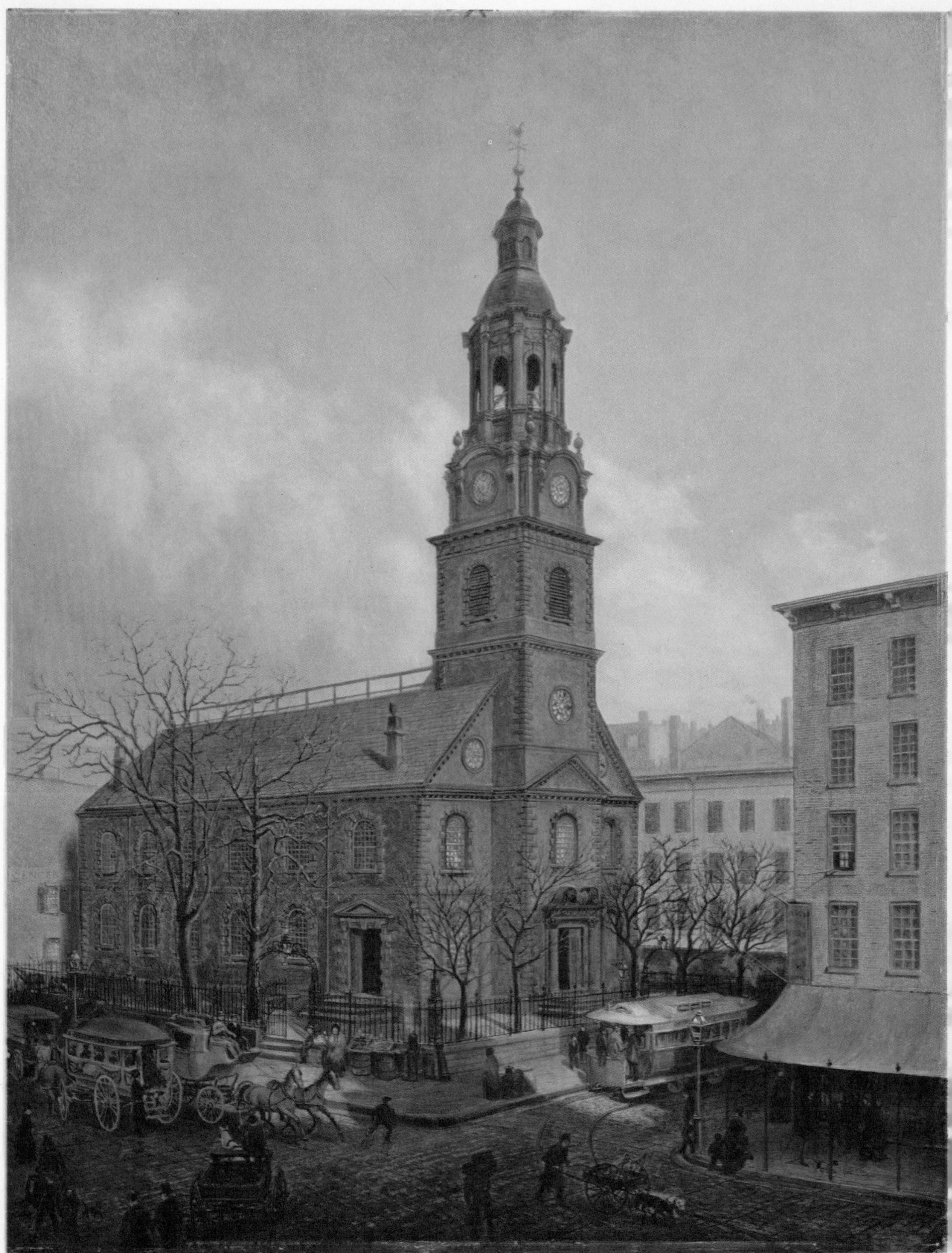

THE OLD NORTH DUTCH CHURCH

OIL PAINTING BY EDWARD L. HENRY IN 1869

COURTESY OF

METROPOLITAN MUSEUM OF ART, NEW YORK

Until 1764, services in the Dutch churches throughout lower Manhattan were conducted solely in the Dutch language but in that year a Reverend Archibald Laidlie, minister of a Scotch church in Flushing, Holland, was called to the Dutch Church in New York to preach in English to those of the congregation who preferred that language. For this purpose the Old North Dutch Church was built, and dedicated May 25, 1769. The property had been willed to the Dutch Church in February, 1723, by John Harpendinck, a member of the congregation who owned a good deal of "Shoemaker's Pasture," and after whom John Street was named. The church was erected upon a portion of this land on the west side of William Street between Ann and Fair (Fulton) Streets.

This structure cost some £7100 to construct and was built of uncut stone. A pedestal and belfry were added in 1820. The original pews and woodwork were exceptional, more beautiful than any previously seen in this country. The main entrance of the Church was on William Street but there were also side entrances on Ann and Fair Streets. The bell, imported from Holland, was hung in the tower June 4, 1770 and a few days before the occupation of New York by the British it was taken down and hidden for the duration of the war, after which it was remounted. The bell from the Church now rests in the Churchyard of the Marble Collegiate Church, on the northwest corner of Fifth Avenue and 29th Street.

The British used the building during the Revolution as a hospital, storehouse, and later as a prison, stripping it of its pews and defacing its walls. It was reopened in December, 1784 and used as a place of worship until 1869, when it was damaged by fire. The old building was finally demolished in the summer of 1875 and its site leased. Here for many years, and until recent times, were the main offices of the Devoe & Raynolds Company, Inc., dealers in paints, varnishes and art supplies. This firm is probably the oldest existing business firm in the city. Though undergoing various changes in names and locations, it has continuously served the public since its founding at Water and Fletcher Streets (where it remained for its first 101 years) in 1754.

This view was painted in 1869, in commemoration of the 100th anniversary of the Church.

St. Paul's Nestlings

IN 1812, the firehouses of Engines 14 and 39 nestled under the protective wing of St. Paul's. A brick wall, into which the firehouses were built, surrounded the churchyard.

Facing the churchyard on Vesey Street, there were only ten houses between the stable in the rear of the Walter Rutherford house at the corner of Broadway and the house of Elizabeth Beekman, at the Church Street corner. Church Street extended only as far south as Partition (Fulton) Street and was forty feet wide, as against ninety feet today. In the illustration, the upper part of St. Peter's, New York's oldest Catholic house of worship, is visible over the roof of the Beekman home. Just to the left of St. Paul's, in the far distance, is the steeple of the Brick Presbyterian Church at Chatham (Park Row) and Beekman Streets.

Franklin Engine Company 39, affectionately referred to as "Old Skiver" and organized in 1812, had as members many prominent citizens and firemen, including the following: Gilbert B. Mott, foreman; John M. Read, assistant foreman; Thomas N. Stanford, secretary; and Uzziah Wenman, later Chief Engineer of the New York Fire Department.

Engine 14 was one of the outstanding companies of the old Department. Abraham Brower was one of the earliest foremen and John P. Roome, a descendant of Peter Roome of Engine 7, was another. Roome was appointed assistant engineer in 1808 and served until 1824.

During the Gulick disturbances* in 1834, the beautifully painted back of "Old Skiver" was shipped to Europe in order to hide it. Chief Engineer John Decker continued his search for it long after it had been returned. He spent one long afternoon in Reuben

Philadelphia-style engine run by No. 14

Bunn's restaurant on Washington Street, questioning Reuben, but the latter, who was a loyal "Skiver," nonchalantly parried his questions although he was sitting on the box in which the treasured back was hidden.

Once, during a race through Chambers Street with No. 5 Engine, one of No. 14's boys fell and was caught in the back by the king-bolt of the machine. Sprawled flat in the street, he shouted in a loud voice, "Go on, Fourteen!" From that day it became the rallying cry of the company.

St. Paul's Chapel, which was completed in 1766, is now New York's oldest church edifice. It is also the oldest public building on Manhattan Island and the only British colonial-built church. Its architect was Thomas McBean, a pupil of Gibbs, who was in turn a pupil of Sir Christopher Wren. The ornamental details were designed by L'Enfant, French architect and Major of Engineers in the Continental Army. The fourteen chandeliers, now electrified, are of Waterford cut glass. Washington's pew is on the north aisle and Governor DeWitt Clinton's pew on the south aisle. In the chancel there is a tablet to Sir John Temple, baronet, the first British consul general appointed to the United States after the Revolutionary War, whose home was in nearby Greenwich Street.

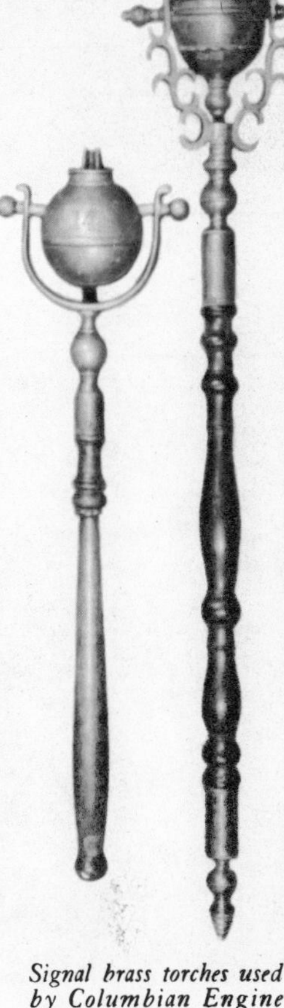

Signal brass torches used by Columbian Engine Company

In the niche above the portico, on the Broadway end of the building, stands the American primitive hand-carved heroic statue of Saint Paul. Little Catherine Havens wrote in 1849: "Maggie (her nurse) says whenever the statue on St. Paul's Church hears the City Hall clock strike twelve, it comes down. I am crazy to see it come down, but we never get there at the right time."

Were Washington to return today, St. Paul's, Fraunces Tavern and the Jumel mansion, where he had headquarters, would be the only buildings familiar

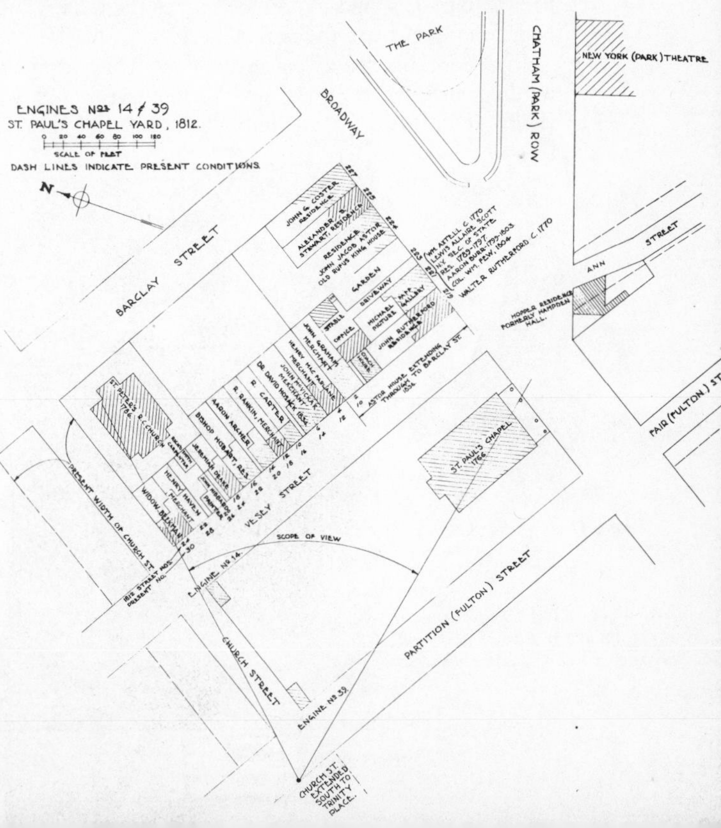

Perhaps the most popular and daring Chief Engineer in the old department, James Gulick's dismissal by the Common Council, in 1835, split the volunteers wide open. Eight hundred firemen marched to City Hall in open rebellion against the dismissal and submitted their resignations from the department. This display of sentiment and regard failed to restore Gulick as Chief Engineer but the firemen later united and nominated and elected their idol Registrar of the City.

Washington passing St. Paul's during his triumphal entry into the city, 1783. Courtesy Museum of the City of New York.

to him in Manhattan. He and his staff are claimed by some to have passed here on November 25, 1783, to take possession of the city, which had been in British hands since the American retreat in 1776. There are a few sentimentalists who on the 25th of November of each year still commemorate the magic of that scene, and a certain editor of *The New York Times* annually visits the Broadway and Ann Street corner, just to await and watch with bated breath the wondrous "Parade," and to doff his cap to the man whose heart and honor are the real America.

In 1799, a fire broke out in the decorating shop and lumber yard of Mr. West, on Washington Street, and rapidly consumed the whole block except five houses. A flaming brand caught on one of the urns on the tower of St. Paul's, and a high wind threatened to destroy not only the chapel but the rest of the city.

As there was no way at that time of going from the inside to the outside of the steeple, several volunteers, including a workman of the church, made their way up over the lightning rods to the danger spot. They chopped the urn down from its base and extinguished the flames by means of water hauled up from below on a cord. These men were later rewarded for their heroism by the Vestry.

There are too many monuments in St. Paul's Chapel and in the churchyard to list them all here. Many New Yorkers are familiar with the tomb and monument of General Richard Montgomery, who was killed in the assault on Quebec, and who is buried under the portico on the Broadway end of the Chapel.

John Wells, the noted jurist and co-editor, with Hamilton, of The Federalist; Thomas Addis Emmet, Irish orator and jurist; Dr. William MacNeven, called the "father of American chemistry"; Christopher Colles, who built the first New York City water works and who conceived the idea of the Erie Canal; a number of British and Colonial Revolutionary officers; the Sieur de Rochefontaine, Instructor of Artillery to the Continental Forces, and many others distinguished in the development of America are immortalized here.

George Templeton Strong and his family attended both Trinity and St. Paul's Chapel regularly. But

John Decker's fire hat

John Decker, last Chief Engineer of the Volunteer Fire Department, February, 1860 to August, 1865

Statue of St. Paul still overlooks Broadway

Strong chafed at what he regarded as the miserable organ playing at St. Paul's and recorded some biting comments on it in his diary during 1841. As he was a great lover of music and a founder of the Philharmonic Society, we can assume that his acid criticism was probably justified.

"February 7, Sunday. Dubious weather. Went to church as usual. Our organist has periodical fits of

St. Paul's steeple saved by a sailor in 1803

insanity and absurdity, during the prevalence of which he plays with as much judgment as a horse. Today he marched us out of church to something between *Yankee Doodle* and the *Overture to the Bronze Horse.*

"April 9, Friday . . . I wished I could get up a musical insurrection at St. Paul's and lynch little Hodges . . .

"April 11, Sunday. Easter—fine weather, though winterish and frosty. Church as usual. Heard my friend, Professor McVickar, this morning. We had very decent music, by the by, for a wonder.

"July 17, Sunday. By the way, I may as well observe that the music at St. Paul's has reached its lowest point—is now the worst possible and may therefore be hoped to improve. I never heard such feeble, lifeless monotony, as it is—and it's especially provoking because little Hodges can play so well when he chooses."

The great repair job done to St. Paul's in 1950 gave hope but no realization that a white steeple and trim might once again grace this beautiful structure. The wooden steeple, which was superimposed on the original square masonry tower in 1794, was then painted white and probably most of the other exterior wood finish as well. It was not until 1840 that the steeple was painted to imitate brownstone, and so it remained until the copper covered steeple was substituted.

In 1805 a high brick wall was built surrounding the churchyard, except for a wrought iron railing with brick piers, which faced Broadway. These brick walls were removed in 1837 and the present cast iron railing installed, with the handsome granite piers at the Broadway corners.

After the burning of Allison's carriage factory in Vesey Street, Engine Company No. 9 inserted in the newspapers a card of thanks to their entertainers, "particularly to the old lady who put her head out of the window and requested us to make less noise."

Charles H. Haswell, who later organized Engine No. 18 uptown, marched with 14 in the parade to honor General Lafayette in 1824, and Owen B. Brennan carried one of the company's torches in the procession. Samuel Y. Coles, James H. McKenny and William Wallace joined the company in 1825. Titus Conklin ("Old Conk"), afterwards bell ringer at the City Hall, joined it in 1826, and John Gaten and Drew Mallet joined in 1828. Henry T. Gratacap, appropriately named fire hat maker, was also a member of No. 14. Gratacap made the first stitched and raised fire hat fronts used by the firemen in New York.

The first "Drover's Inn," kept by Adam Vanderbarrack, was a little above St. Paul's Chapel on the road which is now Broadway. One old lady of Manhattan observed that the land upon which St. Paul's Chapel was built was a beautiful wheat field the year before its construction.

The Washington Market in 1859

Looking south along Broadway from City Hall Park in 1850. Barnum's Museum at the left, St. Paul's at the right and Matthew Brady's Gallery on the southwest corner of Fulton Street. Courtesy of the New York Historical Society

Barnum's Museum

BARNUM'S Museum at the intersection of old Chatham Street (Park Row) and Broadway was a scene of many fabulous events in the history of New York. In early times it was the center of the amusement world and during the heyday of the Astor House was probably the most familiar sight to visitors.

In 1849, Catherine Havens wrote: "We always ask him to take us past Polly Bodine's house. She set fire to a house and burned up ever so many people, and I guess she was hung for it, because there is a wax figure of her in Barnum's Museum.

"Maggie takes us there sometimes, and it is very instructive for there are big glasses to look through, and you can see London and Paris and all over Europe, only the people look like giants and the horses as big as elephants."

Barnum's Museum destroyed by fire, July 13, 1865

Barnum's Museum stood on the site of the original Spring Garden (1712). Barnum purchased the American Museum, which had been constructed there somewhat later, on December 27, 1841. General Tom Thumb made his appearance there in 1842. The building was twice enlarged, once in 1843 and again in 1850. It was set afire by Confederate spies in 1864 but the flames were extinguished before any great damage was done.

The museum, with its marvelous collection of wild life, unbelievable relics, its assortment of strange people and its famous lecture room or theatre, was an institution of interest throughout the country. The famous "egress" story can be matched by Barnum's "omphalopagus" oriental display which amazed the world—the first Siamese twins shown in America.

When a Long Island farmer advertised a cherry-colored cat for sale, Barnum promptly purchased it, sending a special assistant to see that it arrived safely. It was when Barnum found the cat was pitch black that he was supposed to have made his famous remark, "There's a sucker born every minute." Not one to overlook a natural trick, he covered the town with posters inviting everyone to see the cherry-colored cat—which, not surprisingly, came to be referred to as the "Black Crook."

Note: *Barnum was right about a great many things, but as a fire insurance risk, he was about as wrong a prospect as any insurance company could ever hope to have. Every place he ever owned and operated, prior to 1874, was completely destroyed by fire, with enormous building and property losses.*

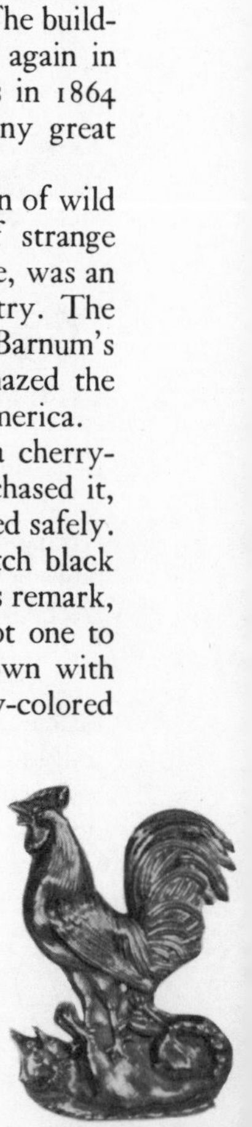

Fireman John Denham's battle with Barnum's Bengal

Broadway looking north from Barnum's, circa 1863

In the 1850's, Barnum put on an elaborate extravaganza, "The Patriots of '76," which called for a large cast of extras—Hessian and Continental soldiers, Indians, etc. Mr. Greenwood, Barnum's manager, recalling the excellent drilling of the Lady Washington Light Guard, the Target Company of Lady Washington Engine No. 40, arranged for nearly all the men in the company to take over these roles, including that of Molly Pitcher.

In the middle of the most exciting act of the opening performance, the City Hall bell sounded an alarm. Bill Racey, the foreman of the Target Company yelled, "Boys, there's a fire in the 7th district." Thirty members of the cast immediately leaped from the stage, rushed up Broadway to get their engine and then raced to the fire, creating one of the most extraordinary sights ever seen. Molly Pitcher was at the head of the rope and a "live Indian" brandished the foreman's trumpet. By the time they reached the fire, they were followed by a curious crowd, who were fascinated at the sight of Hessians, Continentals, Indians and Molly Pitcher, earnestly manning the brakes.

On July 13, 1865, a fire was discovered in the museum at Ann Street.* The firemen were able to open the museum cages and free the birds, which flew about the city to the dismay and wonder of the populace, but the animals were something else again. The fact that many wild beasts were on display created a ticklish problem for the fire-fighters. According to fire department reminiscences, this catastrophe was an epoch in the life of Johnny Denham, a member of Hose Company No. 15, who was known as a quiet, retiring sort of fellow.

During the excitement, a huge Bengal tiger broke loose from its cage and leaped from a second-story window to the street. Some spectators stampeded in terror, others froze helplessly in their tracks. Several police banged away ineffectively with their small pistols, but Fireman Denham grabbed his axe and with a mighty blow sprawled the animal on the street. Cheers went up and Denham "felt a flow of feeling" he had never known before. He ran into the blazing building and carried out the fat lady, who weighed more than 400 pounds. Next, he carried out two children and then made a third trip to bring out the woolly-headed Albino woman. Probably no one was more surprised at his response to the emergency than John Denham and although he seems never again to have caught the public eye, there is no denying that Denham had his day.

George Collyer, who became Manhattan's oldest surviving volunteer fireman, was driving a team with a load of flour in Broad Street when the alarm sounded. Tying his horses to a tree, he ran to meet his engine, the Rooster of Company No. 29, one of the earliest steamers in the department. In the years preceding his death in 1946 at the age of 100, he related in detail his personal observations and experiences in this terrible fire, which was watched by thousands of people.

Fireman Collyer was sent into the building with several other members of Company 29 to release whatever harmless animals they could reach. In a glass-covered cage was a 30-foot python that was beginning to burn, because of the intense heat. It was writhing in pain and Collyer could see it was doomed to destruction. Although he never would have believed that he could be overcome by compassion for such a creature, he was moved to break open the cage and end the snake's suffering with a couple of blows of his axe.

The animals during the fire at Barnum's in 1865

Another museum was opened by Barnum on 14th Street near Irving Place. It was destroyed by fire on the morning of December 24, 1872, and again his extensive and valuable menagerie and practically all other property, valued at more than $300,000, were totally destroyed.

**In September, 1865, Barnum reopened his museum at the Chinese Assembly Rooms on the west side of Broadway, between Spring and Prince Streets. But that place, with its new collection of wild life and curiosities, was completely destroyed on March 3, 1868. The back of this building faced Mercer Street, and Engine 33 of the Paid Department poured streams into the building all day long. As it was extremely cold, the entire building was encased in ice.*

At the height of this fire a huge bear appeared in a second-story window. The firemen tried to save the beast by every means possible but as the flames increased, the bear tumbled backwards into the flames and disappeared.

20
38
HUMANE
24 JAMES KEAN
CLOTHING 24
BOOKS
JAMES McCOMBIE

Iat front
oreman of
ngine 38.

The Bible Savers

Southwark Engine Company No. 38 had a long and illustrious career but its fame, as well as the popular title its men were known by, rested in the success of their operations at the Park Theatre fire, the evening of December 16, 1848. As their house at 28 Ann Street was but a few hundred feet from the fire, 38's engine was the first to reach the scene. They pumped water furiously on the theatre in the hope that at least parts of the hall could be saved, but, soon realizing it was doomed, they immediately transferred their attention to the building at the rear of the theatre which housed the American Bible Society. Their efforts were successful and the building, along with the valuable collection of Bibles owned by the society, was saved.

For this fire-fighting service the society presented the company with a rare and expensive Bible on which was inscribed, "Presented to the Southwark Fire Engine Company No. 38 by the American Bible Society for valuable services rendered in preserving their premises at the burning of the Park Theatre, December 16, 1848." The Bible immediately became a cherished possession of the company and from that time until their disbandment in 1865, 38's boys were known as "the Bible savers."

A popular rendezvous for the members of this company, as well as the men of the Humane Hose who used the same house, was in the basement on the southeast corner of Ann Street and Broadway then occupied by Barnum's Museum. This site, in earlier times, had been occupied in part by the Spring Garden and its tavern, which stood immediately south of the intersection of Broadway and Park Row. The land north of this intersection was part of the vineyard belonging to Governor Bongan. The general locality had always been a favorite amusement and refreshment area.

Toy model of old hose reel.

View shows the house of Southwark Engine Company No. 38 and Humane Hose Company No. 20 at 28 Ann Street in 1846. The building, at that time numbered 16 Ann Street, was three stories high and was built in 1838 by Richard Riker. The two companies were located here from 1843 to 1865. Lower Ann Street between the present William and Gold Streets was once a part of a path known in Dutch days as Van Tienhoven's Lane. Never widened, this part of Ann Street gives a good idea today of the original width of most of lower Manhattan's early thoroughfares.

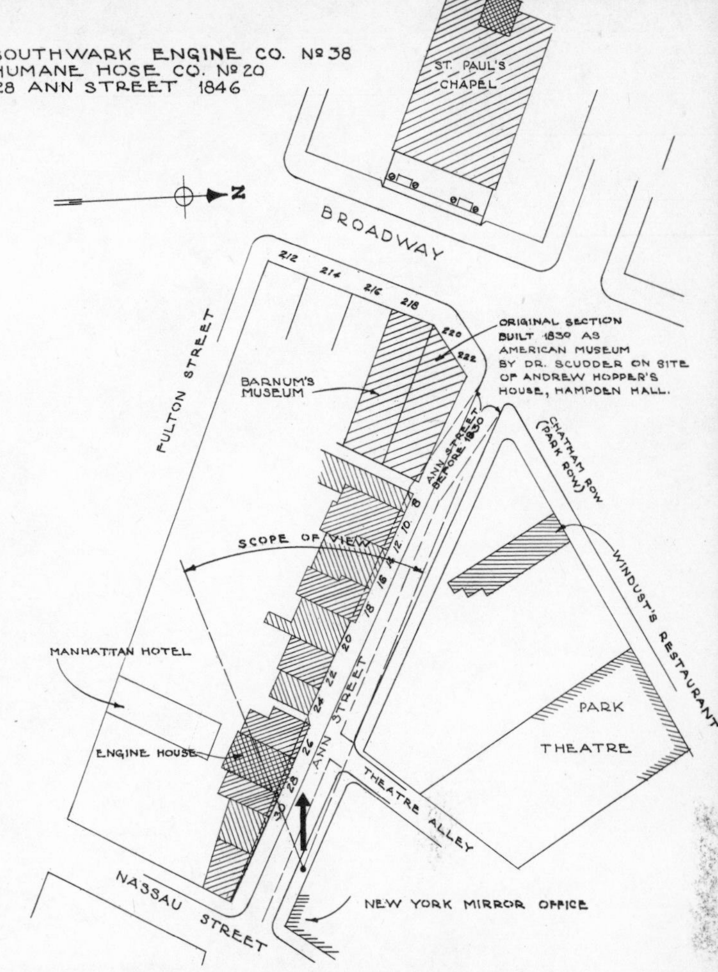

John Greenwood, privateersman and fife-major in the Revolutionary War, is said to have lived for a time in this vicinity, where he practiced the art of dentistry. Greenwood's most distinguished patient was President Washington. Two sets of artificial teeth he made for Washington still exist and are thought to be remarkable examples of dental skill. Greenwood is credited with being the originator of the foot-power drill, of spiral springs to hold the plates of artificial teeth in position, and the use of porcelain in the manufacture of artificial teeth.

In 1894 the late Isaac Mendoza founded a book store at 17 Ann Street and developed it into a local "institution"—the meeting place of many literary personages and noontime researchers. Still operated by his sons, Aaron, Mark and David Mendoza, now at 15 Ann, the shop retains the charm and atmosphere of the type of bookstore so closely associated with old New York.

Uniform button of company member.

Philadelphia-style, double-deck, end-stroke machine used by Southwark Engine Company.

Housed in Thespia

Ladies and gentlemen,
Enlighten'd as you are, you all must know
Our playhouse was burnt down, some time ago,
Without insurance—'Twas a famous blaze,
Fine fun for firemen, but dull sport for plays.
The proudest of our whole dramatic corps
Such warm reception never met before.
It was a woeful night for us and ours;
Worse than dry weather to the fields and flowers.
The evening found us gay as summer's lark,
Happy as sturgeons in the Tappan Sea;

The morning—like the dove from Noah's Ark,
As homeless, houseless, innocent as she,
But—thanks to those who ever have been known
To love the public interest—when their own;
Thanks to the men of talent and of trade,
Who joy in doing well—when they're well paid,
Again our fire-worn mansion is rebuilt,
Inside and outside, neatly carv'd and gilt,
With best of paint and canvas, lath and plaster,
*The Lord bless Beekman and John Jacob Astor.**

THOUGH many early critics bemoaned the lack of a cultural influence in old New York, some of the finest and most stimulating theatre of the day could be seen at the old Park Theatre on Chatham Street (now Park Row). From 1798 to 1848 everything from the Barber of Seville to risque German farces, translated by William Dunlap, could be seen in this theatre which played host to the greatest theatrical talent of its time. Ellen Tree played here as did Fanny and Charles Kemble, Edmund and Charles Kean, Emma Wheatley, Junius Brutus Booth, James Wallack and Tyrone Power—before audiences as receptive and enthusiastic as any on Broadway today. Early New Yorkers were afforded their first and, in many cases, their only glimpse of classical theatre and opera at the old Park, as well as the works of some of America's first playwrights.

Standing at 23 Chatham Street, its rear entrance on the Mews or Theatre Alley which still extends from Ann to Beekman Streets, the house was open every evening at six-thirty, the curtain raised one hour later. The custom generally was to give two pieces—a tragedy and a comedy—sometimes a third was added—a comic song or "a pas seul or pas de deux by danseux."** The pit was provided with board benches without cushions and occupied exclusively by men and boys. The boxes were enclosed in the rear. Privacy was safeguarded by an attendant. In the second tier there was a restaurant and in the third tier a bar and a special section for women.

**From the address of Fitz-Greene Halleck on the reopening of the Park Theatre following the fire of 1820.*
***According to Haswell.*

The Park was one of the most fashionable resorts in the city in its day, despite the hue and cry of several of the city's moralists. Just before an opening of the theatre, one irate gentleman wrote a letter to Mayor Varick and protested against "the corrupting influence of Theatrical exhibitions on the morals and manners of the People," referring to the Park as "a scene of dissipation and licentiousness."

Malachi Fallon—the firemen's fireman

John Howard Payne, author of "Home, Sweet Home," (which, incidentally, was first sung in America at the Park on November 12, 1823) appeared there himself in 1809, for the first time on any stage, as the "Young American Rossius." In 1842, one of the most elaborate balls ever held in old New York was given there in honor of Charles Dickens.

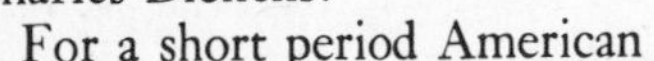

For a short period American Engine Company No. 4 was located in the northeast room of the playhouse. The theatre was consumed by fire in 1820. The house was rebuilt in 1821 but was again destroyed in 1848 when one of the performers carelessly pushed playbills into a low-hanging gas jet. John Gilbert, the actor, spoke the very last words from the stage of the theatre when the alarm of fire was given. It was the final speech of the character, Admiral Kingston, in a play called "Naval Engagements." The theatre was then owned by William Beekman and John Jacob Astor who suffered heavy losses.

It was at this fire that one of Manhattan's handsomest and bravest firemen, Malachi Fallon, covered himself with glory when he dashed into the flaming theatre to rescue the beautiful Mrs. Dyott, an actress who was appearing in the play. The cheers that greeted him when he emerged with his "prize" echoed all over the City Hall Park and established him as the firemen's ideal of a fireman. It was Fallon, years later, along with Sheriff Matt Brennan, another ex-fireman, who was responsible for the capture of John Colt, brother of "Sam'l." Colt had been accused of the murder of a Samuel Adams and was said to have confessed to Fallon who was Warden of the Tombs Prison. Fallon was also one of the group of firemen who left for the gold rush with Dave Broderick and Dave Scannell. Broderick was elected United States Senator from California; Fallon became San Francisco's first Chief of Police; Scannell, after serving as the first Chief Engineer of the San Francisco Fire Department, was elected Sheriff.

Defending the Liberty Pole

For the Jolly Bunch

THE most prominent restaurateur in New York in 1830 was Edward Windust, whose tavern in the basement of 11 Chatham Street was, during that period, the most famous in the city. Windust was unrivalled as a caterer and the atmosphere he created—comparable to that of the London coffee-houses of the Johnson and Goldsmith era—served as a magnet for the members of the literary and theatrical world, as well as for the "liveliest minds in society." After the curtain had fallen on the last act at the Park Theatre, a few doors north, throngs of the curious gathered at the tavern to catch a glimpse of the city's great and near-great who practically made it their home. Edmund Kean, Junius Brutus Booth, the Wallacks and the Kembles were frequent visitors from the world of the theatre, as were Vanderlyn, Jarvis, Trumbull, Morse and Dunlap from the art world, and Zophar Mills, Benjamin Strong and others from the roster of the volunteer firemen.

The records of Nassau (later Valley Forge) Hose Company No. 46 claimed that Windust's Tavern was the place of their origin. This company, which was officially housed at 83 Nassau Street, between Fulton and John Streets, was a "jolly bunch," noted for their speed and spirit. One of their lads, Jimmy Millward, later became Minister to Belgium under President U. S. Grant.

Windust's, sometimes referred to as "the Shakespeare,"* after the earlier renowned tavern of that name at Fulton and Nassau Streets, was opened in the basement of the building in 1824, but as its clientele grew, took over the whole ground floor as well. Its walls were richly adorned with paintings and caricatures of the theatrical celebrities of the day, and its closing, in 1837, left a gap in the city's social life that was not to be filled until the Delmonicos lured the fashionable with their delightful cuisine.

Windust, in company with the Mayor, other officials and firemen, with all of whom he was closely associated, was personally immortalized in the H. R. Robinson print of the Fire of 1835 (see p. 62).

Today as you pass by No. 11 Park Row, Windust's old building, it is hard to imagine that this lonely little survivor was once such a familiar place to the men of wit and fashion. The upper stories are now completely concealed from view by a large advertising sign.

Nearly every building along Chatham Street, from Ann Street to Beekman, was occupied as a tavern or other resort in 1825. The exceptions were Place and Souillard's apothecary at No. 17, George Chilton, the chemist at No. 34, Mrs. Mabbit's grocery store in a small frame house at the Ann Street corner, and Turner's grocery at the Beekman Street corner in a similar building. At 32 Park Row, four doors down from Beekman Street, lived John G. Leake, whose wealth in later years founded the Leake and Watts Orphan House.

In the winter, snow clung to the streets and often

**The "Shakespearean Gallery," where steel engravings of scenes from Shakespeare were sold, was on the floor above the tavern.*

necessitated the transferring of carriage and hack bodies to sets of runners. These rigs, called "booby hacks," were common sights along old Park Row.

The view of 1823 shows the first Park Theatre as it appeared after the fire of 1820, rebuilt, plastered to resemble a brownstone structure, and completely altered in appearance. The Brick Presbyterian Church, in the next block to the north, was one of the best attended in the city. Over the backs of the horses may be seen the "Old White Lecture Room," which stood in the churchyard behind the Brick Presbyterian Church. Above the roof of the lecture room is seen the second building of Tammany Hall (with the flagpole surmounted by a liberty cap) at Frankfort and Nassau Streets. City Hall Park is at the left, with the jail in the distance and the roof of the Free School beyond it at the corner of Tryon Row.

City Hall Park at this time was one of the most attractive spots in the city, the walks and grass-plots being trimly kept and the whole area shaded by groves of elm, poplar, willow and catalpa trees. Several Liberty Poles stood in the Commons, which was the scene of many exciting events in the years before the Revolution. One of them, erected in 1767 by the Sons of Liberty to commemorate the repeal of the Stamp Act, was the object of bitter altercations with the British soldiery.

The neighborhood was the scene of countless civic functions, one of which, the funeral of President William Henry Harrison, was described by George Templeton Strong in his diary. "April 10, 1841. Weather raw, cloudy and unpropitious. Went out at twelve o'clock . . . Chatham Street literally hid with lugubrious drapery. I established myself in Chatham Square, and a fine sight it was to look up at the rising ground towards the park, the houses on each side shrouded with black, the dense mass of people between and in the centre, the procession pouring down, a wide stream of plumes and bayonets and dark banners. It began to pass at a little before one—moving rapidly—headed by the military . . . then the Urn, the General's horse (hypothetical), the "pall bearers," Martin Van Buren, and divers other great men, the civic dignitaries, all the fire companies, about 3,000 men I presume . . . by that time it was half past two and I was tired and it was beginning to snow so I walked down Chatham Street to the Park . . ."

Illustration from an early song sheet showing a gooseneck engine being hauled along Chatham Street

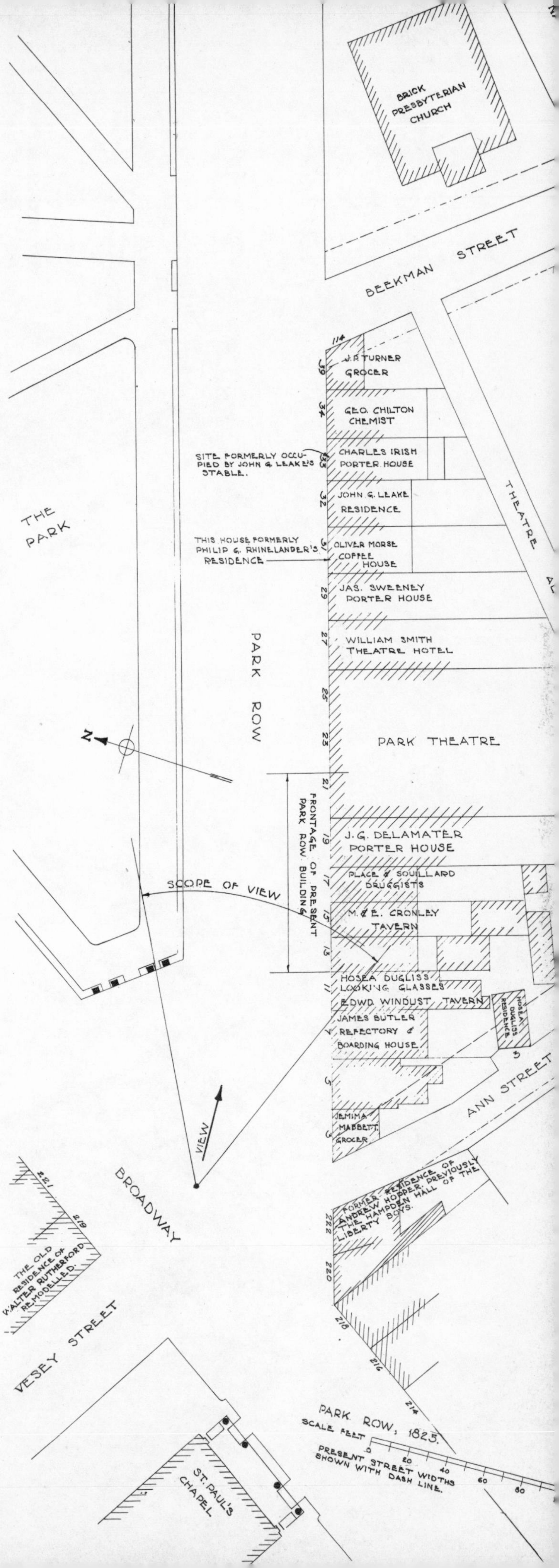

BRICK PRESBYTERIAN CHURCH
BEEKMAN STREET
J.P. TURNER GROCER
GEO. CHILTON CHEMIST
SITE FORMERLY OCCUPIED BY JOHN G. LEAKE'S STABLE.
CHARLES IRISH PORTER HOUSE
JOHN G. LEAKE RESIDENCE
THIS HOUSE FORMERLY PHILIP G. RHINELANDER'S RESIDENCE
OLIVER MORSE COFFEE HOUSE
JAS. SWEENEY PORTER HOUSE
WILLIAM SMITH THEATRE HOTEL
THEATRE
THE PARK
PARK ROW
PARK THEATRE
N
FRONTAGE OF PRESENT PARK ROW BUILDING
SCOPE OF VIEW
J.G. DELAMATER PORTER HOUSE
PLACE & SOUILLARD DRUGGISTS
M. & E. CRONLEY TAVERN
HOSEA DUGLISS LOOKING GLASSES
EDWD. WINDUST TAVERN
HOSEA DUGLISS RESIDENCE
JAMES BUTLER REFECTORY & BOARDING HOUSE
JEMIMA MABBETT GROCER
ANN STREET
VIEW
BROADWAY
FORMER RESIDENCE OF ANDREW HOPPER. PREVIOUSLY THE HAMPDEN HALL OF THE LIBERTY BOYS.
THE OLD RESIDENCE OF WALTER RUTHERFORD REMODELLED.
VESEY STREET
ST. PAUL'S CHAPEL
PARK ROW, 1823.
SCALE FEET 0 20 40 60 80
PRESENT STREET WIDTHS SHOWN WITH DASH LINE.

Seven! Seven!

BEEKMAN Street got its start as a cowpath running through the William Beekman farm. It was known by this name prior to 1730. After St. George's Chapel was consecrated in 1752, and until the British evacuation in 1783, it was referred to variously as New-English-Church Street and Chapel Street, but its original name has prevailed to this day.

From 1656 farmers had the right to drive cattle through it. As late as 1802 a cow gored a man to death in Beekman Street.

Adam Spies, the famous fireman, a member of Engine Company No. 12, lived in this neighborhood.

Hamlet was staged for the first time in New York at the Chapel (Beekman) Street Theatre which was built in 1761.

In May, 1766, a play was presented which, while causing some amusement to the British officers present, was exceedingly offensive to most of the audience. A riot resulted in which the "Sons of Liberty" or others pulled the structure down and hauled most of the parts to the Commons and burned them.

A few days later, "The Sons" seized a press barge, and drew it through the streets to the Commons, where they burned it also.

Company No. 7 (later Lafayette Engine) was organized in the early part of the 18th century as a bucket company, and first located on the former Duke Street or High Street. In the conflagration of 1776 it was Peter Roome, foreman of Engine 7, who struck with his trumpet a Hessian soldier who tried to interfere with the working of his company. He nearly paid for this brave, if rash, act with his life. Nearly beaten to death until rescued by his fellow firemen, he lived to see the Evacuation. His trumpet was preserved by the Roome family and now has a place of honor in the H. V. Smith Museum of The Home Insurance Company. Engine No. 7 moved to Beekman Street in 1798.

The little building which housed Engine 7 in 1798 at No. 114 Beekman Street lasted until 1832 when Beekman Street was widened. About eight feet of the land upon which it stood was utilized to build an addition to the house adjoining on the south which then became the corner house, and was incorporated into Lovejoy's Hotel in later years.

In 1819, the latter building became the property of Henry Eckford, the shipbuilder, and was occupied by his relative Joseph Rodman Drake, the poet.* Eckford added extra stories to the house. The building at the southwest corner of Nassau and Beekman Streets was occupied for many years by Peter Embury as a grocery and residence. Embury was a nephew of the Rev. Philip Embury, the first minister of the Methodist Meeting House in John Street.

* *"Yet I will look upon thy face again,*
My own romantic Bronx."

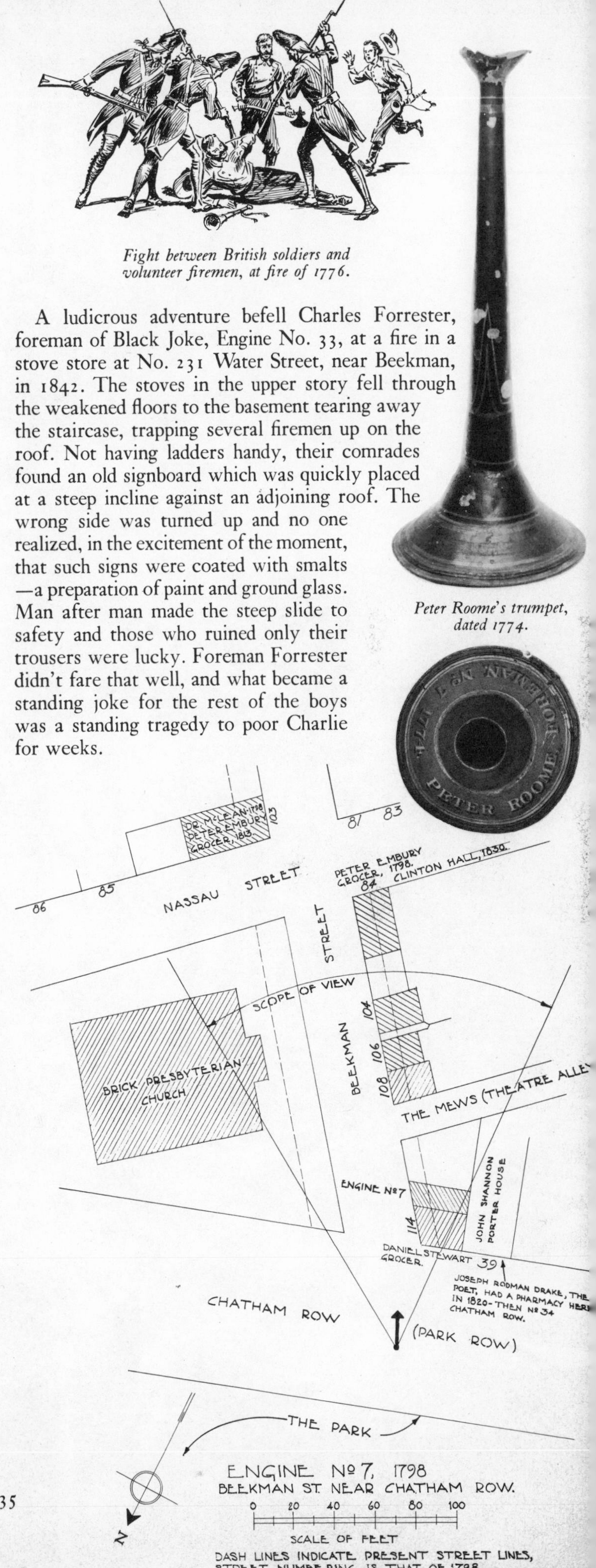

Fight between British soldiers and volunteer firemen, at fire of 1776.

A ludicrous adventure befell Charles Forrester, foreman of Black Joke, Engine No. 33, at a fire in a stove store at No. 231 Water Street, near Beekman, in 1842. The stoves in the upper story fell through the weakened floors to the basement tearing away the staircase, trapping several firemen up on the roof. Not having ladders handy, their comrades found an old signboard which was quickly placed at a steep incline against an adjoining roof. The wrong side was turned up and no one realized, in the excitement of the moment, that such signs were coated with smalts—a preparation of paint and ground glass. Man after man made the steep slide to safety and those who ruined only their trousers were lucky. Foreman Forrester didn't fare that well, and what became a standing joke for the rest of the boys was a standing tragedy to poor Charlie for weeks.

Peter Roome's trumpet, dated 1774.

Buried at St. George's

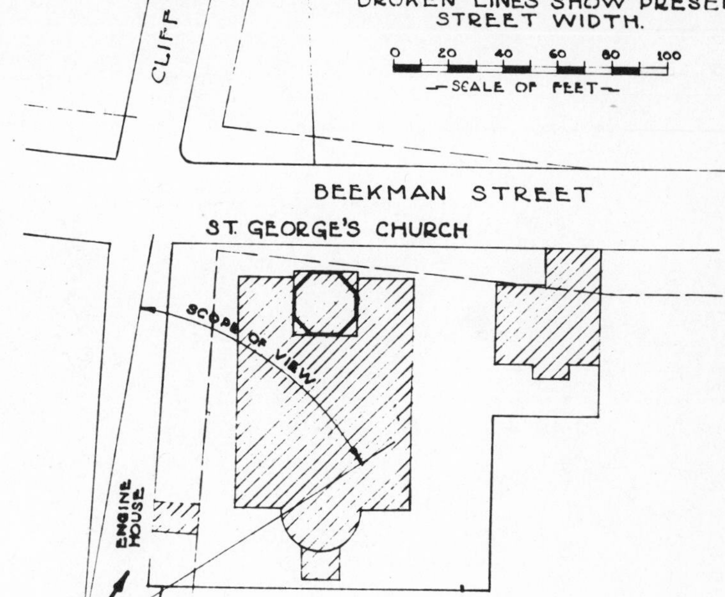

IN 1815, when Engine 7 was quartered in the churchyard of St. George's Chapel, on Cliff Street, its foreman was William L. Mott, merchant; first assistant, Eleazar Lundy, currier; second assistant, Gregory Snether, distiller; clerk, Peter Williams, bookbinder. This district was early associated with the leather and bookbinding trades and later the closely allied paper business.

Tan yards once covered nearly all of the area of the Beekman Swamp or "Kreuplebusch." Oldsters remembered this area as a duck-shooting paradise and the surrounding hills as a favorite place to gather huckleberries.

At the New Year and Christmas festivals, it was the early custom to go out on the ice of Beekman's Swamp for turkey shoots. Each contestant paid for a shot at the turkey's head and neck; if he drew blood, the bird was his for the holiday dinner.

At No. 8 Jacob's Street, there was once a "mineral" spring called "Jacob's Well." Drilling down 130 feet the owners struck a spring which was said to have medicinal properties equaling those of the Saratoga and Congress waters. More than 25,000 persons used it in one year. There was quite a stir, however, when experts discovered that the "mineral" quality of the water emanated from the hides and tan-oak steeping in long-buried and forgotten pits.

Northeast corner of Jacob and Ferry Streets

Where the church stood on Beekman Street was a high mount called Chapel Hill with slopes steep enough for sleighing down to the swamps. There was once an apple orchard on this crest.

On hot summer Sundays in early days, the congregation must have been almost equally bothered by the mosquitoes and the sounds of truant boys playing in the swamp below.

St. George's (1752) was the first chapel built by Trinity Church. The neighborhood at that time was principally aristocratic. George Washington attended a Christmas service during his official residence in the Franklin home in nearby St. George's (Franklin) Square. Robert Nesbit, the Revolutionary War printer, lived at No. 112*, near the church. Washington Irving was baptized here and several generations later J. P. Morgan, Senior, was married in this old church.

The furniture of old St. George's was made from the mahogany topmasts of a ship, the gift of a sea captain church member. The original pulpit, desk and chancel were removed in later years to Christ Church in Manhasset, Long Island.

The mother of Mrs. Catherine C. Havens ("Diary of a Little Girl") was born at 84 Beekman Street in 1801. She said, "Our house in Beekman Street was on the north side, between Cliff and Pearl Streets—a three-story brick house. We moved to Maiden Lane about 1806."

Engine No. 7's house on the Cliff Street side of the church wall was practically buried in the graveyard, a fact which encouraged members of rival companies to make unpleasant remarks.

The chapel, which separated from Trinity Church in 1811, met a great misfortune in January 1814, when it was ravaged by fire. John DeGrauw, a famous old fireman, who was on duty at this fire, commented that a number of women with buckets assisted the men in bringing water to the engines. "Seven," at her own door, never fought more valiantly. As the flames bit deeply into the structure, the steeple which was surmounted by a decorative British crown that had not been removed during the Revolutionary War, was seen ready to topple. Just as the entire tower crashed, DeGrauw heard one of the company remark, "If the crown of Great Britain was never down before, it is down now, at any rate," to the disgust of some Tories who still hung on in the neighborhood. Just before the steeple fell, the old church clock struck three times. The church was rebuilt on the same walls and continued its service until it was abandoned in 1868.

**R. E. Dietz, the designer of the first successful kerosene lamp and manufacturer of many types of fire lamps and signals, lived at No. 66 Beekman Street. In the rear of his house stood the old Shot Tower.*

Note: *The old Beekman Tavern on the northeast corner of Beekman and Gold Streets occupies a building erected about 1820.*

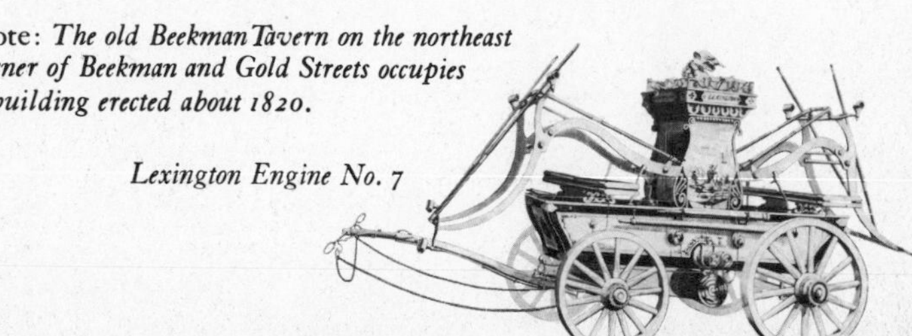

Lexington Engine No. 7

Drawn by W. G. Wall, Engraved, Printed & Coloured by I. Hill
CITY HALL
To the Honble Philip Hone, Mayor of the City of New-York.
This Plate, is Respectfully Inscribed by his obliged Servt. W. G. Wall.
Re-engraved 1919 for the
SOCIETY OF ICONOPHILES
by Sidney L. Smith

CITY HALL

View of 1826

AQUATINT, COLORED. DRAWN BY W. G. WALL.
ENGRAVED, PRINTED AND COLORED BY I. J. HILL.

(This state re-engraved for the Society of Iconophiles by Sydney L. Smith.)

This is said to be the finest engraved colored view of City Hall and represents the building before the cupola was raised in 1830 to accommodate the clock which was installed in 1831. New York City's Hall is considered one of the most perfectly proportioned buildings in the country and undoubtedly reflects the French influence of the designer, J. F. Mangin. Mangin and John McComb, Jr. were the architects.

Ground was broken on April 5 and the corner stone laid May 26, 1803. Although the building was partially occupied July 4, 1811, the project was not completed until 1812. The marble was secured from the Johnson and Stevens Quarry in West Stockbridge, Mass. The brownstone used in the foundations and the rear wall was taken from two New Jersey quarries leased from the Presbyterian Church of Newark. The figure of Justice on the cupola was designed by John Dixey, and inspired the following, from Fitz-Greene Halleck in a poem,* Fanny:

And on our City Hall a Justice stands;
A neater form was never made of board,
Holding majestically in her hands
A pair of steelyards and a wooden sword;
And looking down with complaisant civility—
Emblem of dignity and durability.

In 1848, Catherine Havens wrote: "My mother remembers when the City Hall was being built; and she and Fanny S. used to get pieces of the marble and heat it in their oven and carry it to school in their muffs to keep their hands warm. She loves to tell about her school days and I love to hear her."

A fire bell tower was raised on City Hall roof, May 17, 1830. When Conklin Titus was the bellringer† there, Bill Demilt, a nervy jokester of Engine 14, made a perilous climb up over the lightning rods one evening to invite "Conk" out for a drink. The rest of the boys bet he wouldn't come down.

Bill's grinning face, suddenly appearing out of the blue, almost scared the daylights out of the daydreaming bellringer. When "Conk" recovered his breath, he lost no time in descending (by the stairs) and soon was enjoying a big dose of his favorite nerve allayer, gin and sugar, at Harry Venn's. This was a popular resort for Fourteen's boys and was located at 13 Ann Street. In later years, old "Conk" became the proprietor of this establishment.

Fire destroyed the statue, cupola and roof of City Hall on August 18, 1858, and a new, carved wood figure was erected May 7, 1860. Due to the ravages of time, the latter was taken down July 15, 1887. It was replaced by a copper replica, on November 3rd of the same year.

George A. McAneny, trustee and former president and chairman of the board of the Title Guarantee and Trust Company, now Vice-chairman of the National Trust for Historic Preservation, Trustee of Federal Hall Museum and a director of The Home Insurance Company, often referred to as the "father of American historical conservation," undertook restoration of City Hall to its original design when, as Borough President of Manhattan in 1910, he found the structure to be in an unsafe, dilapidated condition. He received the city's first medal of honor for this and other public services. It is probably due to him, more than to any other man, that City Hall has been preserved for future generations.

**Who was going to look around back? Paint to match the marble of the front and sides now covers this shrewd, economical stroke of the Building Committee who also paid McComb six dollars a day as supervising architect.*

†Old-time firemen claimed they could tell by the sound of City Hall alarm, on what part of the bell-lever old "Conk" held his hand. Incidentally, this stalwart had 21 children. The fact that he wore a wig was one of New York's best-known secrets.

Printing House Square

Formed by the intersection of Nassau Street, Park Row and Spruce Street, and really shaped like a small triangle, the area known as Printing House Square was part of the public pasture in early times. Facing the eastern side of City Hall Park this area was once called City Hall Square though it has been known by its present name since 1861. By that time it had become the undisputed center of newspaper publishing in New York City.

In the days when cows and horses grazed in the vicinity, one of the community's greatest tragedies took place here. Jacob Leisler, lieutenant governor and acting commander of the colony and Major Jacob Milbourne, his son-in-law and chief advisor, died at the hands of the common hangman. Victims of a political plot, Leisler and his deputy were condemned to death by the signature of the Royal Governor, George Sloughter, who was plied with liquor at a wedding in the home of Nicholas Bayard, one of Leisler's bitterest political enemies.

In the deep gloom of the morning of May 16, 1691, the two men were led out to the gallows erected almost exactly where the statue of Benjamin Franklin stands in Printing House Square today and were executed. A driving rain beating upon the tortured faces of Elsie Leisler and her daughter, Mary Milbourne, mingled with their tears of despair. History was later to vindicate the honor of the two innocent men and brand as barbarous murderers the officials and citizens who plotted against them.

By his own choice Leisler was buried in a piece of land which was a part of his garden and the grave was near the southeast corner of Park Row and Spruce Street on ground now occupied by the Tribune Building.

Under the direction of Dr. Spring who officiated there for nearly fifty years, the Brick Presbyterian Church, opened January 1, 1768, was a great influence in the city. The sexton used to stretch chains across the street at meeting time so that the Sabbath peace would not be disturbed by the rumbling of traffic. He was also known as the dog whipper, a title given to church sextons because their duties included chasing stray dogs to prevent their yowling.

In the view at the left, on the southeast corner of Nassau and Spruce stood Martling's Tavern, headquarters of Tammany Hall before the erection of Tammany's own building on Frankfort Street which was finished in 1811. The old firehouse* did not remain long on the church grounds. During its brief existence, however, it was occupied by American Engine Company No. 4, among the first engine companies organized in Manhattan, which was transferred from its house on John Street in front of the Theatre.

This was truly "Newspaper Row." The old *Times* building erected in 1857-8, stood on part of the site of Dr. Spring's Brick Church. The *Sun*, which was

**The building next to the firehouse was the office of the Kine-Pock Institute where the miracle of smallpox serum was pioneered.*

St. Paul's steeple is seen to the right and Trinity's in the far distance under the tree on Nassau Street. The cupola of the Middle Dutch Church at Liberty and Nassau is just visible above the roofs of the houses on the east side of Nassau Street.

The New York World building was destroyed by fire, January 13, 1882. The building was then located at Park Row, Beekman and Nassau Streets.

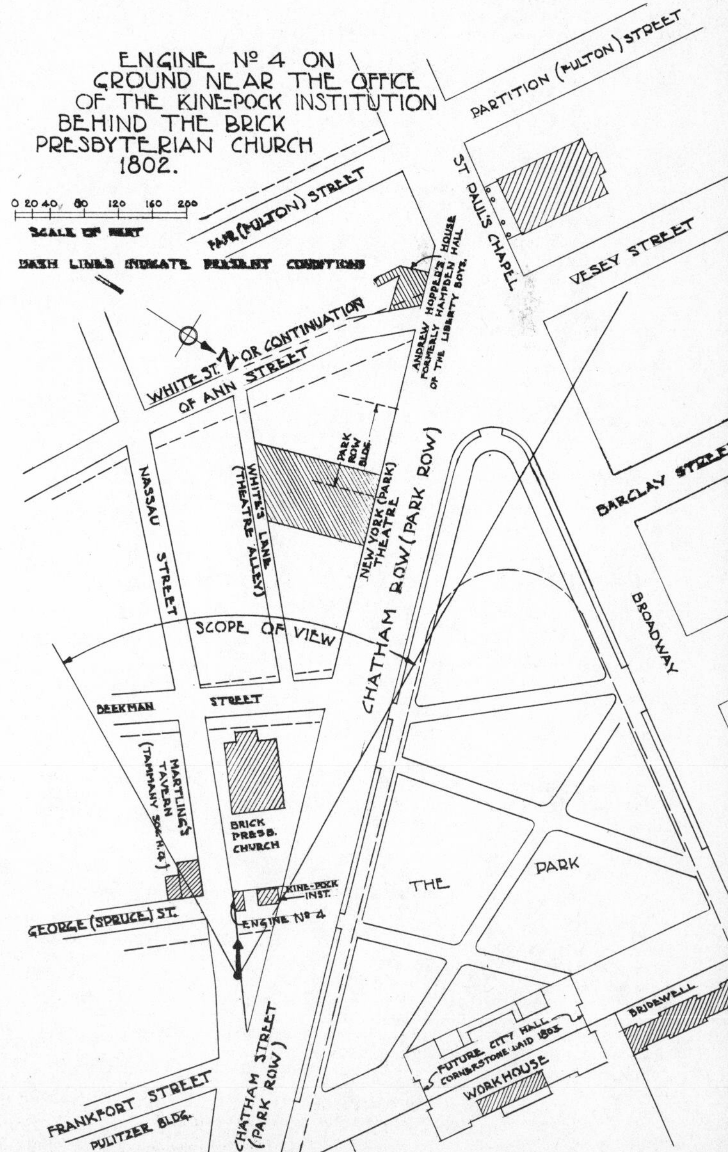

NEW YORK AURORA

NEW YORK, TUESDAY, JANUARY 18, 1842. PRICE, TEN

THE NEW YORK FIREMAN

[*COSTUME OF THE CHIEF ENGINEER*.]

Hark! from the City Hall, the bell,
Fire's dread alarum note is ringing;
With lightning speed, urged by the knell,
The Fireman to the spot is winging

Where the destroying element
In its consuming wrath is raging,
And with the produce of man's toil,
In vengeful flames, a war is waging

See him! undaunted on the spot,
With sinewy arm the engine plying;
Now showers of water, by his toil,
Into the burning mass are flying

His ladder placed against the wall,
He quickly mounts, nor heeds the danger;
Though in the scorching flames he fall,
To craven fear he is a stranger.

Unto the brave in every clime,
And time, as handed down in story,
Honor's been given—none braver deeds
Have done than make a Fireman's glory

A view of Park Row in 1864 showing a section of "Newspaper Row" now occupied by the Brooklyn Bridge. The New York Daily News building was at No. 19 Park Row, the New York Staatz Zeitung at No. 17 and the first quarters of the Associated Press at No. 15 Park Row

organized in 1833, occupied the second Tammany Hall structure mentioned above, and was the city's first penny paper. The name of its founder, Benjamin H. Day, is probably repeated more often than that of any other editor in the profession, for he developed the printing and reproducing screen which is still known as Ben Day. *The Spirit of the Times*, the first sporting paper in the United States, began publication here in 1831.

The *Staats-Zeitung* was located at the corner of Centre Street and Park Row where the Municipal Building now stands, and the *New York World* was on the northeast corner of Beekman Street and Park Row. Joseph Pulitzer's magnificent Pulitzer Building was completed in 1890 on the site of French's Hotel. James Gordon Bennett's *Herald* stood farther down, at the corner of Broadway and Ann Street*, site of the present St. Paul building.

Here, too, was formed the Associated Press, which was founded by the *Journal of Commerce*, the *Courier & Enquirer*, the *Tribune*, *Herald*, the *Sun* and the *Express*, in 1848-9. The *Times* joined in 1851 and the *World* in 1859.

On the southeast corner of Nassau and Spruce Streets, were the offices and salesroom of Currier & Ives. John "Old Time Enough" McDermott, a member of Excelsior Engine Company No. 2 and later its foreman, joined the firm of Currier & Ives in 1854. Born in Manchester, England, he was brought to this country by his parents in 1841 and was a picture-framer by trade. It was at his suggestion and under his direction that the "Life of a Fireman" lithograph series were published. The first four subjects are the most accurate and colorful representations of the operations of the old fire department ever produced.

The series contains many excellent portraits of prominent fire laddies, including those of Chief Engineers Alfred Carson, Harry Howard and John Decker. In "The Night Alarm," Louis Maurer, the artist of the series, depicted Mr. McDermott and his "left-handed Mascot (John F.) Sloper" holding the tongue of the machine. Nathaniel Currier himself was a member of Engine Company No. 2.

Most of the newspaper offices suffered serious fires at one time or another. During a heavy snowstorm on February 5, 1845, at about 4 A.M., a fire broke out in the office of the *New York Tribune*, covering lots No. 158 and 160 Nassau Street. The drifted snow

**Previously located on the northwest corner of Nassau and Fulton Streets.*

had made the streets impassable and it was almost impossible to drag the engines. The few that reached the scene found the hydrants frozen and had to break them open with axes. Although the firemen managed to subdue the blaze*, the loss was great and the adjoining building, at Spruce and Nassau, was also destroyed.

Although the *Tribune's* entire back stock of books was destroyed, Margaret Fuller's "Woman in the Nineteenth Century," was on the presses at another office and was preserved.

The Morse Building, which stood on the corner of Nassau and Beekman Streets, was the first office building erected in the city strictly for general office purposes. The owner was so cautious in his desire for safety that this building was constructed with walls three feet thick. This was fortunate for all of Beekman Street east of this point and perhaps for that entire section of the city, because the building was able to withstand the terrible flames of the *New York World* building fire.

In 1841, No. 126 Nassau Street was the home of Mary Rogers who sold cigars and tobacco in John Anderson's store on Broadway near Duane Street. Mary's beauty and charm attracted many of the prominent men who patronized Anderson's place, among whom were General Winfield Scott, James Gordon Bennett, Fenimore Cooper, Washington Irving and Edgar Allan Poe. Poe, incidentally, was a very familiar figure in this neighborhood for many years and worked in the publication office of the *New York Mirror*, which then stood on the northwest corner of Ann and Nassau Streets.

Mary Rogers was admired and respected by all. Her disappearance and the finding of her horribly mutilated and disfigured body floating in the Hudson River near Hoboken shocked the entire city. In spite of all efforts, this murder mystery was never solved. Edgar Allen Poe based his "Mystery of Marie Roget" on this case, and it is said that the story represented his theory of the crime.

In 1842, New York City, with a population of about 5% of its present total, nevertheless boasted 15 daily, five Sunday and six Saturday newspapers, compared with seven leading dailies and six Sunday papers in 1951. Among them the *New York Aurora*, competing with such famed sheets as Horace Greeley's *Tribune* and James Gordon Bennett's *Herald*, would have been quite inconspicuous were it not for the blazing editorials of its young editor Walt Whitman. The twenty-two year old Whitman edited the *Aurora* for two months before the publishers endeavored to tone down his editorials and he left the paper in disgust. An idea of his style can be gathered from his description of rival publisher Bennett: "A reptile marking his path with slime . . . a midnight ghoul, preying on rottenness and repulsive filth." Later Whitman himself reviled the *Aurora* as "scurrilous and obscene."

The *New York Aurora* was founded in November 1841 by John F. Ropes and Anson Herrick. It was particularly devoted to news of local events—fights, scandals, theatre, etc. The plant and office of the paper were at 162 Nassau Street and were known as *The Great American Newspaper Establishment*. The illustration and poem on the New York fireman appeared in the *Aurora* as part of a series of sketches of New York characters. They are reproduced by permission of Professor Joseph Jay Rubin** who found the files of the *Aurora* in the Paterson, New Jersey, library.

Walt Whitman became editor of the *Aurora* on March 28, 1842. He had already had ten years' journalistic experience, starting as a journeyman printer on the *Long Island Patriot*. Later he had worked on the *Long Island Star* and in 1838 had edited and printed his own paper, *The Long Islander*. A dude and a dandy with a great lust for living, Whitman was referred to as "the pretty pup" by his colleagues.

Charles Dana, Horace Greeley, Joseph Pulitzer and James Gordon Bennett were familiar figures along newspaper row in its lustiest days. Rivers of ink flowed off the presses of the square. But then in the twentieth century came a scramble to move uptown and one paper after another departed. Still carrying on like a hardy perennial on the square, today only the *New York Journal of Commerce* remains, firmly ensconced in the domed World Building. Nearby on North William Street is the German newspaper *Staats-Zeitung*.

The Brooklyn Bridge in 1890

**In an item entitled "The Card," which they published after the fire, the editors expressed their appreciation of these efforts. "We desire to return our heartfelt thanks," they wrote, "to those firemen, who, in defiance of the most furious storm, dragged their engines through streets impassable by ordinary efforts, and desperately, though vainly, struggled to save some portion of our property. Had the fire occurred on any other night of the last two years, these efforts must have been successful."*

***Professor Joseph Jay Rubin and Charles H. Brown recently co-authored a collection of Whitman's* Aurora *articles and sketches under the title, "Walt Whitman of the New York* Aurora, *Editor at Twenty-two."*

My Sister Anna

PETER STUYVESANT was New York's first great man. Almost entirely unsupported by his superiors he was tireless in his determination to do the best he could to fulfill the responsibilities of his leadership. He was a living paradox in that he could be deeply pious or blasphemous, arrogant, yet humble. Unapproachable as civic leader, as a private citizen he was a friendly neighbor and a lover of music, particularly if played by old Mingo, his favorite slave, who often entertained the family with his fiddle.

Peter and his sister, Annake, were the two children of the first marriage of the Rev. Balthazar Stuyvesant. From early childhood in Friesland they were devoted to each other, and although Annake was eleven years younger than Peter, her strong sense of justice and resolute character were always a match for Peter's fiery and impatient spirit.

Anna Stuyvesant

It was to her that Peter turned after receiving the nearly mortal wound in his right leg—a souvenir of his ill-fated military expedition against the Portuguese island of San Martin. The decision to amputate the leg was made during this stay with her. Throughout the long convalescence which followed, she was constant in her care for him.

It is not generally known that Peter and Annake married a brother and sister, the children of Rev. Lazair Bayard, the descendant of an eminent French Huguenot family. Anna was married to Samuel Bayard in the Walloon Church in Amsterdam on November 7, 1638. It was probably about this time that Peter married the lovely Judith Bayard.

Judith accompanied her husband to the New Netherlands, where they arrived May 11, 1647. Annake's husband, on the other hand, died soon after Stuyvesant's departure. She eventually turned to Peter and, in 1656, arrived in Manhattan with her three children. After her arrival she married Nicholas Verlett, an official of the Colony, who received a grant from her brother for the tract called Hobuk (Hoboken) in February, 1663.

In the reminiscences of Sarah R. Van Rensselear, Mrs. Stuyvesant was described as being "gentle and retiring in her manners, but . . . possessed of great firmness of character." She was very beautiful, spoke French, Dutch and English fluently, had a sweet voice and exceptional taste in music, and displayed great artistry in the choice and arrangement of her dress, which was styled in the French manner.

In 1657 Annake succeeded in influencing her tempestuous and strong-headed brother in a matter so delicate that it was a true gauge of her strength and persuasive powers. At that time the Quakers were very severely treated by the Governor. They were banished or thrown into prison on the slightest pretext. "For publicly declaring in the streets," one of them, a man named Robert Hodgson, was led at a cart tail with his arms pinioned, then beaten with a pitched rope until he fell, and at length set to the wheelbarrow to work at hard labor. Filled with compassion for the poor man, Annake intervened in his behalf. Stuyvesant, unable to withstand her prayers and righteous indignation relented. Hodgson's fine was remitted and he was released from prison, but he was banished from the colony. It was owing to Annake's action in this case that, from that time onward, not a single Quaker was so cruelly persecuted in the New Netherlands as Hodgson had been.

His sister's influence on Peter and on the colony was probably much greater than any history book has recorded. She was the personification of that dependable and feminine type of womanhood, who, long before woman's suffrage was ever conceived, could lay aside suffrance long enough to put an end to foolish masculine goings-on.

Peter Stuyvesant appointing the first fire wardens

Seal of Peter Stuyvesant

Petersfield

NOT MUCH is known of the original home of Peter Stuyvesant which stood on the south side of the present 10th Street, 200 feet west of Second Avenue. The residence of Petrus Stuyvesant, great-grandson of the governor, was called Petersfield. It stood on what is now the block bounded by First Avenue, 15th and 16th Streets, and Avenue A, and was located on a promontory (see map) overlooking the river. This pleasing homesite was enclosed by two low meadows through which, on the north, meandered the Crommessie Fly or brook (its source was near Madison Square) and on the south, Stuyvesant's Pond, which in winter often flooded and widened into a larger pond, used for skating. The house was for some time under the care of the son of Petrus, Peter G. Stuyvesant, a great-great-grandson of the Governor, who was president of the Historical Society. By 1831, although the house was still in existence, it was in a state of ruin.

Just beyond "Peter's Field" and mansion, extending up to what later became the site of Bellevue Hospital, was a great bend or bay, which was gradually filled up with earth from the adjacent high grounds, and laid out in streets and city lots. The Stuyvesant flats were converted to meadow and then to building sites. The ancient oaks were uprooted, the hills levelled, and the Indian graves, with their fragments of oyster shell and bits of pottery, scattered.

St. Mark's Church rose on the ground where once the second Dutch Reformed Church had stood, within whose walls lay the Stuyvesant family vault. Here, marked by the original stone, lay the remains of "Petrus Stuyvesant, late Captain General and Governor in Chief of Amsterdam in New-Netherland, now called New-York, and the Dutch West-India Islands. Died in A.D. 1671-2*, aged 80 years. Rest in hope."

Few of the Stuyvesant relics have been preserved for modern eyes. The old country mansion of Governor Peter Stuyvesant, inherited by Nicholas, his great-grandson, with part of its furniture and contents, was burned in 1778. Watson, in his 1828 journal, described portraits of Governor Stuyvesant and his son which he had seen, also Peter's christening shirt, "the very infant shirt of fine Holland, edged with narrow lace, in which the Chief was devoted

*February, 1672.

2 Milestone on the Bowery near 16th Street

The residence of Peter Stuyvesant, great-grandson of the Governor, stood on a site now bounded by First Avenue, 16th Street, Avenue A and 15th Street, overlooking the East River.

Petersfield section of the Stuyvesant estate, circa 1775. Looking east from the Bowery (Fourth Avenue) at a point near the present intersection of 11th Street (see accompanying map).

in baptism and received his christening!" Nicholas told Watson that during the times of Queen Elizabeth, it was the custom to give christening shirts as memorials; later Apostles' spoons were given.

Another custom peculiar to early Dutch families—of which the Stuyvesant clan was no exception—was that practiced by the Dutch patriarch who gave a bundle of goose quills to a son, requesting that he, in turn, give one to each of his male descendants. Watson records an instance of this tradition when he saw one in the possession of James Bogart which had a scroll appended, reading, "this quill given by Petrus Byvanck to James Bogart, in 1789, was a present in 1689, from his grandfather, from Holland."

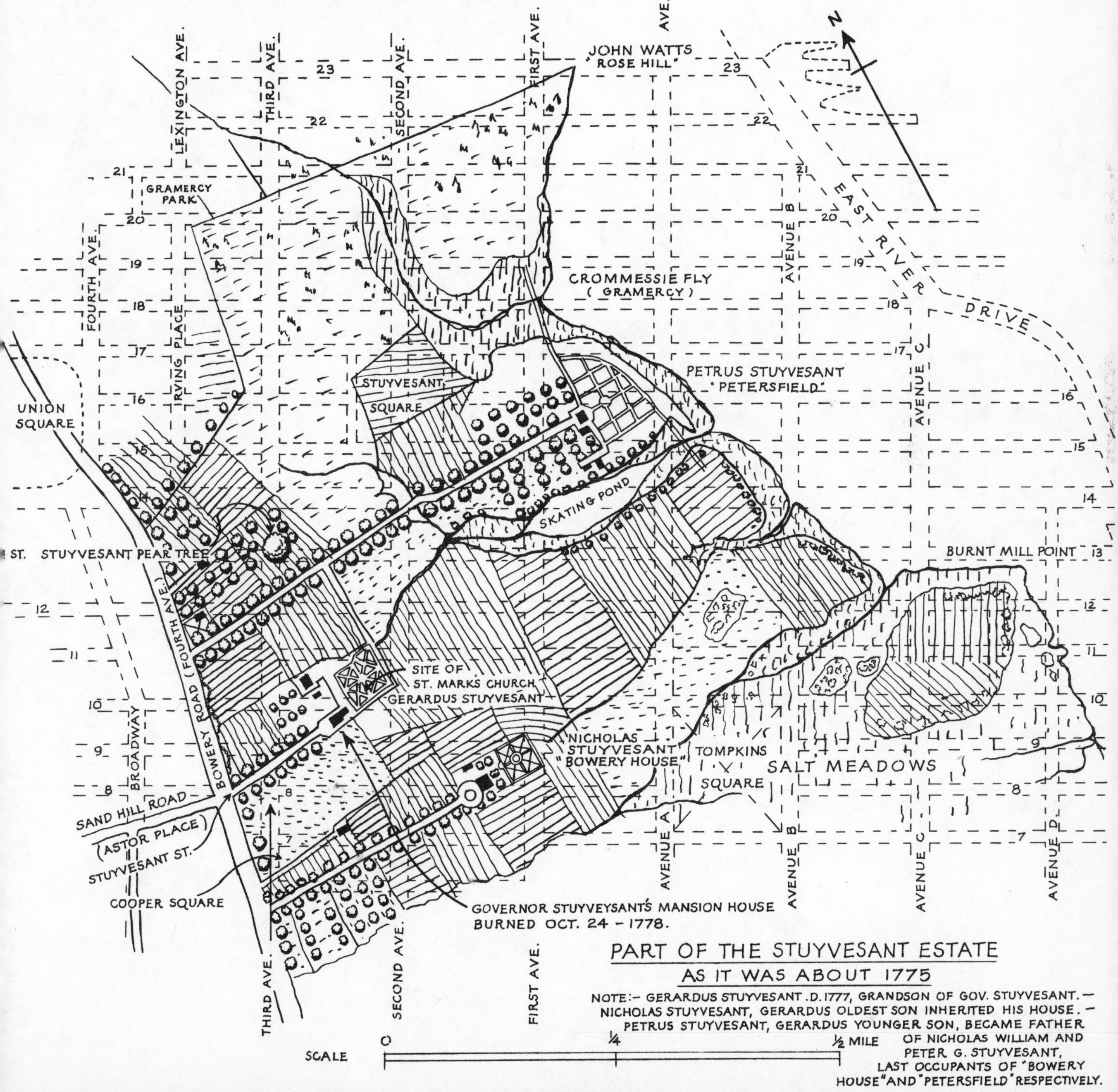

PART OF THE STUYVESANT ESTATE
AS IT WAS ABOUT 1775

NOTE:— GERARDUS STUYVESANT .D.1777, GRANDSON OF GOV. STUYVESANT.— NICHOLAS STUYVESANT, GERARDUS OLDEST SON INHERITED HIS HOUSE.— PETRUS STUYVESANT, GERARDUS YOUNGER SON, BECAME FATHER OF NICHOLAS WILLIAM AND PETER G. STUYVESANT, LAST OCCUPANTS OF "BOWERY HOUSE" AND "PETERSFIELD" RESPECTIVELY.

Stuyvesant's Pear Tree

The Stuyvesant Pear Tree, northeast corner of 13th Street and Third Avenue, in 1863

"Fam'd Relic of the Ancient Time, as on thy form I gaze
My mind reverts to former scenes, to spirit-stirring days:
Guarding their sacred memories, as ashes in an urn,
I muse upon those good old times; and sigh for their return."

ON an otherwise unmemorable day in February, 1867, two vehicles collided at the northeast corner of Third Avenue and Thirteenth Street and destroyed one of the oldest and most interesting relics in New York history. It was a pear tree, brought from Holland by Peter Stuyvesant in 1647 and planted, with others, by his own hand to form an orchard on his estate. For more than two centuries this one survivor was a source of wonder and speculation to the citizens and an inspiration to journalists, diarists and poets.

The old "pere-boome" standing on the Post Road witnessed many strange and stirring events. Beneath its boughs passed many a hero of the troubled times preceding the Revolution. Isaac Sears, a spirituous figure of those days, passed it after his raid on the printing establishment at the foot of Wall Street in which he pied Rivington's type. The bitterly resented Tory editor of the Royal Gazetteer was put out of business by Sears and his company of Connecticut light-horsemen who swooped down on Rivington's shop, destroyed his press, and then, to the tune of Yankee Doodle, carried off every stick of type-fonts to be re-molded into "more understandable words" in the form of bullets.

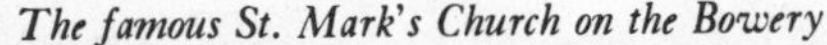

The famous St. Mark's Church on the Bowery

For a century the tree flourished and blossomed; then suddenly the branches began to decay and finally fell off. People supposed it to be dead, but without any artificial measures being taken, the venerable monarch recovered its aged strength and began to sprout hardy new shoots. On May 4, 1820, when the tree was 173 years old, an article in the New York Evening Post stated that "the tree appears to be no more than 30 years old; the fruit ripens the latter part of August, has a rich succulent flavor, and has been known by the name of the spice pear. . . . This is probably the oldest fruit tree in America."

In May of 1847, the tree reached the age of two hundred years, and thus became an object of peculiar regard, though it had long been viewed as an interesting relic. Lossing wrote, "I saw it in 1852, white with blossoms; a patriarch two hundred and five years of age, standing in the midst of strangers, crowned with the hoary honors of age, and clustered with wonderful associations. An iron railing protects it, and it may survive a century longer."

But, alas, Lossing's prediction did not come true, or we might still see the pear tree today, blossoming in an even stranger appearing world and be moved to express our feelings in verse, as Henry Webb Dunshee did in his Address to the Stuyvesant Pear Tree (above).

The scene on the opposite page, looking west toward Fourth Avenue, shows the Octagonal Reservoir of the Manhattan Water Company of which the rotund and cheerful Bob White was cashier. The house to the left was the old Washington Institute. The building later became the first home of St. Vincent's Hospital which, in 1949, in their present home on the northeast corner of 14th Street and Seventh Avenue, celebrated their One Hundredth anniversary.

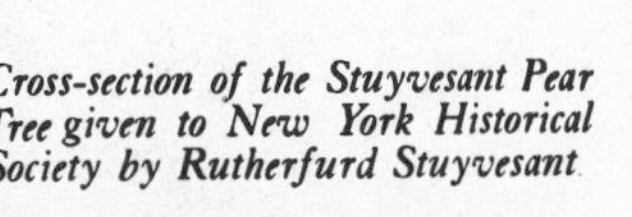

Cross-section of the Stuyvesant Pear Tree given to New York Historical Society by Rutherfurd Stuyvesant

19
LAFAYETTE 19
E.P. CHRYS

On the Bowery

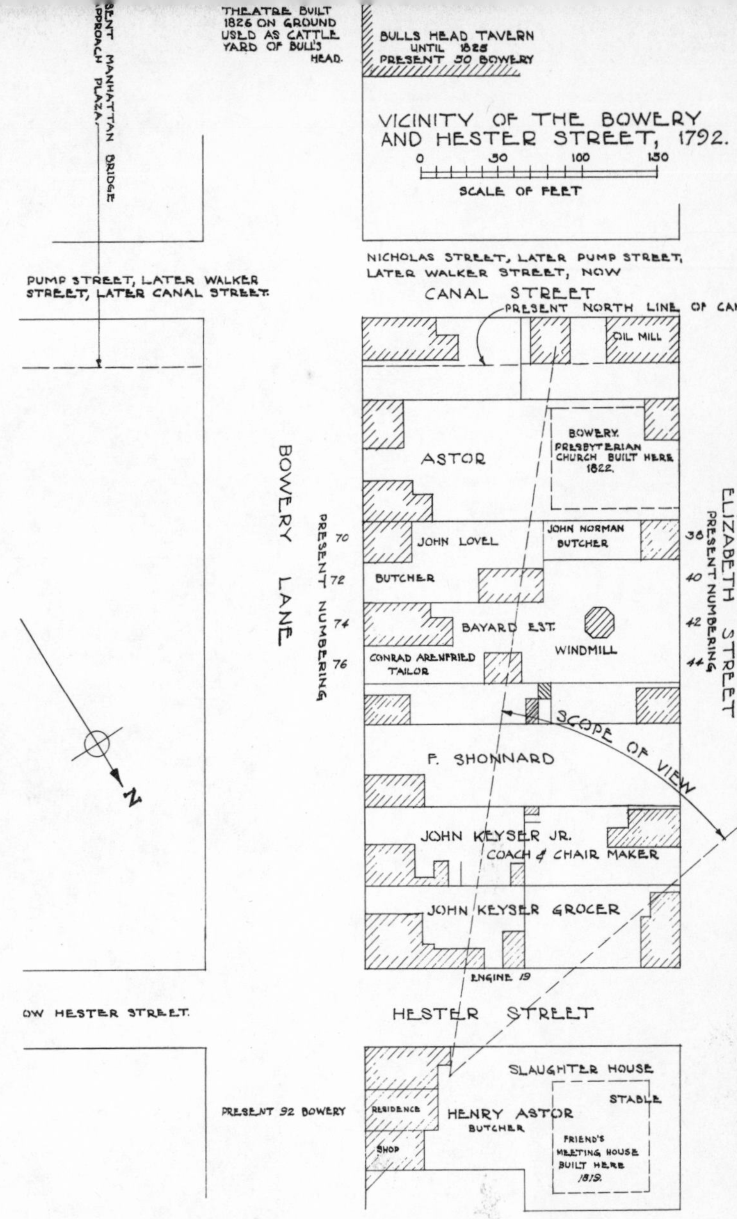

GRACED by a long line of stately poplar trees and the lazy wings of what was probably New York's last windmill, the vicinity of Bowery Lane and Hester Street was one of early Manhattan's most delightful spots. This site, on which the house of Lafayette Engine 19 was constructed in 1792, was acquired from the Bayard estate in 1780 by John Keyser, a grocer on the Bowery Lane, and his son, John Keyser, Jr., a coach and chair maker who built snow runners for the fire engines in 1795.

For three generations the Keysers were merchants on the Bowery. One of them, a prominent Bowery butcher, served as the inspiration for an early slang expression, "I kills for Keyser." Frank Chanfrau used the expression in his interpretation of "Mose the Fire B'hoy," and almost overnight it became "the rage" on the Bowery.

The property directly across the street from Keyser's lot on Hester Street was owned by Henry Astor who lived at 92 Bowery Lane and kept a shop at 94. Astor bought the property from the Bayards in 1785 using the name Henry Ashdore in the conveyance. He came over to America with the Hessian mercenaries during the Revolution and, impressed with the opportunities available here, requested and was given permission to stay.

Bayard's windmill , shown in the view, was built and owned by the Bayards, though they sold their entire Bowery front including the windmill in 1781.

Fire-fighting has always been a hazardous business. On February 6, 1838, fire broke out in a stable on Laurens Street (now South Fifth Avenue) which was owned by Peter Lorillard, early tobacconist who founded the great institution that flourishes to this day. Engine 19 attended the fire and John Buckloh, one of her boys who was holding the pipe, was buried under a falling wall and killed. It was at this fire that the entire company barely escaped being sent to their Maker. Under a constant rain of sparks, within a burning warehouse, the firemen were clambering over a stack of well-filled bags, casually wetting down the burlap underfoot as they moved about their duties. It is well they did, for after the fire it was discovered the bags contained enough gunpowder to blow the entire section and all within it to oblivion.

Engine 19 and Engine 7 waged a minor war over the use of the name Lafayette. Each claimed prior rights to the title, the point being so finely drawn that one company insisted it had met at seven o'clock to adopt the name while the rival company didn't convene until eight.

Lafayette 19 had a very active career and they achieved great prominence by virtue of their bravery in attempting to protect the colored children at the Orphan Asylum fire, July 13, 1861. Under threat of stoning from the rioters who had set out to destroy that building during the Draft Riots, the gallant firemen worked furiously to end the fire while shielding the orphans from the fanatical mob.

The old Bull's Head Tavern, headquarters of the drovers and cattle dealers of the city, was a fixture on the Bowery for many years. It later became the New York Hotel and still later its site was occupied by the Atlantic Gardens. The cattle yard at 46-48 Bowery was the site of the Bowery Theatres which burned and later of the famous Thalia Theatre.

The Old Horse Fair In Chatham Square

When adventurous Dutch souls probed north of the infant settlement of old Manhattan and approached the Fresh Water Pond, they found a well-worn path* leading through the hills lying to the east of that body of water.

It was along this path that the first boweries and farms were settled. As travel along its beauteous sylvan course increased, the old trail was gradually widened and improved, becoming known as the Bowery Lane or Road. The tract of land around the present Chatham Square, later a part of the "Dominie's" grant, was deeded to Thomas Hall, an Englishman, before 1652. Early maps reveal that a considerable rise of ground covered the area which was to become Chatham Square. The hill was steep enough to have diverted the old trail which skirted its eastern base. This ground appealed to Hall as an ideal spot for a tavern. In 1660, Hall and other settlers in the vicinity of the Fresh Water petitioned the Governor for permission to form a village there. The inn was built at the present southwest corner of the Bowery and Doyer Street.**

Although sometimes called the Farmer's Tavern, and once known as the True American, the inn was mostly referred to as the Plow and Harrow from its sign which displayed a painting of those pieces of equipment. At the beginning of the Revolution the old inn and its two barns, then occupied by John Fowler, were used by the Americans as a hospital. After the British occupation of New York in September 1776, during which several skirmishes took place in this vicinity, many of the houses in Chatham Square and on the Bowery Road were sequestered as billets. The 17th Dragoons were quartered along a considerable stretch of the road beginning at No. 1.

Stimulation came to this little rural settlement in 1731, when a smallpox epidemic broke out in lower Manhattan during which the Provincial Assembly found it expedient to retreat to an untainted area. So it was that that august body came up to the healthy "Fresh Water Hills" to conduct their business in the commodious house of Harmanus Rutgers, whose farm extended far to the east towards Corlear's Hook. In 1775 this house was willed to the four daughters of Harmanus' son Hendrick, Catherine Bedlow, Ann Bancker, Elizabeth De Peyster, and Mary Rutgers. Ann lived there until 1790, when the property was sold to the General Society of Mechanics and Tradesmen. The Society found the shape of the lot inconvenient and auctioned it off to Josiah Furman.

As for other thoroughfares in the neighborhood, after access to the original cemetery had been cut through by the building up of the lots on Chatham Square, Jew's Alley became the way leading from Fayette (now Oliver) and Bancker (now Madison) Streets to the Jews Burying Ground. A part of this sacred ground of the early Spanish Jews, maintained by the Congregation Sheareth Israel, still survives. This is probably the oldest existing cemetery, excepting Trinity, in the city today. Pell (once also called Pelham) Street was named after Joshua Pell, who purchased a plot there in 1781. Bayard Street was so

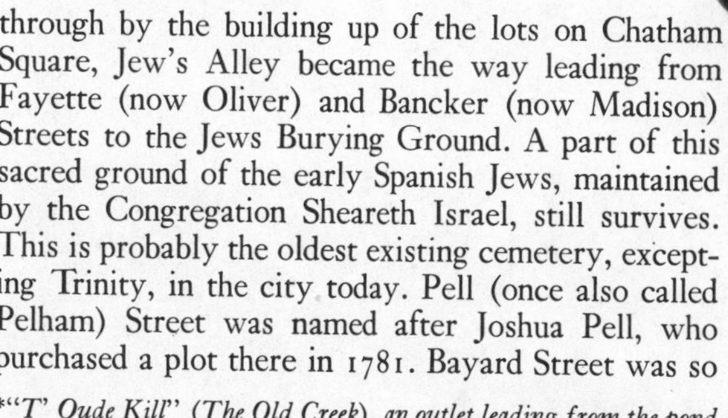

Hitching post

"T' Oude Kill" (The Old Creek), an outlet leading from the pond to the East River crossed the route in the vicinity of the present Roosevelt Street. In wet seasons and high water this intersection could become a morass but the route was dry in contrast to the marshes of the Lispenard meadows to the west. This eastern route was the Weckquasgecks Trail of the Indians which wound a tortuous course through the "howling wilderness" of upper Manhattan all the way to the fording place at the Harlem River.

***In 1665, Adriaen Cornelissen, Hall's husbandman and manager, asked for the abatement of excise tax, "as he is daily asked by those passing by (the Tavrin) for a drink of beer and he can scarcely accommodate them, as he has heretofore found by experience that if he pay the whole tapster's excise, no profit but loss will be realized by the spilling of the beer in carting, loss of time," etc. He obtained relief by being allowed to lay in half a barrel of strong beer weekly.*

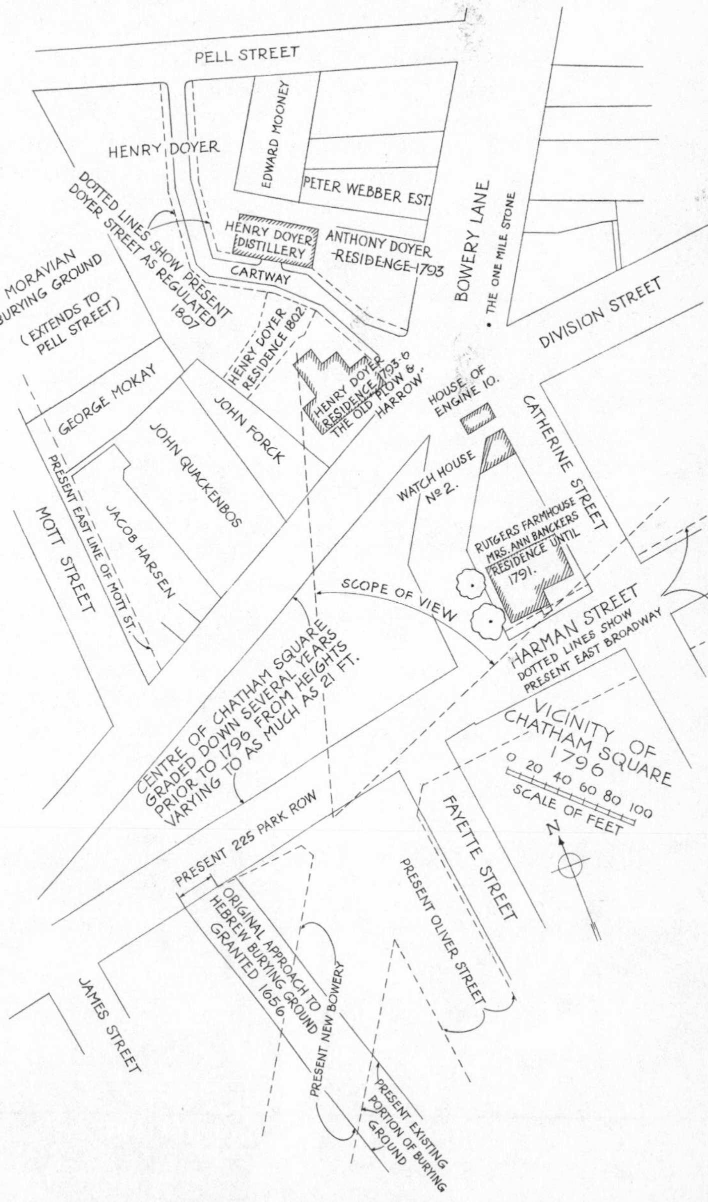

E. Didier's *"Auction Sale in Chatham Square," dated 1843*

designated for Nicholas Bayard, whose large farm included Bayard's Mount, sometimes referred to as Bunker's Hill, the highest ground in that part of the city.

Manhattan once had its own "western" style range* life. By 1671, all horses "at range" were ordered branded and tallied, and two stud horses were "to be kept in commons upon this island." In time, bands of escaped horses bred in the woods and meadows in the upper part of the island. Thousands eventually ran at large and were so numerous and wild as to be dangerous. Many roundups emanated from the Chatham Square area and many a cloud of dust arose there from the branding operations of the Manhattan cowboys.

Bowery Lane was from earliest times associated with racing, trial and sale of horses, and Chatham Square** truly became a "Horse Fair" when it was officially designated the principal place in the city for horse auctions, the ideal spot "for the sale of all horses and other animals to Canter or Gallop . . . provided that Cantering or Galloping shall only be within the limits of the Curb-stone around said Square." At the end of the Revolution all of the King's horses were disposed of here by auction when all the King's men departed from the city.

Henry Doyer bought the Plow and Harrow in 1793 and used it as a residence until he replaced it with a new home built at its rear in 1803. In the meantime he installed a distillery in one of the old "hospital" barns. In 1803-4 he built four houses on Chatham Square on the site of the old inn. One of them, with extra stories added, still remains on the southwest corner of the Bowery and Doyer Street.

Doyer Street was originally the private cartway leading to Doyer's distillery and stables. It was taken over by the city in 1807. Today, on the north side of Doyer Street, opposite Tom Noonan's Bowery Mission, there is an empty lot where the distillery once stood. On this exact spot, corn, squash and other garden truck† as well as flowers have flourished for the past few summers under the care of Howard Wade Kimsey, able superintendent of the Mission.

Around the corner, in a quaint frame structure at No. 6 Bowery, with its backyard adjoining the Doyer Street garden, stands the oldest apothecary shop in the city. Established in 1806, the neat, well-preserved, old-fashioned store of Dr. W. M. Olliffe is still maintained principally as a pharmacy by its present owner —Mr. Herbert Wilkes.††

For some years after the Revolution, Chatham Square continued rural in character, although by 1784 the rise of ground in its center had been gradually cut down to nearly the present level. In 1801 the Bowery was still bordered by farm houses as far north as Broome Street.

After the horse market was banished the square was ordered enclosed with a picket fence and the walks filled with gravel. An appropriation for trees and planting for Chatham Square Park followed in 1812. But as early as 1816 the hustle and bustle of this junction of busy thoroughfares made it necessary for the authorities to order that the park and fence be removed and the area paved.

Shortly after its organization, Niagara Engine Company No. 10 moved into the engine house "stand-

**In 1775 a "rustler," Dick Aldridge, was taken for horse stealing and pardoned. Who's ancestor?*

***For several decades prior to the introduction of the tracks and cars of the Third Avenue Railroad, Third Avenue, in the vicinity of Cooper Square, was used as a fast straightaway half-mile track for trotters. In the post-war period, several circus shows were held on the shores of the Collect at Mount Pitt and mention was made of the remarkable feats of horsemanship performed in 1786 by Thomas Pool, a circus equestrian, on Catiemutz Hill (above Katie Mutz's garden) near the Jewish burying ground.*

†Not a few rare Asiatic vegetables and flowers found their way to furtive germination here. The seeds were apparently carried to America in packing used in cases of imported porcelain.

††Next door is a tattoo shop and an expert black eye fixer.

Two views of a section of the old Jews Burying Ground, the city's second-oldest cemetery

Chatham Square about 1869. Note early funeral proces.

Hat front of William H. Wilson, famous Bowery fire cap maker.

ing in front of the New Watch House" at the head of Chatham Square. No. 10 remained in their original location until 1811, when their house was sold to residents of far-away Harlem. A new firehouse was built on the square in 1812 but when the area was paved it was replaced by a larger house in Fayette Street. While Chatham Engine 15 and Scott Engine 17 occupied this house at different times, in later years it was particularly the home grounds of Engine 15, and during its lustiest days, the runners of the company, known as the "Old Maid's Boys," and numbering several hundred, exemplified the old Bowery tradition of derring-do, tough pride and physical prowess.

The famous old Chatham Theatre just above Roosevelt Street was erected in 1839. At the height of its popularity it was owned by Charley Thorn and Tom Flynn, one of the most successful Irish comedians in theatre history. A heavy drinker, Flynn always met his obligations as businessman and actor.

In the 1840's a temperance crusade swept the city. Led off by the "Washingtonian Battery against Rum" of Engine 20, many engine companies signed the pledge in a body. Inevitably the temperance ladies asked the use of Flynn's theatre to advance the great cause, and with his infallible instinct for drama he not only agreed to their request but promised to address their first gathering himself.

The incredible news that Tom Flynn was to address a temperance meeting spread like wild-fire and to resolute tipplers desperately trying to hold their last line of defense it seemed that they were being betrayed by their best friend and most articulate champion. When the day finally arrived the theatre was packed from pit to dome. The stage was set with a scene from "The Drunkard's Home." On a table near the footlights was a half-filled water pitcher and a tumbler. Flynn stormed upon the stage amidst a thunder of applause. Filling his glass from the pitcher, with great composure the actor drank the contents, cleared his throat, and commenced his address. He was brilliant and voluble. With pathos and humor he pictured the drunkard's steps down to the depths of moral ruin. The audience was profoundly moved. At one moment sobs were heard; at the next, gales of laughter.

After two hours, Flynn reached his peroration with a final burst of eloquence. The audience, awed to deep silence, watched with dismay as the fire of energy suddenly seemed to drain from the actor. He was seen to falter, then fall fainting to the stage. As attendants carried Flynn around the corner to the old New England Hotel, the crowd ebbed away, singing his praises. Some of the temperance folks, however, in nosing around the stage, discovered that instead of water, the pitcher actually contained Old Swan Gin, Tom's favorite. The source of his inspiration suddenly dawned upon them with horror. The story soon got out and the resolutes leaped out of the trenches with a new and tender regard for their hero and enjoyed the first enthusiastic drunk in months. Tom Flynn was never again asked for the use of his theatre and temperance was set back at least five years.

In later days the turbulent Bowery was the most interesting street in the city. Saturday was "gala night." The shops and resorts were brilliantly lighted by turpentine and camphine lamps flaring above the sidewalks which were thronged by jostling pleasure seekers. In the midst of the spiels and confusion, the fruit and roast-chestnut stands and the hot bologna stalls gave forth their pungent odors. Cigar, liquor and souvenir shops plied their wares, and "Cheap Johns" urged the unwary passerby to sacrifice his hard-earned dollars for an eight-bladed pocket knife or for stockings "made in England for the Emperor of Siam."

Discussions are going on with regard to cleaning up the Bowery. Who knows but that once more the old lane may become a tree-lined thoroughfare of beauty?

Olliffe's Pharmacy at No. 6 Bowery—oldest drugstore in New York City.

Directly behind Olliffe's Pharmacy, the lot on Doyer Street where corn is grown by H. W. Kimsey of Tom Noonan's Bowery Mission.

Tom Noonan's Bowery Mission on Doyer Street.

Old frame building at 466½ Pearl Street.

Steve Brodie (at right) in back room of his saloon which stood at 114 Bowery.

C. PARSONS DEL.

LITH. CURRIER & IVES, N.Y.

THE LIFE OF A FIREMAN.

The new era. Steam and Muscle.

NEW YORK PUBLISHED BY CURRIER & IVES, 152 NASSAU ST.

THE LIFE OF A FIREMAN

The New Era. Steam and Muscle

DRAWN ON STONE BY C. PARSONS

LITHOGRAPH PUBLISHED BY CURRIER AND IVES, 1861

COURTESY OF HAROLD V. SMITH, ESQ.

One of Currier and Ives' most famous fire prints, the view shows a stirring example of an early fire scene in old New York—the fire at Murray Street on the northeast corner of Church Street, September 9, 1861. The fire broke out at 11: P.M. and, before the flames were checked had consumed seven buildings in the neighborhood. The City Hall is shown in the distance.

The engine in the center is thought to be the "White Ghost," Lady Washington Engine Company No. 40, which was made up at that time, of a group of young men from the Center Market district. The machine was of the piano crane-neck style and had two 8½" cylinders and a 9" stroke. It was built by W. H. Torbass in 1856.

The steam engine on the right was made by A. B. Latta of Cincinnati, Ohio, who built the "Uncle Joe Ross," Cincinnati's first steam fire engine. This engine, a so-called self-propelled machine was one of the several early machines paid for and presented to New York City's firefighters by the fire insurance companies. The engine was cumbersome and heavy to haul but once in position produced a powerful stream of water. In February, 1855, there was a test between this steamer and the Exempt Company's "Mankiller," a hand-pump engine. The hand-pumper threw a higher stream but after her men had dropped the brakes in exhaustion the steamer continued to sustain a heavy stream of water. This insured its success as the engine of the future even though held in resentment and even ridicule by some of the city's firemen.

The steamer at the left is "The Elephant," the Lee & Larned machine used by Manhattan Engine Company No. 8 and the first practical mobile steam fire engine put into regular service in New York City. A gift from the city's fire insurance companies in 1859, the steamer, even though weighing 5600 pounds, was considerably lighter than earlier steam engines used elsewhere because of the substitution of steel and brass for iron in its construction.

The fireman in the center, shouting directions, is John Decker, hero of the Draft Riots of 1863 and the last Chief Engineer of the New York Volunteer Fire Department. Decker served from 1860 to 1865, when the department was disbanded, and was one of the most popular firemen in the city. He later became president of the Veteran Firemen's Association.

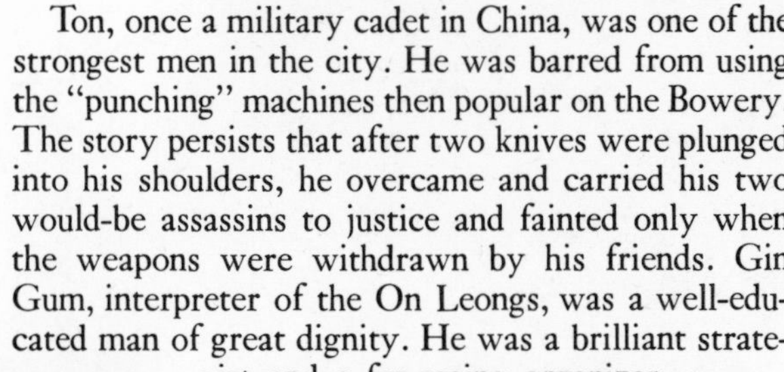

18 Mott Street, about 1900

華 Chinatown 埠

LYING between old Five Points and the Bowery is New York's Chinatown. Within this small island of six blocks, in the great sea of humanity that forms Manhattan, live approximately 7500 Chinese-Americans. This area, however, becomes a gathering place on weekends and holidays for many of Greater New York's total Chinese-American population of more than 30,000.

Lŭk Yee (green coat)

As time goes, Chinatown is comparatively recent in origin. In 1858, a Cantonese named Ah Ken, the first Chinese to settle here, opened a small cigar store in Park Row and made his home in Mott Street. By all accounts, he was an industrious, respectable man. But a decade later, Wah Kee, the second arrival, discovered that while he could do a good business in fruits, vegetables and oriental curios, he could acquire much larger profits by operating a gambling den and opium-smoking dive above his store at 13 Pell Street. As word of Wah Kee's riches spread abroad, new arrivals began to follow his example, riffraff of every nationality were attracted by the promise of easy money and the neighborhood swiftly began to deteriorate.

The Chinese Theatre, 5-7 Doyer Street. Now occupied by Tom Noonan's Bowery Mission.

The population of Chinatown mushroomed suddenly. As late as 1872 there were only 12 Chinese in the area but by 1880 the number had increased 700. During the next thirty years, the population grew to nearly 15,000.

The Chinese have always been great "joiners," having since early days in America, formed family societies as well as associations based on the districts of South China from which most of them came. The first association formed in New York was the Gee Kung or Chinese Freemasons.

The tongs, which originated not in China but in America, were organized first as sort of vigilante associations and later as business organizations. The tong wars broke out in 1899 and almost immediately gave the district a sensational and notorious name. The Hip Sings of Pell Street and the On Leongs of Mott Street were the two great New York rivals and at one time a member of either tong virtually signed his death warrant if he ventured into the territory of his enemies. Tom Lee, "Mayor" of Chinatown, and a deputy sheriff, and Mark Duck, a brave little man and a gambler who would wager his total wealth on the odd or even number of seeds in an orange, were credited with being the heads of the rival tongs for many years. It is the author's belief that the real tong leaders behind them were men such as Ton Bock Woo of the Hip Sings and Gin Gum of the On Leongs.

Ton, once a military cadet in China, was one of the strongest men in the city. He was barred from using the "punching" machines then popular on the Bowery. The story persists that after two knives were plunged into his shoulders, he overcame and carried his two would-be assassins to justice and fainted only when the weapons were withdrawn by his friends. Gin Gum, interpreter of the On Leongs, was a well-educated man of great dignity. He was a brilliant strategist and a far-seeing organizer.

The "tong war" era was one of quiet-footed assassinations* and open battle. Fifty Chinese were killed at the famous Bloody Angle in Doyer Street in one fight. Murderous hails of bullets swept Doyer and Mott Streets on a number of occasions. The tongs repeatedly signed truces and often broke them. Peace once was obtained by the direct intervention of the great Chinese Minister of State, Li Hung Chang, at the request of the U. S. Government. Tong wars are now in the past and Chinatown has become one of the quietest and safest parts of the teeming city.

Today, Chinatown is merely picturesque. Like the Bowery, the fires of emotion have burned out. About the only ones who ever shudder when walking through the narrow streets are the passing firemen, who envision what could happen here, in one of the city's worst firetraps.

If you visualize the Chinese-American as the living counterpart of the inscrutable, mysterious depths of Asia, think again. These solid American citizens are enterprising merchants and sound businessmen. They are shrewd, honest and courteous. Many of their antecedents in America were prospectors for gold and silver in the Rocky Mountains, side-by-side with yours. Today, Chinatown, with its tourist attractions, its piquant, delicious food, its strange vegetables and herbs, dried meats and sea food, tinkling novelties and colorful processions is a tourist mecca and a neighborhood of responsible parents and respectful children. Odd to say, this much maligned district has little or no juvenile delinquency and enjoys the lowest crime rates found among any segment of the population.

**Hatchets used by the tong fighters or Boo How Doy, were nothing like the crude hand axes usually depicted. They were very finely made, usually with a metal handle and metal head, delicate and small-headed like a tiny Indian tomahawk. "Executioners" could carry them neatly rolled in a newspaper tucked under an arm and could thus strike by stealth even in broad daylight with little chance of detection. The paper did not resist the razor-sharp cutting edge, and at the same time concealed any trace of blood.*

One of the most famous of the gaudy figures who haunted Chinatown and the Bowery in the days of their tawdry splendor was Chuck Connors (The Insect), who was glamorized as "The sage of Doyer Street" and the "Bowery Philosopher" by contemporary newsmen. He originally was a fireman on the 3rd Avenue Elevated Railroad, when they ran steam locomotives. Later, he lived by running benefits for himself at Tammany Hall, in the name of the Chuck Connors Club.

Vicinity of Five Points in 1827, a "bull-baiting, rip-roaring hell"

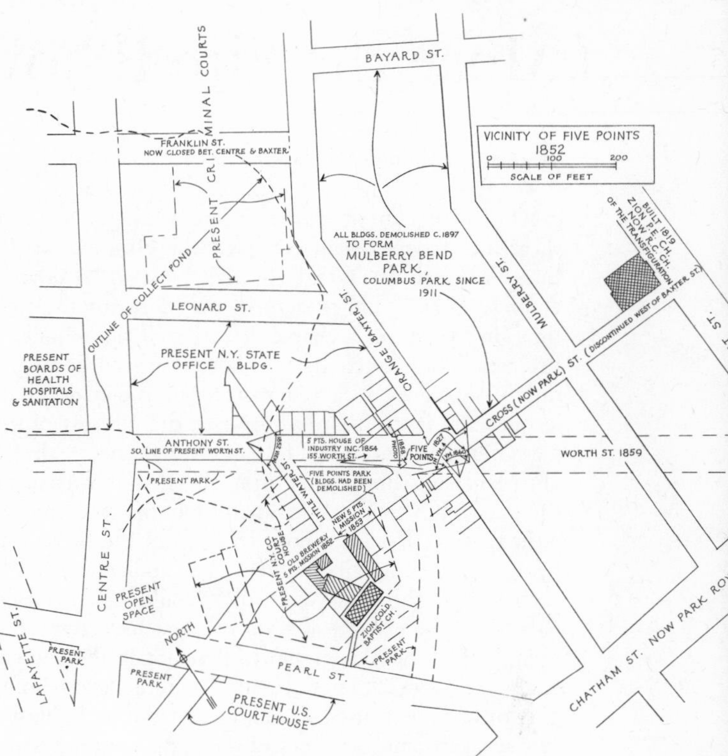

Five Points

FIVE points was formed by the intersection of Orange (Baxter), Cross (Park) and Anthony (Worth) Streets. In 1820, it was the poor people's Coney Island—but it soon became much more, and much worse. It was a section inhabited by toughs of the worst sort and by derelicts—outcasts of both sexes, and all ages. Philip Hone refers to "swarms of ragged, bare-footed, unbreeched little tatterdemalions, free-born Americans . . ."

At dusk, the Hot Corn Girls appeared, carrying their baskets of roasted ears while singing

Hot Corn! Hot Corn!
Here's your lily white corn.
All you that's got money—
Poor me that's got none—
Come buy my lily hot corn
And let me go home.

The earnings of the Hot Corn Girls were in direct proportion to their beauty and charm, and many a savage fight was waged over them.

In a notoriously bad section, the worst was the Old Brewery. This began its career respectably, as Coulter's Brewery in 1792, and was noted for the quality of its brew. In 1837, however, it had outgrown its usefulness and was transformed into a dwelling place for the thugs of Five Points.

It was a five-story building, rotting and dilapidated. Along one wall an alley led to a single large room in which more than seventy-five men and women of assorted nationalities and races lived together. This was the Den of Thieves. The name was appropriate. Along the other wall ran another filthy lane called Murderer's Alley, worse than the first.

Upstairs there were about 75 other chambers, housing more than 1,000 people—men, women and children. The section was a warren, with underground passages and murderous cul-de-sacs, into which the police dared venture only in large numbers, for the Old Brewery for a period of more than fifteen years averaged a murder a night. Not all the notorious characters were men. We know of Margaret Hall, known as "Gentle Maggie," and Elizabeth Jennings, "Lizzie the Dove," who settled their differences over the man of their choice with flashing knives.

Five Points was too tough, too unlawful, too unsavory to last, even in the New York of a century ago. The Old Brewery was finally razed, the last of the gangs destroyed. Today it bears little resemblance to the bull-baiting, rip-roaring hell it was in 1850. One bit remains—the little house running through from Worth to Park Streets, which was there then and is there now. The old Five Points Mission has also survived. Now located uptown and dedicated to helping needy children, it recently celebrated its 100th anniversary.

"Leather-head"

ld photograph of the Five Points rea made by E. Anthony in 1864.

Rear of an old frame house at 173 Worth Street

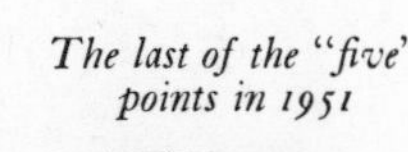
The last of the "five" points in 1951

The Old Brewery as it appeared, December 1852

Mose the Fire B'hoy

On the evening of February 15, 1848, at Mitchell's Olympic Theatre, a benefit performance was given for Benjamin Baker, stage manager for the theatrical group which occupied the Olympic at the time. A sketch was written by Baker, called "A Glance at New York," and presented, probably for the first time on any stage, a characterization of the volunteer fireman. The sketch itself, poorly written and only mildly amusing, was reminiscent of the many burlesques and farces that had aroused the interest of the playgoer that season and few in the audience realized, until halfway through the first act, that it was to be the toast of New York for many months to come.

In that act the farm boy, fresh from the hills, was being swindled by city slickers. The audience, used to such fare as this time-worn plot, could easily anticipate the "lines" and actions of the characters. But then, just as the play began to flounder in the ancient cliches, a new character appeared and like a refreshing gust of cool air roused the audience from its lethargy. His savage bluntness shocked them, his courage while racing to a fire and fighting it, thrilled them, and his dialect, that of the contemporary Bowery boy, both fascinated and horrified them.

Completely dominating the stage, Frank Chanfrau, as Mose the Fireboy, gave a memorable performance. He had the tone, the look and the manner of a real Bowery boy who ran with the engines, risking life and limb without pay, and his remarkable portrayal was rewarded by a tremendous ovation.

Mitchell's Olympic Theatre, where Mose was born.

Last scene from "New York As It Is"—1848

The piece was at once revised and the part of Mose amplified. An accompanying role was added, that of his "gal" Liz, charmingly interpreted by Miss Mary Taylor.

The play met with hosannas from most of the aisle-sitters of that era, a great deal of the praise being lavished on Chanfrau. Of his performance one critic exclaimed, "His characterization was perfect in every detail. He understood the phraseology and mannerisms that characterized the famous 'B'hoy dat ran wid der mersheen.' He affected the soap-locks of that period; the black 'plug' hat, the red double-breasted shirt, black trousers thrust into boots, the 'long-nine' cigar set between lips in a cocky upward tilt, the 'gallus' high-heeled boots and the general jerky motions and speech of a bully-boy of old Bowery Lane." Mose, the success of the season, ran seventy nights—an extraordinary record considering the fact that the population of New York at the time numbered no more than 400,000.

On the opposite side of the critical ledger, the dignified Anthony Child, editor of The Fireman's Journal, took Chanfrau to task for his personation. Though he praised Chanfrau as an actor, he sharply criticized the part of Mose. He said, "Its effects upon the fire department are serious, in the estimation of those who are not acquainted with its members, as they set every fireman down as a character degrading to youth." In spite of Mr. Child's remarks the firemen of the day flocked in droves to the old Olympic Theatre and their laughter and tears mingled in chorus with those of other citizens of New York.

Mose and his gal Liz

Not generally known until later was the fact that Benjamin Baker modeled the character of Mose after a New York fireman named Moses Humphreys, a member of Lady Washington Engine 40. Moses was touted as the most powerful man in the Volunteer Fire Department as well as being held as the city's most feared street-fighter. Humphreys, or "Old Mose" as he was called by his legion of admirers, was king of the Bowery and Five Points until he met his match in a street battle with Henry Chanfrau, a member of Peterson Engine Company No. 15. Ironically Hen was an elder brother of the young man who was destined to immortalize Humphreys' name and character. Frank, then a lad of twelve, was an interested observer of the fracas.

The fight, which took place in the vicinity of Pell Street and the Bowery on a Sunday afternoon in 1836, was one of the most ferocious brawls ever witnessed on the streets of old New York and it furnished conversation for many years afterwards. Not only the two gladiators and their companies but hundreds of other firemen and bystanders were involved in a brutal knockdown, drag-out struggle that ended in wild disorder and near panic. Humphreys eventually was unmercifully beaten by Hen, and the rest of Lady Washington's boys were badly defeated by members of the Peterson Engine Company.

Old Mose, his pride wounded beyond repair after this crushing fall from physical fame, fled to Honolulu and was never seen in these parts again. His legend, however, continued to grow, chiefly through the imagination and glibness of the volunteer firemen who created a prodigious character, the pride of old New York, whose stature and feats vied with that "country hick," Paul Bunyan. Frank Chanfrau as well as his older brothers, Peter, Joseph, and the mighty Henry, each ran with Peterson Engine 15 or the "Old Maid" in his time. This company, one of the toughest and most efficient of the old New York volunteer fire companies, was Chanfrau's "dramatic" school, a fact that was fully attested by the amazing accuracy and realism that made his role so outstanding.

The Bowery Boys guarding a hydrant

Frank Chanfrau was born in 1824 in "The Old Tree House," a frame dwelling which stood at the corner of the Bowery and Pell Street. The father of the Chanfrau boys, a French sailor, came to America on a ship that had brought Lafayette here on an early visit.

After trying an apprenticeship at carpentry and the ship-joiners' trade Frank found his way to the stage of the Bowery Theatre as a supernumerary. While there he attracted the attention of Edwin Forrest and other theatrical celebrities by his excellent imitations. He eventually made the rounds of several New York theatres in a number of different roles until he adopted the character of Mose which gained him fame and carried him to almost every theatrical town in America.

After his great success at the Olympic, other theatres were anxious to share his popularity so for a time he acted the play twice nightly, once at the Olympic and again with a companion piece called "New York As It Is" at the famous old Chatham Theatre of which he was manager.

Some of the later Mose plays were "Mose in California," "Mose in a Muss," "Mose's Visit to Philadelphia," and "Mose in China." Other actors, including John E. Owens and Junius Brutus Booth, Jr., began appearing in similar plays, but Chanfrau, who originated the type, was unique in popular favor.

A nostalgic remembrance of the New York Bowery boy* and Chanfrau's characterization was expressed by "Florry" (J. Frank Kernan) in his "Reminiscences of the Fire Laddies." He said, "I remember his red shirt, his fireman's hat with its high enscribed ensign in front advertising the number of the machine with which he ran, and its prodigious rear extension of brim; his soap-locks, his trousers in his boots, and his brass speaking trumpet . . . Mose was a coarse creature with an abominable dialect . . . and he had other habits of speech and conduct unfitting him for refined society; but he would plunge into a burning house and bring out in his arms helpless women and children, and stand on top of a ladder, with the flames all around him, enacting exploits of the most prodigious peril and valor; and the people loved him and went to see him, thronging the theatre wherever he appeared."

**The story of Thackeray and the Bowery boy has, of course, been told many times, but may be worth repeating. Walking through the Bowery, a friend suggested to Thackeray that he get into conversation with one of the "B'hoys." Thackeray went up to one who was loitering on a corner and said, "I want to go to the Bowery."*

The reply was, "Well, sonny, you can go as long as you don't stay too long."

Cherry Hill

IN 1754, New York lay in fear of attack—not by the British, for the Revolution was almost twenty years in the future, but by the French and Indians. On Cherry Hill, which is the section now overshadowed by the Brooklyn Bridge, palisades and blockhouses were erected in defense. At that time, the East River encroached much further upon the city. The palisades, which were made of cedar logs fourteen feet long and ten inches in diameter, extended from Cherry Street across to Windmill Hill and over to Dominie's Hook, on the North River. The attack never materialized.

In later years, another enemy, the British, fortified a line from Corlear's Hook to Bunker's Hill, near old City Hall, while their Hessians were encamped on another hill near Corlear's Hook. At this time, Cherry Hill was the finest residential district in the city. Indeed, when George Washington was inaugurated President, he maintained residence in the Franklin house* at the corner of Franklin Square. In 1817, the same house was occupied by DeWitt Clinton. Nearby, at 5 Cherry Street, was the home of John Hancock. Adjoining the Hancock mansion, at 7 Cherry Street, was the first house in New York to be supplied with illuminating gas. At Pearl Street, a few doors from Dover, stood "the most beautiful house in America," the Walton House. Constructed of yellow Holland brick, it had a double pitched roof covered with tiles and a double row of balustrades. At one time, its gardens extended down to the river. In the 17th century, Walton's Shipyards stood here. In 1794, the Bank of New York opened in the Walton House, and across the street, at 159 Queen Street, was the home of Isaac Roosevelt, third president of the Bank.

At 23 Cherry Street stood a tavern, The Well, which had its own claim to distinction. A favorite of army and navy officers, it was patronized by the officers of American privateers during the War of 1812 and it was to serve their hearty appetites that the beefsteak party originated here.

Another popular attraction in this neighborhood was Bowen's Waxworks, at 74 Water Street. The figures of many of the prominent men of the period were on display, done in wax and appropriately costumed. On September 14, 1789, General Washington, Martha and the Custis children visited the waxworks and were privileged to see a full-length figure of the President himself inside a canopy, dressed in Revolutionary War uniform, "a Fame, also in wax, extending over his head crowning him with a wreath of laurels."

After the turn of the 19th century, however,

**The palatial house of Walter Franklin, merchant and brother of Chief Engineer Thomas Franklin. Morris Franklin, Walter's son, was foreman of Engine 24, and became president of the N. Y. Life Insurance Co. in 1848.*

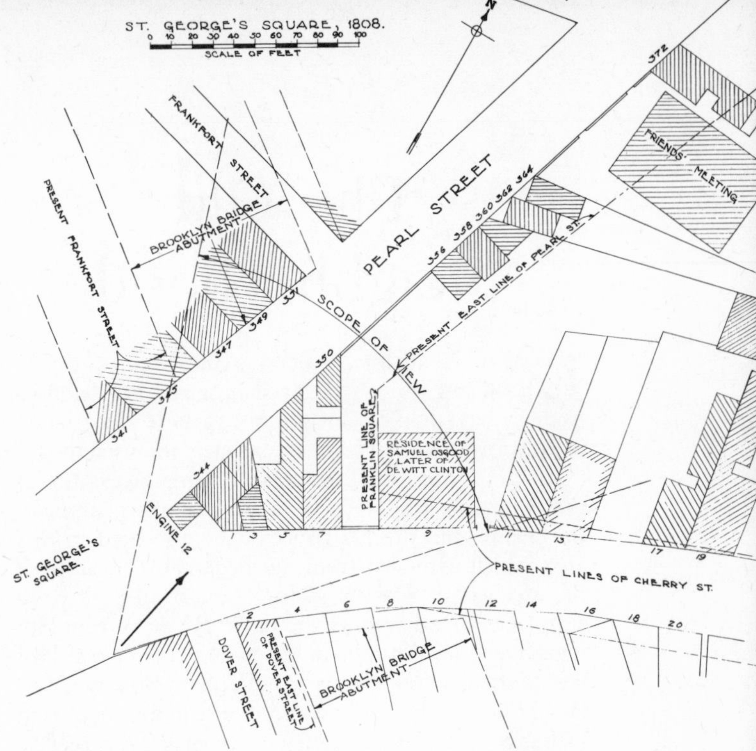

Cherry Hill soon passed its fashionable prime. The hungry hordes of Europe immigrating to the free new world settled on the fringes of Cherry Hill and exerted irresistible pressure on the aristocratic inhabitants, forcing them northward. By 1840 the mansions had given way to long rows of tenements second in ill-repute only to the horrible hovels of Five Points, to the north. Most unsavory of all, probably, was a dive maintained by a giant Negress known as The Turtle. She weighed over 350 pounds and it is said that her nickname was aptly chosen. The type of establishment she ran is left to the imagination of the reader. Cherry Hill had seen the last of gracious living now.

To go back, however, to better days—Cherry Hill was protected by Engine Company 12, which was located in Franklin Square, first known as St. George's Square. In 1832, the Company moved its headquarters to Rose Street, near Frankfort—and later to William Street, near Duane.

Franklin Square was the home of some noted printing and publishing establishments and was the scene of the terrible fire which broke out in the Harper & Brothers print shop on December 10, 1853. (Harper's was built on the site of the Roosevelt Sugar Refinery.) Spreading with incredible rapidity, the fire destroyed the entire building and attracted thousands of watchers from all over the city. Miraculously, over 600 people who were in the building escaped unscathed. The loss in this fire amounted to over $1,000,000, of which only $200,000 was covered by insurance. Fletcher Harper, watching his life's work go up in the flames, turned to Adam Pentz and said "If I could have my grandchild back, all this would mean nothing to me." (His grandchild had died several months previously.) The four Harper brothers met that night and decided to rebuild. The great house celebrated its centennial in 1950.

Nearby, at 338 Pearl Street, was the home of the famous weekly Police Gazette, which has become a byword in American pink-page literature. Published by Richard Fox, the Police Gazette catered to the male audience and at one time was standard equipment in every barber shop. The building still stands and traces of its earlier days can still be seen in the elaborate fire-escapes facing Dover Street. In intricate metal designs on the fire-escapes can be seen the heads of foxes (after the publisher) and heroic boxing scenes.

Emblem of the East River Savings Bank

Another quaint link with the past goes back even further. Every year, the owner is required to pay to the city a fee in peppercorns. This traces directly to a lease on the property in 1701! The building is now owned and occupied by the Thomas W. Dunn Co.

William M. Tweed was born at 24 Cherry Street and his father's chair-making shop was at No. 5. Captain Samuel Chester Reid, the man who, with one ship, stopped a British fleet long enough to prevent reenforcements from reaching New Orleans in 1812, lived at 27 Cherry.

Cherry Hill has gone through the cycle that is so typical of New York. First rural, then residential, it was the highly fashionable section at one time, gradually declining to genteel middle-class and finally slums. Now it shows signs of climbing to a new dignity again.

A view of Market and Cherry Streets in 1859 showing the flour store of S. Valentine, a charter trustee of the East River Savings Bank which was founded on Cherry Street. The Mariner's House is at the left of the store.

The Old Neighborhood

EAST of Chatham Street was the "old neighborhood" to those who lived in it and the "slums" to those who didn't. This was the earliest "East Side" of New York, as that term is sentimentally used today.

In early times the plantations; farms or estates of the Rutgers*, the Roosevelts, the Gouverneurs, the Beekmans, and the De Lanceys were located in this area which extended from the Beekman's Swamp all the way east to Corlear's Hook, and from Park Row and East Broadway to the East River. Originally there were several hills in the vicinity. Monkey Hill was a small knob near North William Street. Cow-foot Hill rose above the swamps a little distance from Hague Street in the westerly curve of Pearl Street... old folks related after the Revolution that they well remembered Indians making and selling baskets while camping at Cow-foot Hill. Cherry or Peck's Hill lay further east. There were several other eminences in the area ending with Gouverneur's Hill near Corlear's Hook.

Today, Henry Street, which has a number of well-kept old houses in the vicinity of the Settlement near Scammell and Gouverneur Streets, still gives some hint of the pleasant, almost rural, character of this section.**

The promontory of Corlear's Hook, known to the Indians as Nechtanc and to the British as Crown Point, was once part of a farm first owned by Jacob Van Corlear in Governor Van Twiller's time (1639). A party of frightened Indians sought sanctuary there from their northern enemies, the Mohawks, in 1643. Keift, a nervous and befuddled governor, ordered them attacked and they were disgracefully massacred at night by an expedition under the command of a notorious henchman of the governor's. Thomas Beekman bought the property in 1652.

The Hook was always believed to be the hiding place of vast riches, the loot of both Blackbeard and Kidd but it was prospected for generations without success. The high ground there was a favorite picnic ground and the beach below a favorite swimming and fishing place. In the Revolution the point was fortified and for some time the Hessians under General Knyphausen were camped nearby. Baptists from the city once used Corlear's beach as a place for immersion rites. On July 25, 1780, H.M.S. *Blonde* got off course and ran on the rocks at Corlear's. As time went on the Hook became a notorious place. Refugees from the indignation of outraged citizens and from the law swelled the criminal population there to sizable proportions. The number of "blondes" and their victims who got off course and were wrecked on the rocks of Corlear's Hook has never been estimated.

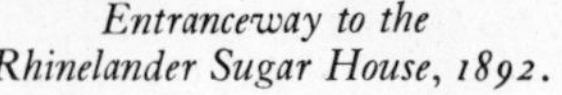

Entranceway to the Rhinelander Sugar House, 1892.

The Sugar House, Rose and Duan used as a prison during the Rev

In the so-called uprising of Negro slaves in 1741, a white preacher named John Huston was considered a leading instigator and was hanged in chains upon a gibbet erected at the southeast corner of Henry Rutgers' farm which is now covered by the Smith housing development. The gruesome remains of the victim, looking with sightless eyes out over the broad expanse of the East River, had a lasting effect on the entire vicinity and Huston's ghost was supposed to have haunted this spot for decades.

Engine Company No. 33, "Old Bombazula," was located at the Grand Street Market in 1813. Seven years later the company moved to a little one-story frame house on the top of Gouverneur Hill. When they moved again to a new firehouse at Henry and Gouverneur Streets in 1828, one of the Gouverneur ladies offered a golden trumpet to the company if they would adopt a new name, "Lady Gouverneur," in honor of her family which once owned a large tract of land in this neighborhood.

As many of the lads of No. 33 Engine were ship workers, their affection centered upon a certain privateer, the *Black Joke*. During the War of 1812, this fast and sturdy little craft, a champion moulded by their own hands, performed many daring feats extremely disconcerting to the mighty fleets of England, and its gallant commander made his home

Hook and ladder racing down Dover Street. Second boy on the right is believed to be Al Smith, who was an ardent buff as a young man.

**Below Henry Rutgers' farm once stood a race track to which folks of the 18th century made their way on an old road which followed the course of Bancker (now Madison) Street and its continuation, Bedlow Street.*

***Mechanics Alley, a relic of the old shipbuilding days, runs from Water Street to the river, nestling closely along the south side of Manhattan Bridge.*

Edward Brooks of Brooks Bros. was a member of Eagle Hose No. 1.

port headquarters in Cherry Street where they often saw him. Moreover, the title of Lady Gouverneur suggested the kind of thing the boys in the yards as well as the gunners of the *Black Joke* had been opposed to. So when a vote was taken the new name became "Black Joke," and the company continued to use the same old tin trumpets. This fire company started the first target company in the city and from this origi-

Original window of Sugar House. It appears today in walls of building at southwest intersection of Rose and Duane Streets.

nated the Black Joke Guard.

In the presidential campaign in 1844, Black Joke turned out without orders for a Polk and Dallas parade. A few days later a parade of Whigs led by Tom Hyer, the prizefighter, and others of his ilk, many mounted on horses, came up East Broadway into the Fourth District which was Engine 33's home ground, carrying a bell belonging to the Allaires. Then someone was irresistibly tempted to ring the alarm for the Fourth District. As usual, Black Joke turned out fast and soon saw the reason for the tap. When they swung around to return, some of the horses shied and others got tangled up in the ropes. The resulting stampede practically broke up the procession. A short while after, 33's engine was turned into the Corporation Yard "tongue first" or suspended, and Black Joke Company of the "East Side" was a thing of the past.

New York's lower East Side was a training ground for hordes from other lands. If the life amid the teeming tenements was rough, those who ran its gauntlet were strong. From the old neighborhood came some of America's famous men. Alfred E. Smith, Cardinal Hayes, subway-builder Sam Rosoff, Walter Winchell,

Engine panel showing the rescue of young Samuel Tindale at the Hague Street explosion, February 4, 1850.

Irving Berlin, David Sarnoff, James J. Walker, and Jimmy Durante are only a few examples of the many from the East Side who rose to fame.

Vast changes have taken place there in the last decade as the modern housing developments—the Governor Alfred E. Smith Houses, the Lillian Wald Houses, the Vladeck Houses and the Jacob Riis houses—now cover much of its original area.

To poor people engaged in a daily struggle to keep bread on the table, the great events of the past have less significance than the problems and small victories of everyday life. While the ghosts of history walk unseen in the old neighborhood, East Siders watch with great interest and warm sympathy the drama of life as it unfolds in their crowded streets. They, too, have their tragedy, triumphs and minor miracles. One such is the story of little Vera Costa.

Anna was on her way to Mr. White's bakery on Catherine Street to buy the loaves of bread that made such fine "heroes," those mammoth sandwiches her father and brothers carried for lunch each day. From the fourth floor fire escape of 79 Oliver Street the deep brown eyes of eight-year-old Vera watched her friend Anna who sometimes "minded" her. A neighbor called, "Vera, look out you don't fall."

"If I do, I'll climb right up again." With the childish boast still fresh on her lips the little girl stepped back, clutched the air in terror for an instant, then fell screaming through the fire escape opening from the fourth floor. Instinctively, she grasped a wooden safety gate on the second floor which came off and fell beneath her in such a way as to stop her on the first floor landing and break the force of her fall.

The gas-man was the first to reach the child and taking her limp form in his arms he placed her in his truck and drove to the Gold Street (Beekman Street) Hospital. Although cut and bruised, Vera had miraculously escaped any serious injuries but she was kept over night for observation. It was her first night away from home but more than by fright or by homesickness she was disturbed by the knowledge it was Thursday, the night for macaroni, and she had missed hers.

Next day as no one had cab fare her sixteen-year-old brother went to the hospital and proudly carried Vera home to a tearful but joyous welcome.

Mariners Temple on Catherine Street

On the southeast corner of Henry and Scammell still stands St. Augustine's (All Saints Church), which was built in 1828.

The boldness of their spirit can be seen in their faces—Hose Company No. 2 at 5 Duane Street

LIVE 44 OAK

The Terrible Turk

In early times, parts of the shoreland along the East River from Rivington to 10th Streets were broken up and segregated from the mainland by marshes and inlets. The largest of these, a rise of land completely surrounded by water during high tide, acquired the name of Manhattan Island. The Island was primarily a shipyard settlement as many of its denizens built, sailed or owned most of the ships produced by the East River shipyards.*

In August, 1824, the master shipbuilders of the Dry Dock organized their own fire company for the protection of the shipyards in the vicinity. Made up of ship calkers, painters, carpenters and mechanics, the company was quartered in a small frame house, built by themselves, in Columbia Street, near Houston, the river at that time reaching almost to Goerck Street. They were not attached to the Fire Department for four years but in 1828 they received the number 44, and had a brick house, one story, with a peaked roof, built for them in Houston Street, between Lewis and Manhattan Streets. In respect to the live oaks from which the ships' "knees" were made the company adopted the name of "Live Oak" Engine Company, a title they held for many years.

In 1830, Henry Eckford, the shipbuilder, took a crew of ship carpenters with him to Constantinople to work on a contract he had there. Several of the men were members of No. 44. Unfortunately, Eckford died within a short time after reaching Turkey, and the men returned home. One of them, a lad named Russell, had let his beard and hair grow long and was dubbed the "Old Turk" by his companions. The name soon was extended to the whole company and remained with

Live Oak's side-lamp

Live Oak Engine No. 44

**Even before shipyards were ever dreamed of this East River shore was the favorite careening place for the early ships of the Dutch and British and it was near here that Captain Peter Warren brought the Frigate Launceston whose raids upon shipping brought much wealth to New York and much popularity to this brave officer in the period between 1741 and 1746.*

ENGINE NO. 44 HOUSTON ST.
BETWEEN MANHATTAN & LEWIS STS
1845

SCALE OF FEET
DASH LINES SHOW PRESENT CONDITIONS EXCEPTING AS OTHERWISE NOTED

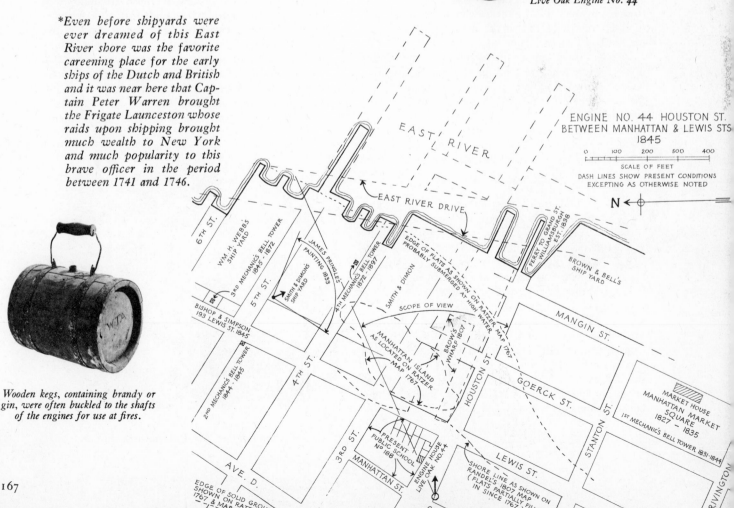

Wooden kegs, containing brandy or gin, were often buckled to the shafts of the engines for use at fires.

The Smith and Dimon Shipyard on the Hudson River at the foot of 4th Street. The Rainbow, the Sea Witch and other famous clipper ships were built there.

them from that time on, though rival companies often referred to them as "the Terrible Turks" because of their street-fighting prowess. On the back panel of their engine, they had a painting of a female figure with outstretched arms. Edward Penny, Jr., was foreman of the company in 1836, and was elected an assistant engineer of the Department in 1837. It was from Penny and his brothers who were rough and tumble street fighters that the name "Old Turk and the Six Bad Pennys" came into use.

John S. Green, one of the members, was run over in 1851 while going to a fire in Houston Street, near Ridge, and killed. The company gave a huge ball for the benefit of his family. William H. Van Ness, the engine builder, was also a member of this company.

The company was always bitterly opposed to steam engines in the fire department, and fought hard against their introduction. Five of its members went to the war in 1861 with Ellsworth's Zouaves.

The "Terrible Turks" painting their engine panel

"Old Turk" had another side and was also affectionately known as the "singing engine," as a number of members were noted for their fine voices. William H. Webb and his boys would often find occasion for a good round of song but the best voice belonged to Frank Walton, a tall, handsome lad who was known as "the Minstrel Boy of the Saw-Pit."

When gold was discovered in California, Walton, as well as others from the New York Fire Department, caught the fever. Walton sang his way to San Francisco early in 1850 by a clipper ship around the "Horn." He joined the procession to the mountains, but like most of the dreamers, found nothing but hardship and an empty poke.

A number of New Yorkers who saw him on the trail or in the mines, were surprised by his old faded red shirt with the big No. 44 still showing on the chest. Things went from bad to worse until one cold winter's night in wild, wild Nevada "Old Turk's" mockingbird slept his last sleep, alone on the trail with his tattered old firecoat wrapped around him.

Haswell has reported that Old Turk's engine was destroyed in the great shipyard fire but a check of the records proves that it was Black Joke's (Engine Company No. 33) goose-neck engine that was destroyed.

James R. Steers and his brother George, famous shipbuilders, who were enthusiastic members of Live Oak Engine Company No. 44, received special comment in an article entitled, "The Old Shipbuilders of

Frank Walton, Old Turk's mockingbird.

The death of "the Minstrel Boy of the Saw-Pit."

James R. Steers, designer of the yacht, America.

The yacht, America, built in 1851 for John C. Stevens, by James and George Steers, members of the Live Oak Engine Company

New York" which appeared in Harper's Magazine, July, 1882, as follows:

"One of the most brilliant successes of the clipper era was the yacht built in 1851 for John C. Stevens and several other gentlemen, who desired to secure a vessel which would win the Queen's Cup at the annual regatta of the London Royal Yacht Club. She cost about $23,000.

"This yacht, *America*, schooner of 170 tons, was designed by George Steers of James R. & George Steers and built by his firm for John C. and Edwin A. Stevens, George L. Schuyler, J. Beekman Finley, and Hamilton Wilkes. There were fifteen starters, ranging from 47 to 392 tons, and the *America* not only won by some 25 minutes, but proved to be much the faster vessel on all points of sailing.* So marvelous was the performance of the *America* that there were many who believed there was some propelling machinery on board. In illustration of this opinion Lord Yarborough visited her and after looking all through between decks, boldly asked the sailing-master, Brown, who was in charge, to lift the hatch in the cockpit, in order that he might be fully advised upon the question of the alleged existence of a propelling machine in her stern.

"The subject so interested that officer that he obtained from the authorities an order to build two war vessels, the Shark and the Grampus, after the same model. Steers and Thomas also furnished plans for the construction of an immense ship-house and an inclined plane, by means of which they were successful in hauling up the frigate, *Congress*, for repairs. In 1824 the two shipbuilders came to New York and built at the foot of Tenth Street, on the East River, the first ship-railway ever seen in the United States; it consisted of rails laid on an inclined plane, upon which a cradle was run for the purpose of drawing vessels up out of the water in order to repair them; and in consideration of their enterprise, the Legislature granted to the railway company a charter for a bank, to last 'as long as grass grows and water runs.' Thus was founded the Eleventh Ward Bank, now the Dry Dock Bank. The only other institution that ever received such a charter was the Manhattan Company."

Henry Steers, grandson of the Henry Steers who built the Shark and Grampus, later distinguished himself by building some of the largest vessels of the Pacific Mail Steamship Company, and some of the largest steamboats ever seen on Long Island Sound.

The house of Black Joke Engine No. 33 in 1820. This company, a bitter rival of the Terrible Turks, was located on Gouverneur Hill at this time.

**The race was held off Cowes, England, on August 22, 1851. Its 100th anniversary was marked by a notable exhibit of the State Street Trust Company in Boston, under the tender care of the Honorable Charles Francis Adams, Allan Forbes and Philip Potter.*

RUTGERS HOSE No 2

"Spinning Wheels"

IN 1852 this old east side neighborhood still retained a rural or small village character. In the view at the left the house in the distance, between Jefferson and Clinton Streets, facing Rutgers Place, was the Hendrick Rutgers mansion built of Holland brick in 1754. Hendrick was the son of Harmanus Rutgers whose old farmhouse stood at the head of Chatham Square. Hendrick's son, Colonel Henry Rutgers, who in time inherited the place, died in 1830 leaving the greater part of his farm, including the mansion, to William B. Crosby* who was living here in 1852. It was at Crosby's instigation that Monroe Street was cut through the two blocks bounded by Jefferson, Cherry, Clinton and Madison Streets which surrounded the grounds of the mansion. This portion of Monroe Street was named Rutgers Place. The house was remodelled at this time, two wings being added and its entrance changed to the north side. After Crosby's death in 1865 the house was sold. Finally hemmed in by the encroachments of the slums, it was demolished in 1875.

In this vicinity were located two of the finest hose companies in the Volunteer Department. In the accompanying scene are shown Oceana Hose No. 36 racing down Jefferson Street and Rutgers Hose Company No. 26 hauling a two-wheeled "jumper" from their house at 166 Monroe Street, to take after the "Ocean Wave."** These companies were great rivals and had many exciting races and encounters.

Rutgers Hose organized in 1839, served until the Department was disbanded in 1865. Their new machine, acquired soon after this race, was a very attractive four-wheeled carriage. The running gear was red, with gilt stripes. The reel was red, ornamented with beautiful gilt carving of intertwining olive and oak branches. Over the arch on each side of the reel was a miniature equestrian statue of Washington. On the front box was a representation of the Rutgers mansion and the motto of the company: "The Noblest Motive is the Public Good."

On the back tool box was a painted portrait of Colonel Rutgers; on the side panels were small sketches of a girl coyly peeping through a lattice. All painted areas were set off with rich scroll work; the lifters (cranks to turn the reel) were made to represent sea horses; the running lamps were the "neatest in the city." This carriage was built by Van Ness and painted by Moriarty.

Oceana Hose was organized March 5, 1845 at a meeting held in the home of Frances O'Connor at 94 Madison Street and shortly afterwards was quartered at 189 Madison Street. Painted by Edward Weir, their carriage, also called the "Old Red Gal," was one of the most elaborate in the city. The panels were extremely beautiful. On one, "Oceana" was shown with attendant nymphs rising from the sea. On another was a view of the High Bridge at Harlem and a group of Indians; on the back the Park and Bowling Green fountains were pictured, and on the front was a painting of a little girl washing her feet at a hydrant.

John H. Waydell joined Oceana's boys in 1847. "Old Boy Waydell," as he was affectionately referred to in later years, had quite a time with his new friends. One of them jovially bet another that the new member would miss the first working fire. Several nights passed without a fire occurring anywhere in their district. Waydell, who had heard of the wager, vowed that nothing could cause him to miss that fire if he were alive. Now all this had been going on close to the time of his approaching marriage. As luck would

Parade hose cart

U.S. TOBACCO INSPECTION WAREHOUSE NO 2.
MONROE ST.
CLINTON ST.
WM. DENNISTOUN WOOD YARD
IRON FOUNDRY
A. DURYEE
CHERRY ST.
DURYEE & SONS
3 STORY BRICK RESIDENCES BUILT 1834-5.
RUTGERS PLACE
W. B. CROSBY RESIDENCE
SCOPE OF VIEW
JEFFERSON ST.
3 STORY BRICK HOUSE BUILT 1840
LUMBER YARD
COAL YARD
RUTGERS HOSE
MONROE ST.
N.

RUTGERS HOSE CO. No 26
1852
MONROE ST. NEAR JEFFERSON
0 20 40 60 80 100 120 140
SCALE OF FEET.

**Crosby was the great-nephew of Colonel Rutgers*
***"Life on the Ocean Wave"—see appendix.*

have it, Waydell actually was enroute to his wedding, listening to the happy tones of his wedding bells, when the loud, harsh clang of the fire alarms rudely drowned them out. The situation called for a quick decision and the persistant signal urged that he make it instantly. Away he went—straight to the fire—and was among the first to arrive at the blaze. His passing of this unexpectedly severe test of his loyalty to duty, fortunately followed by the gracious forgiveness of his disappointed bride, elevated him to a high place in the estimation of his comrades.

Waydell ran to a fire in the winter of 1852, helping inside the ropes of Engine No. 19. His running mate tripped, fell and was killed under the wheels of the machine. From that time on, Waydell lost heart for active fire-fighting after having served a long and useful career.

A number of prominent New Yorkers were members of Oceana Hose Company, including Richard B. Ferris, president of the Bank of New York; Alonzo Slote and his likable brother Daniel, who has been mentioned as being the living inspiration of Mark Twain's character "Dan" in "Innocents Abroad." Also, William D. Wade, John R. Platt, William A. Woodhull, each of whom at different times were elected to the presidency of the New York Fire Department.

William Durand Wade died a few days before his thirtieth birthday and was possibly the youngest president the Fire Department ever had. He was much admired and a tablet in his memory was erected

The beautiful hose reel of Oceana No. 36

in the old Volunteer Firemen's Hall at 155 Mercer Street.

Woodhull was the recipient of a handsome rosewood desk which was presented to him upon his election as president. This relic may now be seen in the H. V. Smith Museum of The Home Insurance Company where it is carefully preserved for posterity.

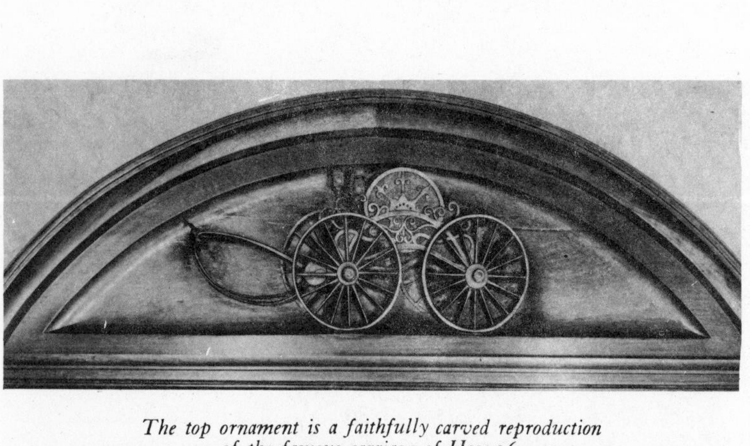

The top ornament is a faithfully carved reproduction of the famous carriage of Hose 36.

Engraved silver plate on drop lid of secretary

The hand-carved rosewood desk which was presented to William A. Woodhull by members of Oceana 36 upon his leaving the company for the presidency of the Fire Department.

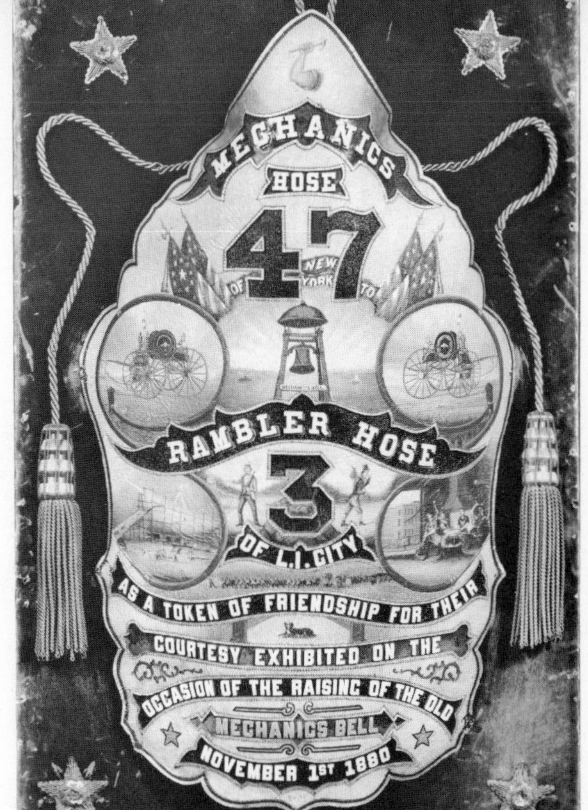

The Mechanics Bell

PERHAPS the most famous alarm in the city was the old Mechanics Bell which was located near the East River shipyards for many years. The bell was erected in 1831 by a group of mechanics who had a "labor parley" with the local shipbuilders, winning the right to a ten-hour working day in the shipyards. In commemoration of their victory those workmen, who were also volunteer firemen, contributed the bell which served as a fire signal until 1873 when it was taken down. After being stored for several years the bell was recast and hung in a new tower at the foot of 4th Street in 1880. The bell is pictured on the beautiful leather testimonial shield which was presented by Mechanics Hose No. 47 to their friends of Rambler Hose No. 3, as "a token of friendship for their courtesy on the raising of the old Mechanics Bell, November, 1880."

John Quigg

This shield also shows the Mechanics Hose carriage, "Mechanics Own," on a cobble-stone pavement, faced by Rambler Hose on a country road. Below, on the left, is a scene showing carpenters and sawyers at work in Webb's old shipyard. Facing this is the marine smithy with William Young's tool store and the mechanics headquarters visible through the open door of the forge building. Below is a vivid illustration of a stirring race among Hose No. 47 and Engine No. 44 and Hook and Ladder Truck No. 13. John Quigg was in command of the hose cart which can be seen passing the "engine." Beneath can be seen the house of Mechanics Hose, in East 4th Street, and "Major," one of the most famous of the old fire dogs. This huge Newfoundland served with the company until the department disbanded and then joined Engine Company No. 3 of Elizabeth, New Jersey, where he died some time later. It is said that Major knew the stroke of the fire-bells as well as any member and that he took great delight in the races between the Mechanics and other companies.

This handsome presentation piece, designed and painted by John Quigg, last foreman of Mechanics Hose, and master painter and decorator, is one of the finest pieces of craftsmanship of its type extant. It is now a part of the H. V. Smith Museum collection.

Daniel Kelley, one-time foreman of Mechanics Hose No. 47, was a man of great strength and reckless courage. Shortly after he was elected foreman, the leader of a rival machine attempted to strike Kelley with his trumpet while they were "rubbing hubs" in racing to a fire in Columbia Street. Kelley seized the opposing foreman, threw him across his knee and spanked him like a schoolboy, amid the roars of the crowd. After serving many years in the Volunteer Department, Kelley suddenly resigned and disappeared. Years later a former fireman visited a monastery at Hoboken and, by chance, was referred to a Brother Bonaventure. The visitor was being conducted through the house by the Brother when the latter suddenly turned and pressing the hand of the visitor said, "How do you do, Joe! Do you remember me?"

Brother Bonaventure was Danny Kelley, the Mechanics' old-time foreman, and the visitor was one of his former comrades. Kelley explained to the amazed visitor that he joined the order many years before in an attempt to save his soul by helping to extinguish the fires of the Devil.

"Mechanics Own"

The Mechanics Bell

A marine smithy

Webb's Shipyard

THE OLD HOME FACING DOWN CHESTNUT STREET

NO. 7 MADISON STREET – 1950

KODACHROME BY HARRY COLLINS

In all of Manhattan today it would be a formidable task to find a dozen structures which survive from the eighteenth century; but, scattered throughout the great metropolis in areas which had early origins, are uncounted numbers of buildings of a somewhat later period in various stages of perfection, disfigurement, or alteration. This is particularly so in Greenwich Village, bits of Chelsea, Bloomingdale, and on the lower East Side extending from City Hall through Chatham Square, Chinatown, and out toward Corlear's Hook. Here and there, among the rather sad survivals, you will find a sparkling little gem of a home which, in the tender hands of its owners, has never suffered from the social cycles that have changed the neighborhood in which these have stood throughout the years.

Such a home is that of Dr. Bernard Poggioli, at the present No. 7, (old No. 3) Madison Street. The kindly and imaginative doctor has continued to maintain this architecturally simple but attractive structure which, while not actually dated, probably was erected in the period between 1800 and 1820.

This part of Madison Street was once called Bancker Street after a man named Bancker who had married one of Henry Rutgers' daughters. The thoroughfare followed an old cowpath, which was cut through this part of the country in 1755. The eastern part was named Bedlow in honor of the husband of another Rutgers' daughter. In earliest times, there seemed to have been a definite connection between what is now the lower end of Duane and Bancker Streets. In 1812, folks petitioned the Common Council for permission to change the name of Bedlow, due to the unsavory character of a few of its inhabitants at the lower end. It was at this time that the entire street was named Bancker. The name was again changed in 1826 to the present Madison Street.

The rallying cry of Eagle Engine 13 often rang out in this neighborhood and the boys of Engine 13, housed in nearby No. 5 Duane Street, were, from 1857 to 1864, the pets of the neighborhood.

On February 4, 1850, this old home was almost shaken from its foundations by one of the worst and most unexpected disasters ever to strike New York. In Hague Street, not 500 feet away as a sparrow would fly, occurred an explosion, in which a 100-lb., high pressure boiler in a machine shop burst with such violence that it almost lifted the five-story building off the ground and transformed the entire structure, with its heavy machinery, into a ruined burning mass.

It was on a Monday morning, and the employees of the shop had just assembled to commence operations for the day. Sixty-seven were killed, over fifty injured and only a handful of the survivors escaped disfigurement or permanent crippling. The records of daring and heroism, in the efforts of the firemen on that day, would fill volumes.

In time, No. 7 came into the possession of Patrick Divver, an alderman who became a judge of the supreme court. The judge had a brother, Danny, who worked as a tanner in the "Swamp." Dan was a tall, handsome figure of a man who was a well-known and popular member of Eagle Engine 13. This youth was one of the first to volunteer for active service in the 11th New York Regiment, upon the outbreak of the Civil War. He was unanimously elected a 2nd Lieutenant. During the training period, prior to embarkation for Virginia, he was an inspiration and example to his fellow soldiers.

The eagerness for action displayed by his company was soon satisfied, because this comparatively green outfit was thrown into the thickest of the first bitter battle at Bull Run and had to stand the brunt of a series of reckless Southern charges.

The battle cry of old Engine 13, "Get Down, Old Hague!" were the last words ever heard by a number of the boys in blue and grey. During a temporary lull in the desperate action, Danny Divver was found dying of a dozen wounds, and he was carried off the field into a wheelwright shop by Surgeon Gray of the regiment. He died soon after, and his body was never again found or identified, due to the fact that the Southerners made another sally and took the ground upon which he lay.

No. 7 Madison Street still looks down what is left of Chestnut Street—so-called because of the many Chestnut trees which were once abundant in this vicinity. This narrow and short thoroughfare, which once connected Madison and Oak Street, before it was bisected by the cutting-through of the New Bowery and Chambers Street, is now the shortest street in the city.

At the present No. 3 Madison Street lives Mrs. Teresa Hart in the house where she was born, and who on a summer's evening likes to reminisce with her life-long friends in the neighborhood about the good old times before the housing developments.

The successful experiment of John Fitch's steamboat took place on the Collect, the summer of 1796

Fresh Water

How enviable was the first white man who, cautiously making his way north from the lower beaches through the woods and meadows, came upon the Collect*, Manhattan's largest natural fresh water pond. With what wonder and delight he must have beheld its lovely setting. There it lay in its pristine beauty, placidly reflecting the wilderness around its shores. From its sides rose short sheer cliffs and noble hills covered with groves of hickory and chestnut trees.

The waters of the pond were of unusual purity, fed mainly by gigantic springs which bubbled from its great depths. Two crystal streams carried off its waters, one into the East River and the other into the marshes that lay toward the north and west.

Hunting and fishing parties from various Indian tribes sometimes camped beside its deep cool waters in search of the shell fish and other sea food which they craved. One of these tribes, probably the fierce "Manhattans" described by Henry Hudson, was known to have built "wikwams" along the western shore and settled there. These tribes devoted little time to cultivating the soil and lived chiefly by hunting and fishing. The early settlers never encountered them. The existence of their village was known only by the huge shell deposits strewn over the lake's western promontory.

Dutch used the Pond for sk

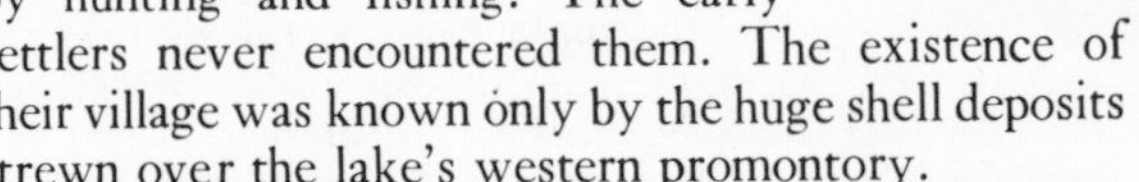

The early Dutch settlers named the hill to the west "Kaltchhook," literally "Lime Shell Point," and the pond itself, which they called Der Kolck , gradually came to be known as the Collect or Fresh Water Pond.

A drawing of the Collect as it appeared about 1742 was made by David Grim. On the west side of the Little Collect was a garden, separated from the big Collect by an island-like knoll. Once a real island, the Dutch built a powder magazine to keep their powder

*Originally named by the Dutch, "Der Kolck," which meant, "The Rippling Water." It was also referred to as the "Versche" or "Fresh" water.

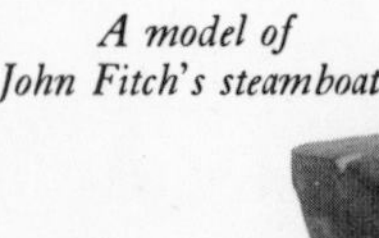

A model of John Fitch's steamboat

Engine Company No. 3 which was located by the Collect in 1745

out of harm's way. About seventy-five years later, a man named Ramney who had lived in Cross or Magazine Street reported that he had dug up the remains of the old magazine, and he could see evidence that water sometimes had enclosed it.

Fish were so abundant in Fresh Water Pond that for over a hundred years the early settlers were allowed to drop their rods and lines freely, without any restrictions. But when some fishermen began to use nets, the Common Council decided to put an end to such goings on and ordered that "if any person or persons whatsoever do from henceforth presume to put, place or cast into the pond . . . any hoop-net, draw-net, purse-net, casting-net, cod-net, bley net, or any other net or nets whatsoever . . . every person so offending against the tenor of this law shall, for every offense, forfeit and pay the sum of twenty shillings current money."

In winter salt water often came up from the North River through the Lispenard meadows to the Collect

The Tombs Prison was built on the site of the Collect Pond

and raised the ice. Benjamin Strong of Engine 13 and W. J. De Grauw of Engine 16 each wrote in 1785 that as boys they had skated from the East River through the Old Kill, making their way over the Little Collect around the island where the old powder magazine used to be, then skating over the main lake and

The Collect circa 1787

The house of Supply Engine No. 1 on Franklin Street

on down over the meadows to the Hudson.

And in 1850, little Catherine Havens recorded in her diary: "My grandfather had ships that went to Holland and he brought skates home to his children and they used to skate on the Canal that is now Canal Street and on the Pond where the Tombs is now, and my mother says that the poor people used to get a rib of beef and polish it and drive holes in it and fasten it on their shoes to skate on."

The land surrounding the Collect was in dispute for over a century. In 1791 it finally became the property of the Corporation. By this time, the growth of the city had formed a considerable settlement around the borders of the lake and the Corporation immediately staked off boundaries and ordered surveys with a view to filling in the pond and laying out streets.

J. F. Mangin, a French architect who had won fame in the city opposed this idea. He suggested that a dock or basin be made in the deep water of the Collect for a harbor, which would communicate with the North and East Rivers by a canal forty feet in width and thus bring the city's shipping into the heart of its

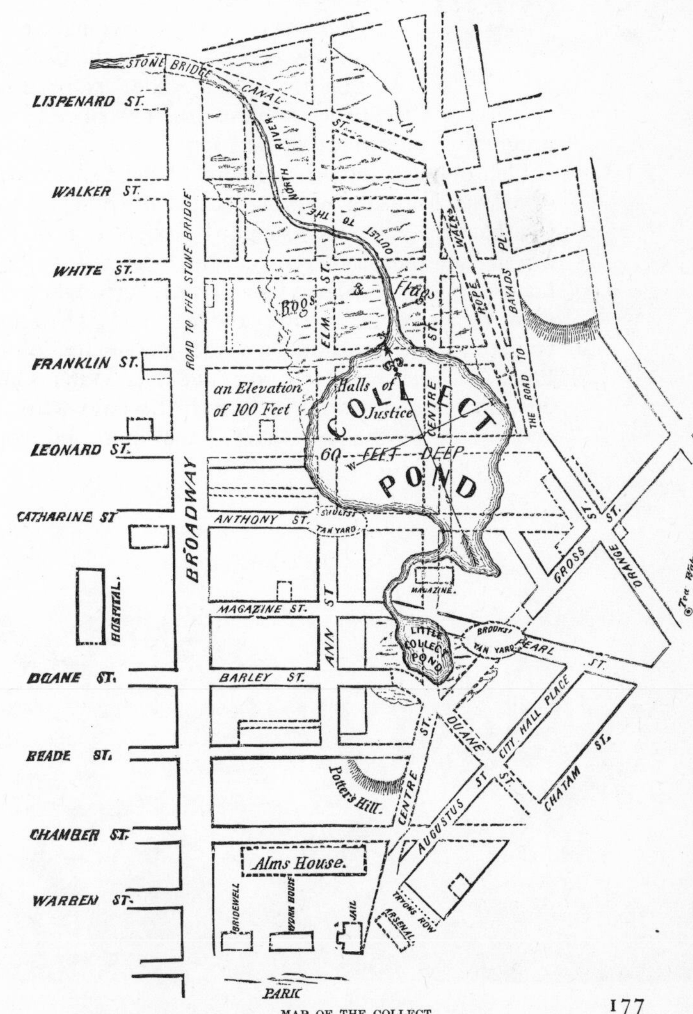

MAP OF THE COLLECT.

The Tea Water Pump at Roosevelt and Chatham Streets, one of the city's principal fountains.

commercial activity. If New York had remained a small town there could have been no better plan but by the time the Corporation got through discussing the idea, the city had already grown to proportions that far exceeded the Collect's capacity as a harbor.

After many years of wrangling, it was finally decided to fill in the Collect. Within the next decade, its beauties were quite obliterated, a sacrifice to the progress of mechanical civilization.

One of the principal fountains of the Collect was called the Teawater, a spring located at Roosevelt and Chatham Streets which supplied a population of ten or twelve thousand people with water for their favorite beverage. David Grim remembered when the carmen first began to deliver the tea water. The fountain later came to be known as the famous Tea Water Pump. A very popular resort, the Tea Water Garden, once stood nearby to furnish the boys with an antidote for all the pure water. But as more and more business houses came to be built in the area, the Garden too disappeared, for in 1829, Watson wrote, "I found the once celebrated Tea Water Pump, long covered up and disused—again in use—but unknown—in the liquor store of a Mr. Fagan, 126 Chatham Street, I drank of it to revive recollection."

During the ten years that the Collect was being filled up, the formerly lovely spot became a nuisance and an eyesore. The city had decided to fill the pond with the cleanest and best earth that could be obtained. Part of Hangman's Hill and Catiemutz Hill were taken off and a large crest of the famous Bayard's Mount was also used to partially fill the body of water. However, a tannery and slaughterhouse had been built at the water's edge and from them dead animals, and every species of rubbish and offal were also thrown into the Collect. In some places, the parts filled in were so low that houses were filled in up to the second story. Some of the springs and fountains were left open and in 1811 the council was still considering whether it would not be expedient to leave them.

In that same year the first outdoor circus was held

Stone Bridge Tavern, Canal Street and Broadway

Knapp's Tea Water Pump was on Tenth Avenue near 14th Street

Currier's view (the artist was L. Maurer) of the race between Engine Company No. 21 and Hose Company 60 in 1854. Matthew Brennan, foreman of 21 is shown urging on his men who are just passing the hose reel. The cupola and fire bell tower of City Hall can be seen in the background. Courtesy of the H. V. Smith Museum.

on the site of the Collect. The stage was built in the open air and there was no ring. Instead of charging admission, the managers passed a tambourine around and the audience dropped into it as many coins as they pleased. The circus later moved to the horse training track at Broadway and Prince, afterwards the site of Niblo's Garden and Theatre.

By 1816 an ordinance had been passed for the regulating and paving of Collect (Centre) Street, and the histories of the period tell us that "an humble class of buildings were erected on the site of Collect Pond." In later years the famous Tombs Prison was built there.

arly ngine ntern

The carriage of American Hose Company No. 19

The "Kissing Bridge" at Canal Street

By the Reservoir

"THE whole affair was as flat as Manhattan water," George Templeton Strong commented in his diary on February 22, 1841; Catherine Havens quoted her mother as saying that this same water "was brackish and not very pleasant to drink." Until 1842 when Croton water was furnished to New Yorkers, one source of the water to which citizens referred so disparagingly was the reservoir on Chambers Street owned by the Manhattan Company, forerunner of the bank of that name, and built on a plot they bought in 1800. The supply was obtained from a well in Reade Street by a sun and planet wheel steam engine and then driven into the reservoir and distributed in some streets through wooden pipes. The reservoir was surmounted by the reclining bronze figure of Aquarius, the water bearer. This statue was said to have been removed to the bank's quarters in Wall Street and subsequently lost. The reservoir was demolished in 1842.

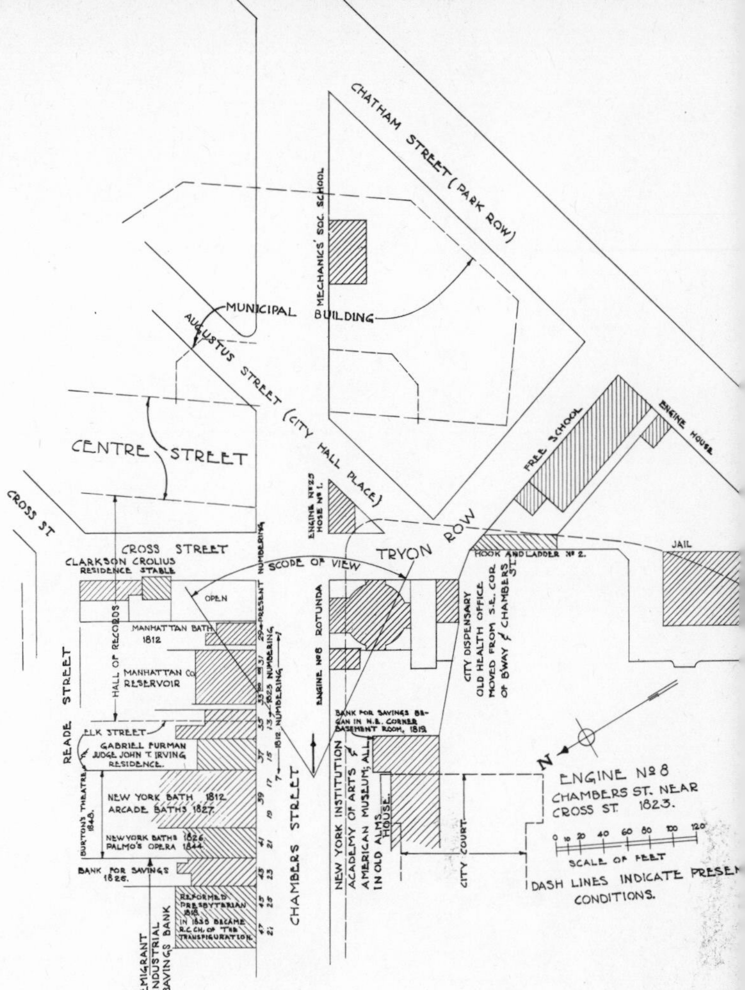

House and carriage of Pearl Hose No. 28

At one time most New Yorkers obtained their drinking water from wooden pumps located throughout the city. The most pure and copious of these was the famous Tea Water Pump at Chatham and Roosevelt Streets. According to New York housewives, its water was unsurpassed for making tea. Another flow of good water came from a spring near Bethune Street. Known as Knapp's Tea Water Pump it was peddled in carts at two cents a pail by the enterprising Mr. Knapp. Many houses and stores had cisterns in the yards for collecting rain water from the roofs. This was dipped up by buckets for use in washing. Public cisterns to supply the fire engines were never sufficient as a dependable source.

The Arcade Baths, located conveniently near a source of water, were in time enlarged by the addition of the adjoining structure, also a public bath. Later the combined buildings were connected with a place of amusement, known successively as Palmo's Opera House and Burton's Theatre. To the west of the Arcade Baths was the first bank building erected by the Bank for Savings.

The famous Pearl Hose Company No. 28, among the handsomest run in the city, was one of the machines housed in the firehouse shown. Pearl Hose was largely made up of merchants, one of whom, Edward W. Wilhelm, owed his life to the fact that when he was holding the nozzle at the Jenning's fire, the length of hose prevented his entering the building which later collapsed.

The Vigilant, a double-deck, end-stroke machine located in the area of the Reservoir for a time

In Tryon Row

ON Tryon Row in proximity to the Rotunda were quartered several fire companies among whom considerable rivalry existed. It is recorded that "in January, 1831, three attempts were made to set fire to the house of Engine 21." Hook and Ladder Companies 1 and 2 were each located in Tryon Row at different times in their early years. The famous Mutual Baseball Club was named after the Mutual Hook and Ladder Company No. 1 and was organized

Foreman and men of Hook and Ladder Company No. 2, located in Tryon Row.

in their house. Their ball grounds were at the Elysian Fields in Hoboken and they played the Atlantic, Eagle, Empire and Gotham Clubs during the years 1859 and 1860. No. 24, Jackson Engine, once had its headquarters at No. 2 Hook and Ladder House.

Engine No. 25 (Merchants Cataract) had been situated at the Broadway end of the Park near the Bridewell, but in 1821 the Common Council resolved to remove it and to build a one-story brick building "on the piece of ground opposite the Rotunda."

For some years Protection Engine No. 22 was stationed at Chambers and Centre Streets and its membership was made up mainly of merchants. The company ran an old machine affectionately referred to as "Old Junk," but it was a large and powerful engine. It was destroyed by the great saltpeter explosion at Broad Street and Exchange Place in 1835.

Floyd S. Gregg, familiarly known as "Old Trap Door" Gregg, and later a fire warden, was one of the outstanding members of this company. Gregg was instrumental in enforcing a law requiring merchants to keep their basement hatchways closed at night. Always a menace to the life and limb of firemen and pedestrians, cellar openings were a serious fire hazard.

Two-wheeled jumper, style of hose carriage first used in New York City

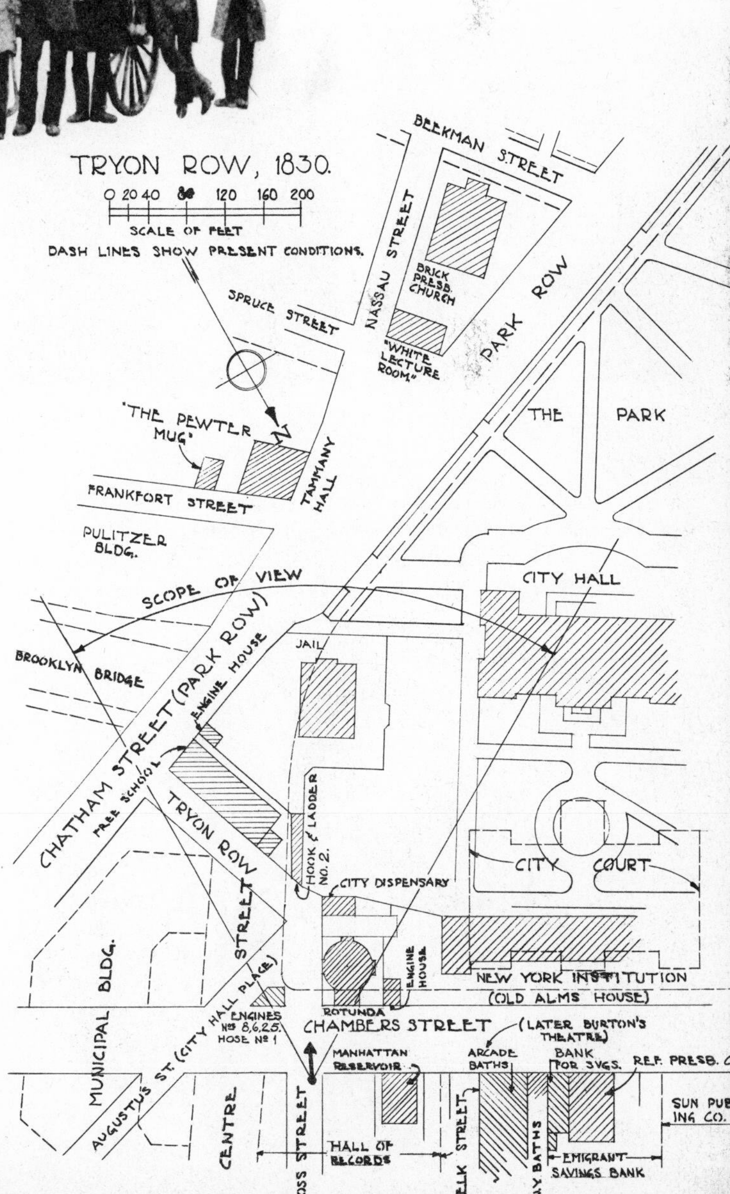

The view looks directly down Cross Street to Tryon Row which extended into Chatham Street. Cross Street was eliminated when Centre Street was opened shortly after 1835. The street intersecting Tryon Row and Cross Street in 1830 was originally known as Augustus Street. In later times it was called City Hall Place. When the present Municipal Building was constructed, circa 1910, its southerly end was eliminated and the portion remaining to the north re-named Cardinal Place. Most of the buildings in the area were connected with various functions of the city.

Center of Art

The Rotunda on Chambers Street—once the city's center of art

WHEN John Vanderlyn was an obscure country boy, who liked to amuse himself by drawing charcoal sketches on the stable door, he never dreamed that one of the most unhappy figures in American history would be instrumental in his rise to artistic eminence, nor that he was destined to become the founder of the first important art center in the United States.

John Vanderlyn

By one of the happy accidents which seem to occur so often in the lives of famous men, Aaron Burr stopped one night in the town where young John lived, and noticing the boy sketching, was so struck with his talent that he promised to help him become an artist. "When you are ready to begin your studies, come to visit me at my home at Richmond Hill," he told Vanderlyn. Beyond that, his only instruction to the astonished lad was that he must not forget to bring along a clean shirt.

William Dunlap

John Wesley Jarvis

The promise remained as fresh in the memory of the man who made it as it did in the heart of the excited boy. It was not long before Vanderlyn left the farm behind and made the long trek to New York. When he reached Burr's home and demanded admittance, the servants refused to believe his story and angrily told him to go away. Vanderlyn, however, remained adamant and was so insistent that he was finally admitted to the presence of Aaron Burr and his daughter.

Burr welcomed him and became his patron. He sent him to study, first with the American painter, Archibald Robertson, and later at the atelier of the great French classicist, David, in Paris. Thus Vanderlyn became one of the first Americans to be influenced by a style and atmosphere other than that of England, where an entire generation of American artists were being molded under the guidance of Benjamin West and others.

After his return to America, Vanderlyn rapidly won renown as a portrait painter. Among the notable men who sat for him were Governor Yates, General Taylor and Andrew Jackson, whose portrait he completed in 1821.

He obtained a lease to ground in City Hall Park through the assistance of Aaron Burr, and in 1817, the circular brick building called the Rotunda, the center of art of which he had long dreamed and planned, was built. During the following three years, panoramic views of Paris, of the Palace and Gardens at Versailles, and of the attack of the allied forces on Paris were displayed there, but the enterprise was a complete failure financially. The unfortunate artist was forced to surrender his property to the city, and the galleries which had formerly housed artistic and historic treasures became successively the home of the Court of Sessions, the Naturalization Office and the Post Office. In March, 1845, the New York Gallery of Fine Arts was given permission to occupy the Rotunda.

Engine panel from the Hope Engine

Before its demise, the works of Thomas Cole. John Trumbull, S. F. Morse, John Wesley Jarvis, A. B. Durand, Waldo, Jewett, Frazee and William Dunlap were exhibited in the Rotunda, along with the few old masters which local citizens had purchased from European galleries.

William Dunlap was a leading figure in the art circles of the period. Although he lost an eye early in boyhood, he had an astonishing career, not only as an artist, but also as editor, author, playwright and historian. He was, for a time, the lessee and manager of the Park Theatre. He was a major figure in the establishment of both the American Academy of Fine Arts and the National Academy of Design. His book, *The History of the Arts of Design in America*, is today regarded as the primary source book on early American art and artists.

Hand-painted fire bucket

One of New York's earliest artists, Evert Duyckinck, who arrived here from Holland in 1638, painted the city arms on fire buckets. He was the founder of a whole dynasty of artists, and through four generations his sons and their sons won fame as craftsmen in limning, painting, varnishing, gilding, glazing and silvering of glass, and as painters and teachers of painting.

The number of early American artists who at some time in their careers worked as sign, coach, fire-bucket, and especially fire-engine painters is quite astonishing. Vanderlyn was no exception, and was responsible for decorating a number of fire-engine panels.

The story of John Vanderlyn and Aaron Burr did not end with the building of the Rotunda. Never by word or action did the artist evade his debt of gratitude to the fallen idol who had befriended him. When Aaron Burr was a poverty-stricken and broken man, spending his last years in exile, he appealed to Vanderlyn to fulfill his last wish. The artist assented. He brought the forgotten old man home, so that he might die in his native land.

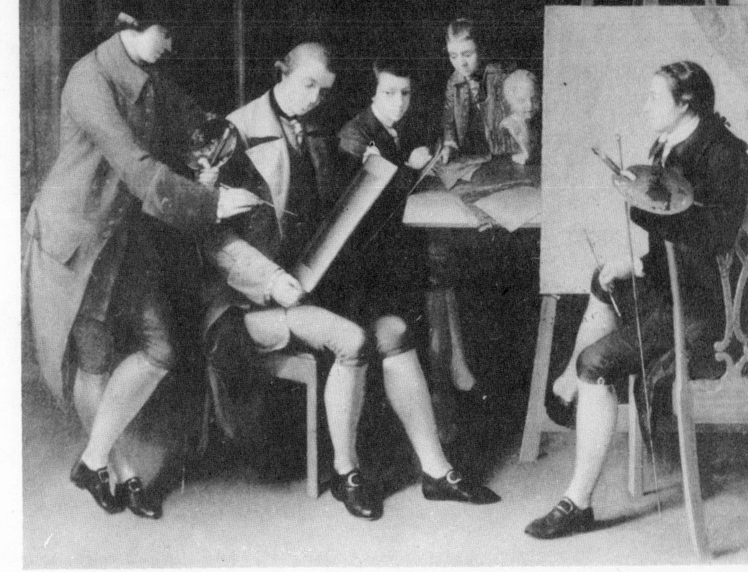

The American School, painted by Matthew Pratt in 1765. It shows Benjamin West in his studio with several of his young American pupils. West is seen standing at left correcting a drawing held by Pratt

Some of the country's foremost artists decorated buckets, fire engines and painted engine panels

A beautiful painting of a young Indian brave and his maid, in an idyllic pose and setting, by the artist Quidor, who was one of the most sought-after panel painters of his day, achieved the widest notoriety. When Engine Company No. 14 was disbanded for fighting, the painting, which had formed its back panel disappeared and the combined efforts of the fire and police departments and special detectives failed to find a clue to its whereabouts. When No. 14 was reinstated, the painting mysteriously reappeared, and to all questions the fire laddies merely replied that there had been a supernatural force at work and that such forces had "best not be tampered with."

The same site today showing the handsome Hall of Records building in the foreground, the Emigrant Industrial Savings Bank (the building with painted advertising sign) and the Stewart Building. A section of the City Courthouse appears at the left.

The New Look

In the view of 1827, Cross Street is seen at its intersection with Chambers Street, with the Rotunda dominating the foreground at the southwest corner. When the new buildings on the Bellevue Hospital site superseded it after 1812, the Almshouse, west of the Rotunda, was converted to private use and became known as the New York Institution. This building, demolished in 1861 to make way for the Tweed Court House, was once occupied by John Scudder's American Museum. The Bank for Savings opened in the northeast basement of the building in 1819.

The Reformed Presbyterian Church, built in 1818, is the brick building with gable to the street. It was occupied in later years by the Roman Catholic Church of the Transfiguration, which today occupies the church building at Park and Mott Streets, originally built as a Lutheran Church. At No. 51 (the second house west of the church) once the home of the Irish Emigrant Society, the Emigrant Industrial Savings Bank began operations in 1851. The present limits of the bank building plot rest on the sites of the two early banks—their own on the west and the Bank for Savings on the east.

A. T. Stewart acquired the low houses at the Broadway end of the block and demolished them about 1850 to provide space for the extension of his store. He built the first unit on the site of the burned Washington Hall and eventually extended the store into Chambers Street as far as the Emigrant Bank site. Catherine Havens noted in her diary that Mr. Stewart "is making a palace of a store" on Broadway.

Washington Hall, erected in 1810, which appears in the view above the roof of the Reformed Presbyterian Church, was used as a hotel and became the headquarters of the Federalist Party. At the extreme right of the view is the dome of the Broadway Tabernacle, where William Lloyd Garrison, Wendell Phillips and other famed abolitionist orators pleaded with New Yorkers to join with them to destroy the hated institution of slavery.

By the Bridewell

THE firehouse of Merchants Engine Company No. 25, organized in 1801, was located near the Bridewell Prison from 1832 to 1834. Close to Broadway on the plaza which now crosses in front of City Hall once stood the Windmill on the Commons which was built in 1662. Even then the stones and ironwork of a still earlier mill were used in its construction. In 1689 it was struck by lightning and destroyed. It was rebuilt but by 1723 it was just a memory.

The barracks built for use in the French and Indian War also housed British troops during the Revolution. Except for farms, Manhattan beyond this point was largely undeveloped and in its natural state. The Collect or Fresh Water Lake was only a little way north of here. In those days a Potter's Field lay along Broadway just above the present line of Chambers Street.

In that old Potter's Field the epitaph of a patriotic Englishman, a Mr. Taylor, once attracted much attention:

He loved his country and that spot of earth
Which gave a Milton, Hampden, Bradshaw birth—
But when that country—dead to all but gain,
Bow'd her base neck and hugg'd the oppressor's chain,
Lothing the abject scene, he droop'd and sigh'd—
Cross'd the wild waves, and her^ untimely died.

During the Revolution this area was the scene of much military activity, especially that involving prisoners confined in the Bridewell* and other buildings on the Commons. In 1828 the daughter of Mr. Peter Grimm stated that she remembered seeing the British barracks enclosed by a high fence extending from Broadway to Chatham Street along the present Chambers Street. This enclosure had a gate at each end, the one on the Chatham Street side being called Tryon's Gate from which Tryon Row was named.

Firecap, circa 1812.

The sadistic British provost marshal, William Cunningham, was infamous for his cruelty to American prisoners of war. He was quick to execute prisoners—and just as quick to continue drawing their rations. One night Cunningham, who performed his executions under cover of darkness to avoid public disfavor, led a condemned member of Washington's life-guard to Hangman's Hill, just north of the barracks. When they arrived, the bodies of two men executed the previous night still swung above the open grave that had been prepared for them

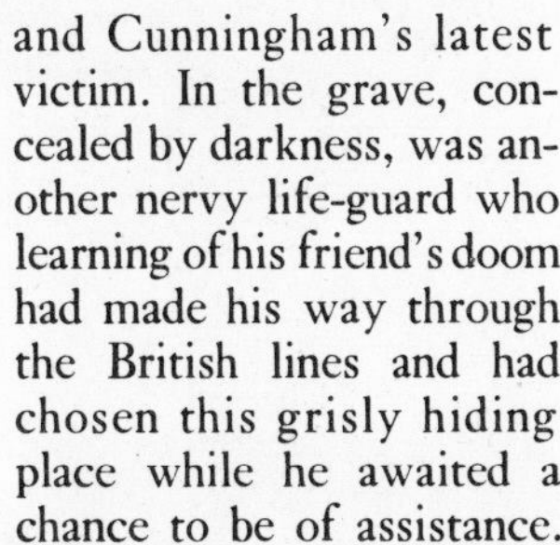

and Cunningham's latest victim. In the grave, concealed by darkness, was another nervy life-guard who learning of his friend's doom had made his way through the British lines and had chosen this grisly hiding place while he awaited a chance to be of assistance.

By an unexpected stroke of luck, when the hangman cut down the bodies of the two men, one of them fell into the grave where the would-be rescuer was lurking and suggested an idea for his friend's release.

"Windmill on the Commons," built in 1662.

As his hour approached the condemned life-guard who had been granted three minutes for prayer knelt beside the open grave. To his amazement he heard his friend's voice instructing him to turn his back when the three minutes were up. His shackles would then be cut and he could escape. At the expiration of his three minutes of grace, just as the hangman stepped forward, the prisoner turned his back as instructed. At that moment up from the grave rose a figure which the hangman, Cunningham,** and his attendants assumed was the ghost of the man whose body had just fallen in. Wild panic broke out and in the midst of the uproar the reprieved soldier and his rescuer were able to flee to safety.

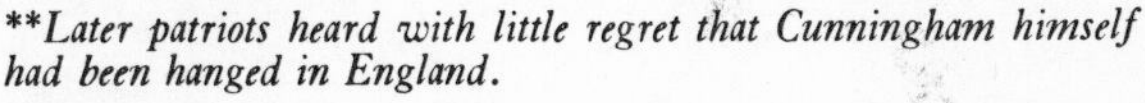

****Later patriots heard with little regret that Cunningham himself had been hanged in England.*

The mascot

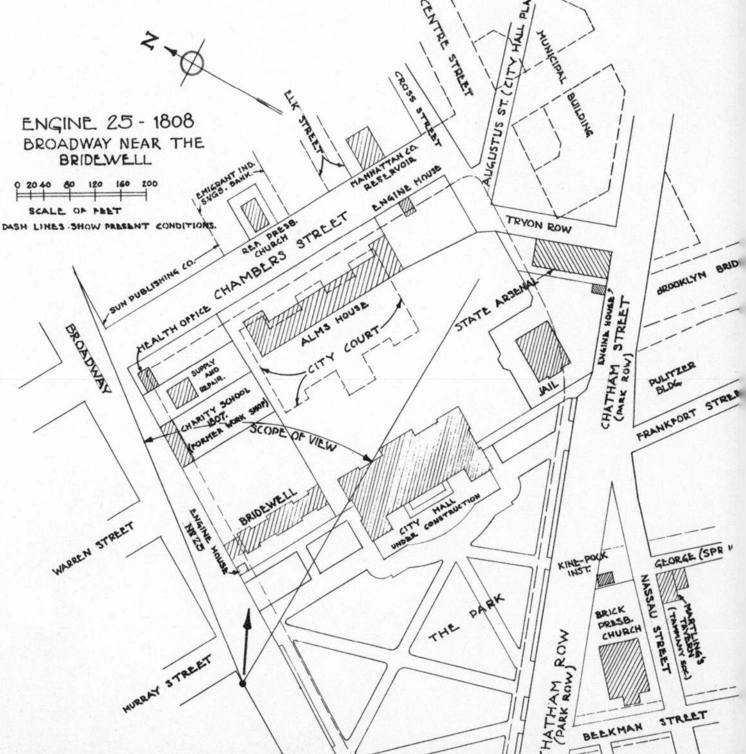

**The jail near the Bridewell for a time held Alexander MacDougall who was later to be a general in the Revolutionary Army. MacDougall was imprisoned because he wrote a public notice protesting the action of the General Assembly in voting supplies to the British troops. While in jail he was visited by forty-five "Sons of Liberty" (the number was talismanic on account of the Scottish uprising of "Forty-five") who came to applaud his action. Later forty-five female "Sons of Liberty" led by Mrs. Malcolm (wife of the General) also came to demonstrate their support.*

Corporation Yard

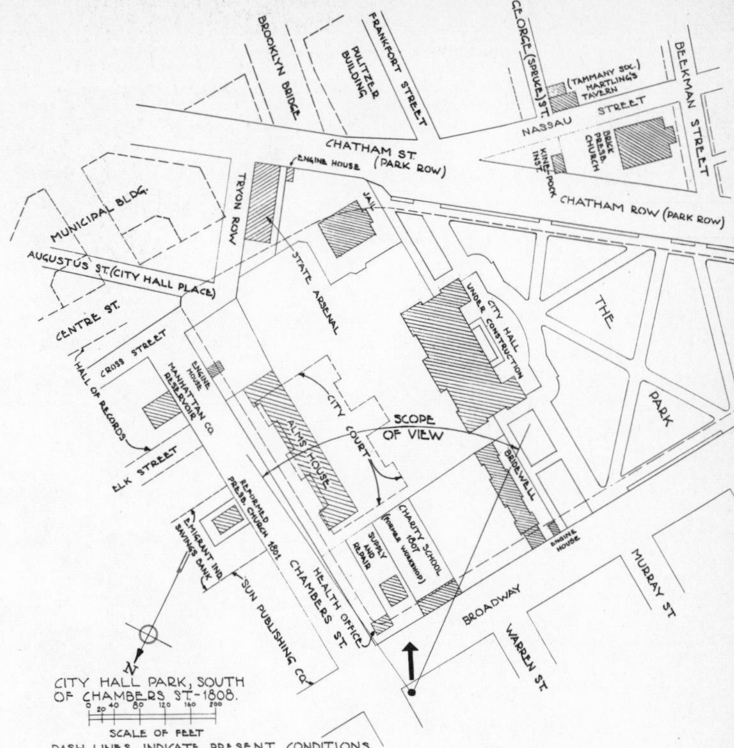

During the first part of the nineteenth century, damaged fire equipment was given first aid treatment and restored to fire-fighting condition in the Corporation Yard—the name given to the storage and repair yard of the city's Department of Supply and Repair—then located in City Hall Park.

It may seem strange today to think of this as a location for such a structure but in times past the Park was the site of many buildings owned by the city and housed various municipal departments. The Health Office was quartered in the frame building at the left of the Corporation Yard in this scene. At the right was New York's first public school. This modest beginning of the city's great public school system originally served as a work shop* before being used for the study of the three R's. The building was offered to the Free School Society by the City Corporation as temporary accommodation while the State Arsenal (Chatham Street and Tryon Row) was being adapted for use as a school. But in those days, too, "temporary" was an elastic word, and the classes of the Charity School were held there from 1806 to 1808.

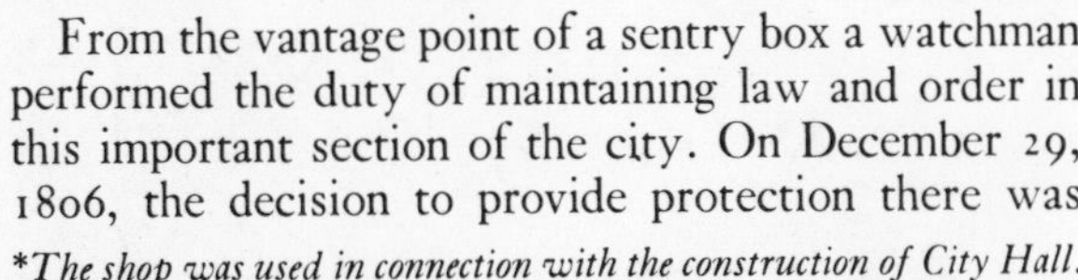

Early advertisement

From the vantage point of a sentry box a watchman performed the duty of maintaining law and order in this important section of the city. On December 29, 1806, the decision to provide protection there was taken by the City Council when it was "ordered that a watch-box be placed at and adjoining the fence of the Health Office in Chambers Street and that a watchman be then stationed."

**The shop was used in connection with the construction of City Hall.*

Interior details of a goose-neck engine, showing rocking beam and capstan

The city's needy were sheltered in the Almshouse, built in 1797 at the rear of the Old Poor House. This long building, which was later used as the New York Institution, was destroyed by fire in 1854. The present City Court House was built on the site in 1861-67.

Below the Charity School on Broadway stood Bridewell Prison where many patriots had been jailed during the Revolution when British troops occupied New York.

In the distance, can be seen the Gaol or Provost-Gaol, where Major Cunningham, the Provost Marshal during the occupation of the city, held sway and was personally responsible for the torture of hundreds of American prisoners. Cunningham was the most despised British officer in the city and the whole town rejoiced on Evacuation Day, November 25, 1783, when he was chased down Murray Street by a broomstick-wielding Mrs. Day, who prevented him from hauling down the American flag she had proudly fastened to a pole in her front yard. The courageous woman is said to have dealt him such a blow on his nose with her broomstick that "the torrent of gore made Cunningham's front as red as his back." This scuffle, from which Mrs. Day emerged victorious, was the most one-sided battle of the Revolution.

Note: *Between the jail and the Bridewell were the beginnings of a building that New Yorkers hoped would be the finest of its kind when it was completed—the new City Hall. And their hopes were justified; the building then being constructed would continue to dominate the scene almost 150 years later.*

Various firemen's tools, spanners and axes

Drawn & Etched by T. Horner

Aquatinted by J. Hill

BROADWAY, NEW-YORK.

Shewing each Building from the Hygeian Depot corner of Canal Street to beyond Niblo's Garden.

Published by JOSEPH STANLEY & Co.

BROADWAY, NEW YORK, 1834

Looking North from Canal St. to a point Beyond Niblo's Garden at Prince St.

AQUATINT, COLORED, BY J. HILL — DRAWN AND ETCHED BY T. HORNER

PUBLISHED BY JOSEPH STANLEY AND CO. — COURTESY OF THE HOME INSURANCE COMPANY

This view is particularly interesting, since it gives a good idea of the vehicles, and especially the street trades, of the period. The building near the center foreground, on the northeast corner of Canal Street, was a branch of the British College of Health, and was known as the Hygeian Depot. The building with the gable roof, seen a little higher up Broadway, between Howard and Grand Streets, was Tattersall's well-known horse and carriage mart, riding school and livery stable, established in 1829 and named after the similar establishment in London. In 1846 it was the largest horse market in the United States.

Niblo's Garden, in the distance on the corner of Prince Street was for many years one of the most popular places of entertainment in the city. The site on which Niblo's was built had been an area known as the "Stadium"; a place used as a drill ground for Militia officers during the War of 1812, and occupied by a circus for many years.

Later two brick buildings, one of which was occupied by James Fenimore Cooper, were built next to Niblo's on the Broadway side.

William Niblo, who had previously been proprietor of the Bank Coffee House in Pine Street, moved to this location in 1828.

The old circus building stood in the center of the garden for some time, and Mr. Niblo used it to stage theatrical performances which were so popular that he was soon required to build a more pretentious theatre.

Catherine Havens recalls the famous resort in her diary: "I will now tell about the Ravels. They act in a theater, called Niblo's Theater and it is on corner of Broadway and Prince Street. My biggest own brother goes there with some of his friends to see the plays and he said he would take me to see the Ravels. But when my father found out about it he would not let me go. He said he did not think it was right for Christians to go to the theater. I went out on our front balcony and walked back and forth and cried so much I hurt my eyes."

Miss Havens' father proved to be right for other than moral reasons, as it was not too long afterwards, September 18, 1846, that the entire theatre was destroyed by fire, during an engagement of the tight-rope performers, the Ravels. It was at this fire that fireman Thomas Boesé held the nozzle so near the fire that his clothes caught fire. Another pipeman turned a hose on him to catch the flames and the cold water paralyzed Boesé on the spot.

At the Seat of Learning

So here's a health to the Tiger's friends,
May they be near or far,
And three times three for the Tiger's boys,
And one good old Tig-a-a-a-r.

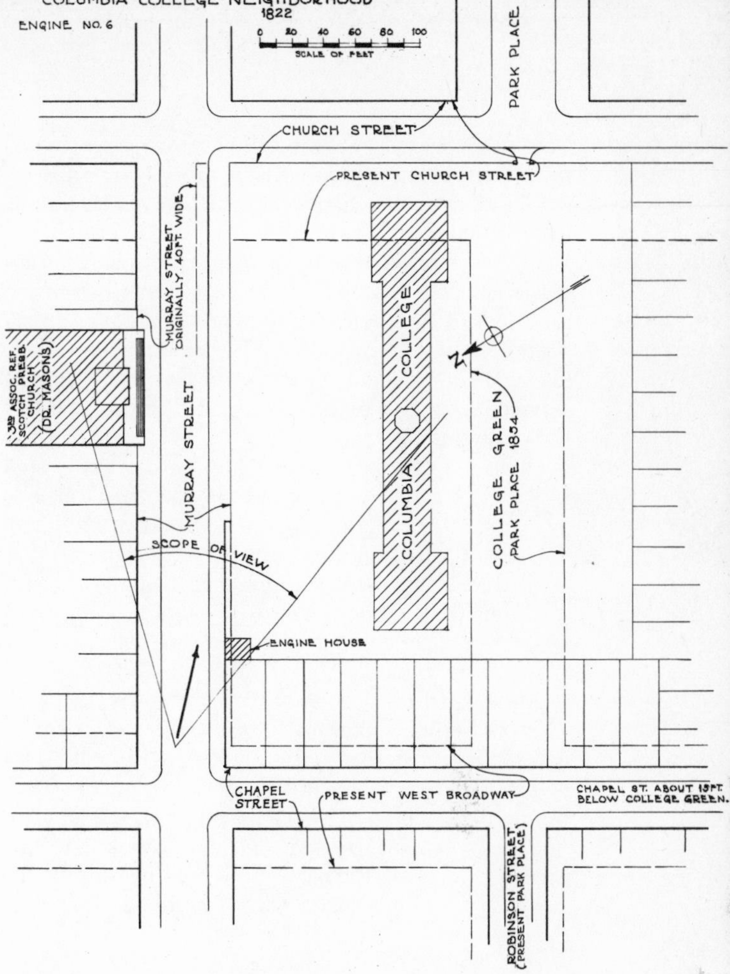

King's College, renamed Columbia in 1784, was the scene of many stirring events and riots of protest on the part of "rebel" students during the days leading to war. Years after, in 1845, in the excavation of the cellar for the house of John C. Stevens in the old college grounds, two pieces of cannon were brought to light. It was believed that these were captured and hidden in 1775 by Alexander Hamilton and the Liberty Boys. Before the city was lost following the reverses on Long Island the college was used by the American Army as a military hospital.

In 1822, the date of the accompanying view, Columbia College maintained a pew in St. Paul's for the use of her students and in that year her medical library was begun. The church opposite the "campus" was the Third Associate Reformed Presbyterian Church, which was built in 1812 on the lots now numbered 41-47 Murray Street. It was a handsome brownstone structure. Its rector was the celebrated Dr. John M. Mason. He had been rector of the Scotch Presbyterian Church in Cedar Street for thirty years, resigning in 1810 to become Provost of Columbia College. This church, incidentally, was removed in 1842 and rebuilt stone by stone on 8th Street at the head of Lafayette Place, on the site of the present Wanamaker store. In later years it became Saint Ann's Roman Catholic Church, and ended its days as a theatre.

Tweed's hat front and trumpet.
V. Smith Museum)

The college itself moved in 1857 to the old buildings of the New York Deaf and Dumb Asylum at Madison and Park Avenues, from 49th to 50th Streets.

Neptune Engine Company No. 6 was organized in 1756, with quarters in Crown (Liberty) Street, near Kip (Nassau) Street. Like several other early fire companies they started as a bucket company while their engine was under construction.

Jonas Addoms was one of the most noted members of No. 6 during the Revolutionary War. His regiment was the first to march into the city at the evacuation of the British troops in 1783 and he personally touched off the first salute of thirteen guns at Fort George in honor of the occasion. He helped to reorganize his fire company in the same year when the company moved to Murray Street to the house built in the wall of Columbia College grounds, where they remained until 1832. The nickname of "Bean Soup," by which this company was known, gives some hint of the festive spirit of its members. They were, from first to last, great feeders, joiners and organizers, as attested in their later years by the famous Americus Club, the target companies,

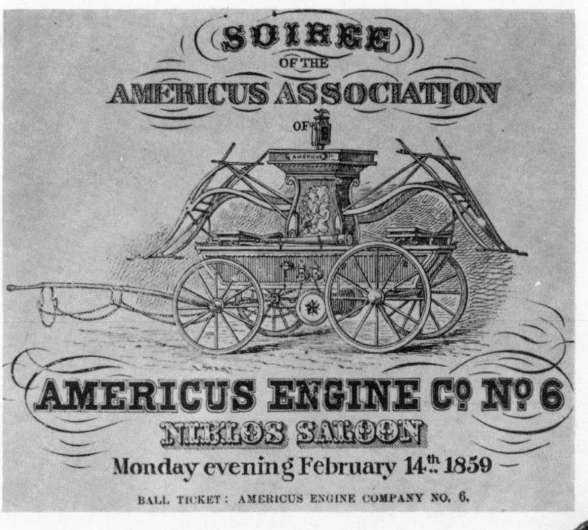

Firemen's Ball Ticket issued by Americus.

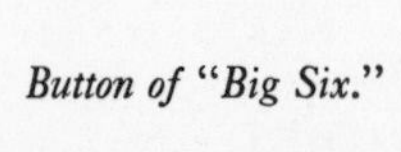

Button of "Big Six."

Foreman's Trumpet of "Big Six."

"Cudney Guards," "Wm. M. Tweed Guards" and the "Young Americus Guards."

One of their members in the early days was a lame man named Tom Flender. Flender used a crutch and had the reputation of being the "most long-winded runner in the department." All he asked for, when inside the ropes of the engine was "plenty of room to swing his crutch in." Lame or not, Flender was a good fireman and he was later rewarded for his many years of service to the city by being appointed to the Board of Fire Commissioners.

Hudson Engine No. 1 and No. 6 were early and bitter rivals, but any fight or race between Engine 6 and Engine No. 8 was sure to attract city-wide attention. When the word was passed that these two companies were running to a fire, hundreds along the route would drop whatever they were doing to watch or even to help Six or Eight "raise the hill."

No. 6 also carried on such a bitter rivalry with Clinton Engine Company No. 41 that "Old Stag" had to be transferred to the upper districts of Manhattan. That company was finally disbanded in 1836 for fighting with Engine Companies 21, 23 and 36. Engine No. 6 was reorganized in 1848 as "Americus" and received the name "Big Six" because of the size of their "Philadelphia-style" double-deck engine,

"Big Six," by E. R. Campbell.

AMERICUS
6
WMT

At the sound of Centre market,
And the Island Bell's first stroke,
We rolled the Champion of new york,
The second old Black Joke.

As up the Bowery we did go,
Upon the iron rails
We quickly saw old Forty one,
Up near the old Hay Scales.

We stopped, then turned our Engine around,
Our crowd being very slim,
And when we started to come home,
They tried to run us in.

The Tigers quickly saw their game,
And soon they were about,
And as the Stag boys piled in them
The Tigers laid them out

Now here's success to old Americus,
And may they never lag
But live to whip the fighters,
Of Forty one old Stag

We ne'er forget that Old Black Joke,
Was Champion to the last,
We take her place with credit now,
And never can be passed.

So here's a health to the Tigers friends,
May they be near or far,
And three times three for the Tigers boys,
And one good old Tig-a-a-r.

Early photo of college grounds by Columbia President Nathaniel Fish Moore.

which weighed over forty-two hundred pounds. This engine ran over two members of the company who tripped on their rope in front of Lord and Taylor's store while racing to a fire through Grand Street in 1855. One of these men was so severely injured by the engine passing over his chest that he spent eight months in a hospital. Miraculously, however, the man lived to rejoin his company.

An outstanding member of later days was Joseph Johnson, the artist, who was responsible for Big Six's Engine being the most elaborate in the city. They received their new "Philadelphia-style" engine in 1851. Johnson's work on the machine was so fine that it was placed on exhibition at the American Institute Fair at Castle Garden, easily winning the prize diploma. It was he who painted many of the engine panels which are so valuable today, as well as designing many of the handsome invitation cards for the various firemen's balls. Thanks are due to Johnson also for the only portraits we have of Chief Engineers Gulick and Anderson. His brother, David Johnson, a member of Peterson Engine Company No. 15 and also a painter, was famous for his landscapes and his painting of the Harper's Fire.

In addition to their valuable and often valorous fire-fighting service to the metropolis, the city's volunteer firemen were a mighty political power. No fewer than nine mayors of New York were elected either

Interior of Americus engine house.

The belted crest of the Tiger "clan."

as active firemen or officially sponsored by the closely organized fire companies.

This was the reason behind William M. Tweed's unrelenting effort to get himself elected a foreman of a fire company. He joined several companies before he was successful in 1850, with Engine 6, in achieving the first important step in his political career. Tweed was the son of a chair manufacturer. His popularity was undeniable and at a huge banquet at his home Tweed was given an elaborate watch and chain, and, of greater importance, a preview of the enthusiastic support of the rank and file of the populace. It was probably this response that prompted Tweed to form the famous Americus Club, a Tammany bulwark in its early days, which helped to get out the so-called fireman vote.

While the history of Boss Tweed's "success" and his infamous ring is too well known to repeat here, his beloved "Tiger," which was painted on the body panel of Engine Company No. 6's machine was turned against him by Thomas Nast, the artist. The savage cartoons of Nast, who used the tiger to symbolize Tammany Hall and the Tweed regime, did much to turn the public against him.

In January, 1854, the company moved to a brownstone house on Henry Street, near Gouverneur Street. This house was a beautiful one and considered the most sumptuously furnished and equipped engine house in the city.

Firehouse of Americus Engine Company No. 6 on Henry Street, the most elaborate in New York City in the days of the Volunteers

ALES AND PORTER 321
23

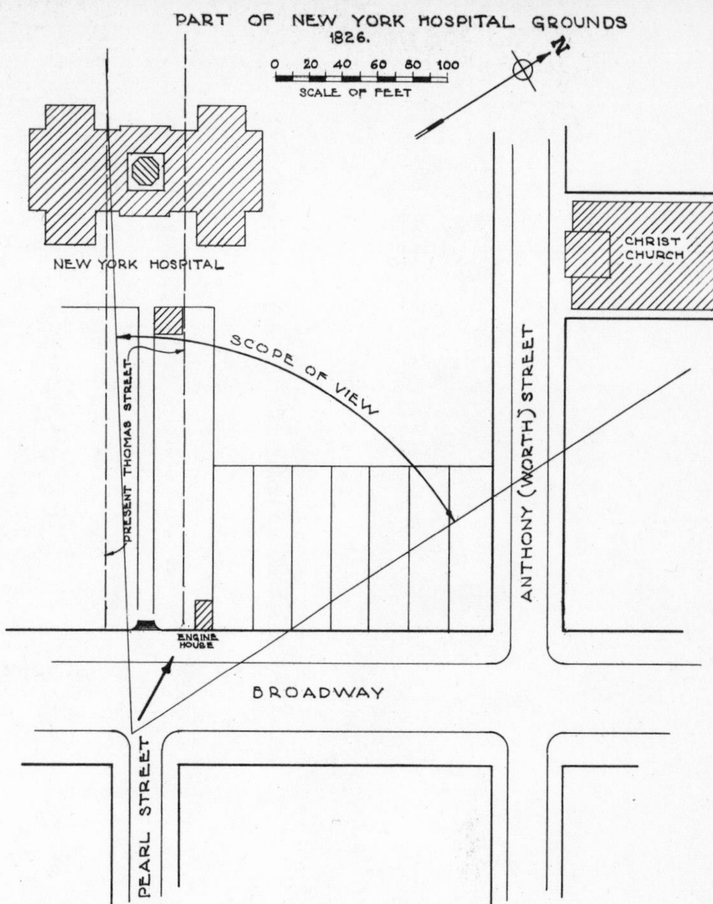

The Hospital Grounds

IN 1828, old Abram Brower recollected that in his youth the vicinity of Broadway and Worth Street was open country, far "out of town." The blackberries which grew wild there were so abundant "as never to have been sold."

The cornerstone of the New York Hospital was laid in 1773 and the completed building was ready for occupancy when the Revolution began. During the war it was used as a military hospital and earthen breastworks were thrown around it. When peace was restored the institution reverted to public use and continued to serve the citizens of the city until 1870.*

In April, 1788, it was the scene of the so-called Doctors' Riot which was instigated by the careless handling of cadavers by medical students and the wild rumor that private graves were being desecrated to secure anatomical specimens. An infuriated crowd rushed to the hospital, and grew to such a proportion that the militia had to be called along with a group of armed citizens led by Governor Clinton, Mayor Duane and Baron Von Steuben. The governor was about to use armed force on the mob when Von Steuben intervened and, in the process of dissuading him, was struck in the forehead by a stone. The flying missile not only knocked the Baron down but, according to one eye witness, changed his whole perspective on the proceedings. Four of the mob were killed and a score more severely injured when the militia was forced to fire on them.

United States Engine Company No. 23 was reputed to be one of the most efficient and harmonious engine companies in the city. They were the proud possessors of a silver-plated goose-neck engine which they housed just within the Broadway entrance to the hospital. Their renown was in great part due to the fact that the company was composed of a number of Manhattan's outstanding business men. Probably the most famous was John Ryker, who assumed the duties of Chief Engineer of the Volunteer Fire Department in 1836, succeeding the firemen's idol, James Gulick.

Another member, Hugh Gallagher, was among the casualties at the Jenning's clothing store fire. He was found pinned against a wall by a heavy safe which had fallen from the floor above. The day after the fire, thousands of persons gathered in the hospital yard to inquire about the injured and to pay tribute to the outstanding heroism of their firefighters.

On the southwest corner of Broadway and Anthony Street stood one of New York's lowest establishments of refreshment and sin, widely known as "The Finish." An appropriately named institution of infamy and degradation it was a place to be avoided by the fire laddies, who had trouble enough with its celebrants from time to time.**

Catherine Lane, now a narrow, nearly forgotten street, lying a few paces north of Worth Street and extending down to Lafayette, was a picturesque place where mint girls, hot corn and strawberry vendors sold their wares. In later years this was the battleground for one of the worst brawls in New York history. The relatively new gangs of the Five Points attempted to flex the muscles of their growing power by a challenge to the Bowery "Bhoys." The latter, at first contemptuous, finally took the time to accommodate them. More than five hundred hoodlums fought up and down Catherine Lane in one of the wildest rough and tumble fights ever seen.

Frank Moss reported that while he stood watching from a distance, he saw one hundred policemen, marching in precise formation up Broadway, turn down the lane. It was almost impossible to see what happened next because of the brick-bats, clubs, hats and other missiles filling the air, but it was only a short while before the limping, bleeding, half-naked survivors of this brave official sortie retreated out of the lane and made their escape as best they could. The two gangs, which had called a truce long enough to "clean the coppers," then resumed their battle, which eventually ended in complete victory for the "Bhoys" from the old Bowery.

**The present Trimble Place, once a wagon road leading from Duane Street into the Hospital grounds, was named for George T. Trimble and his son Merritt, both of whom served as presidents of the New York Hospital. In 1817 George T. Trimble lived in a boarding house opposite the Franklin Mansion (Washington's residence) in Cherry Street.*

***The Ranelagh Gardens, one of New York's favorite resorts, was opened by John Jones in 1765. This famous place, established in the former mansion and grounds of Anthony Rutgers, stood at 232-236 Church Street.*

The House Near the Opera

"WHERE there is danger there you will find Hope" was just a way of saying, when the fiery demon threatened, that there you would find the boys of Hope Engine Company No. 31. This lively group of "firesparks," organized in 1805, was assigned a small house on the southwest corner of Church and Leonard Streets, in a section of the city once called "Frogtown." Whether this appellation came from the amphibious croakers of the not-too-distant Lispenard Meadows to the north or from the presence of the Eglise Du St. Esprit, center of worship for the French colony of the city, on the Franklin Street corner, now can be only guessed at. Nearby stood the Dutch Reformed Church, the Italian Opera and the Zion African Methodist Episcopal Church. When the latter was enlarged in 1820 to occupy the corner, Hope's house was moved to a space at the rear of the church. After 1834 the company was located in Chapel Street (West Broadway) near Beach.*

The night of November 18, 1833, was an important one in the history of New York music. With a competent performance of Rossini's "La Gazza Ladra," the new Italian Opera House, the most resplendent theatre in the country, opened to an enthusiastic audience of music lovers. Many fine representations followed during the next six months, but the quality of the music far surpassed the box office, and the enterprise failed.

The building, leased to James Wallack, the father of Lester Wallack, was renamed the National Theatre and opened again on a more profitable basis in 1836. Both Charles Kean and Edwin Forrest appeared there in a number of plays.

The neighborhood grew to be a most peculiar one, consisting of a weird combination of thespians, saints and sinners. Huddled close by the churches and the theatre were an unusual number of "boarding houses" and even more notorious resorts of vice. It seemed inevitable that Providence should strike the blow of retribution that it did in 1839.

It was on a quiet Monday afternoon late in September that huge columns of smoke rolling over the city gave the sign that a big fire was in the making. The word soon spread that the National Theatre was burning. A general alarm brought engines racing from all parts of the city. Thousands of curious spectators followed. Soon the entire vicinity was one of almost indescribable confusion.

The theatre burned so fiercely that the flames quickly spread to the adjoining French Church and leaped Leonard Street to ignite the Zion Church. Both the Dutch Reformed Church and school and many houses were caught in the spreading flames. Throughout, the firemen worked in a steady, calm manner and they showed great heroism in a number of rescues and in the saving of much property. The despair-stricken French; the despondent Negroes; chattering half-clad women; actors; horn-toting musicians; excited citizens—all combined in this drama of fire.

The contents snatched from the various buildings and hastily tossed into the street presented a scene of ludicrous contrasts. Stage properties and church furniture; bibles, prayerbooks and hymnals; tomes of Shakespeare, and the librettos of operas were strangely mixed with the gaudy contents of a number of the indefinitely described houses in this curious neighborhood.

When the fire was brought under control the theatre was a total ruin. James Wallack lost uninsured property valued at twenty-five thousand dollars. Empty walls remained of the Dutch Reformed Church, which was insured for ten thousand dollars. The French Church and contents were severely damaged and numerous other buildings and dwellings partially

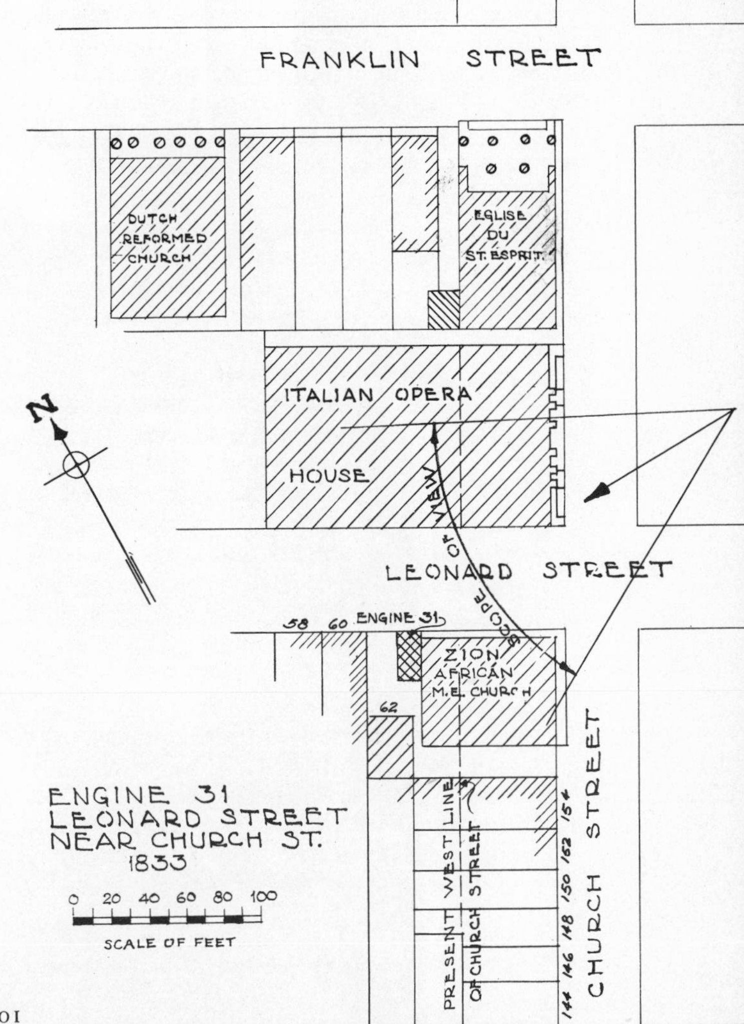

**This firehouse, probably the second oldest in the city, was built in 1823 and is still standing at 246 West Broadway. It now is in commercial use.*

destroyed. Only the Zion African Methodist Episcopal Church was fully insured.

The National Theatre was soon rebuilt and the other scars of the vicinity disappeared. Sin, not so easily dislodged, flourished more openly than ever.

Fate did not hesitate to strike a second time. Another fire, in 1841, finished the theatre forever. George Templeton Strong recorded the event in his diary on May 29th of that year.

"Still very warm and tonight it's raining," wrote Strong. . . . "The National Theatre was burnt up in about twenty minutes this morning: set on fire unquestionably. The walls have mostly tumbled down and from the way they tumbled, one would think the builder had used gum arabic for mortar. One side pitched into Verren's yard and did no other damage than smashing his necessary—another tumbled on a new and very magnificent temple of Venus—kept, I believe, by that respectable person, Mrs. Brown—and demolished one-half of it—and one unfortunate young lady*—who had set up in business a day or two before and whose life fell a sacrifice to a modest reluctance to appear in public in deshabille. It's said that the first individual who emerged from the establishment when the alarm was given was that indefatigable pipelayer, Mr. Glentworth. I don't believe they'll rebuild the theatre in a hurry."

Having taken two strikes Providence apparently did not risk a third, at least in such a spectacular way. In any event Madam Brown and the unchastened coterie of "Frogtown" continued their nefarious ways to later and unrecorded fates. We know from Mr. Strong again, that at least Madam Brown and her own

The Italian Opera House at Church and Leonard Streets

little flock of pigeons retained their health and good spirits until November, 1841.

On Tuesday, November 30, he wrote, "It's said that there were divers very curious characters who contrived to find their way into Mrs. (Valentine) Mott's ballrooms at her late (Prince de) Joinville

**Ellen Roberts, age 17; the redoubtable Zophar Mills, Assistant Engineer, recovered her body the next morning.*

Adriatic Engine No. 31

rout—among the rest the notable Miss or Mrs. or Madam Julia Brown with some of her sisterhood—sent there, 'tis said, by some malicious acquaintance of the Mott's whom they had cut, in their magnificence, on returning from Paris, and who took this mode of revenge for not having been invited."

★

That the death of the unfortunate young woman, Ellen Roberts, alias Margaret Yagers, became a case "celebre" was evidenced not only by the sympathetic and kindly regard of a man of George Templeton Strong's position and high religious standards, but by the prominence given it by *The New York Fireman.*** This newspaper, a three-penny weekly, appeared for the first time on May 1, 1841, published by D. Lewis Northup and Co. at 19½ Ann Street. It was issued from the same office as that of the *New York Pioneer*, a weekly which had previously catered to the interest of firemen. Both were probably printed by William Applegate of Franklin Hall, at 17 Ann Street, at the corner of Theatre Alley.

A new publisher, A. W. Noney, took over for two issues, May 29th and the one of June 5, 1841, in which appeared the following remarkable account of the death of Ellen Roberts. Noney was then succeeded by W. H. Smith and Wm. H. Land.

In keeping with the journalistic style of those days, *The Fireman* published a truly remarkable melodramatic account of the girl's background and demise, complete with villain, virtue and the wages of sin—all set in a 19th century morality piece.

WRITTEN FOR THE FIREMAN

A Biographical Sketch of the Life of Ellen Roberts, alias Margaret Yagers

"Ellen Roberts, who was killed by the falling of the rear walls of the New National Opera House, was born May 16th, 1824.

"She was the only daughter of a respected and wealthy widow lady residing in Philadelphia, and moved in the first circles of society;—fashioned and endowed by the lavish hand of Nature to adorn even the most exalted sphere of that society, she shone a bright particular orb amid its galaxy of stars, and vied with each in radiance. Here was a beauty that not only

***Edward Alexander Holmes, later editor-in-chief and owner of the* New York Dispatch, *was at this time a typesticker for the* "Fireman." *Members of "Big Six" please note that he was a compositor in good standing (and later became president) of Typographical Union No. 6.*

captivated the eye, but enchained the sense. There was an air of voluptuous though graceful elegance in the exquisite contour of her form and features, and the slight shade of pensive thoughtfulness which sat upon her fair brow, added a sweeter charm to the magic of her loveliness. Her soft blue eyes, half hidden by their drooping lashes, beamed with intelligence and truth, and their calm sweet light softened and subdued the sternest heart.

"Miss Roberts was, as we have said before, moving in the highest circles in Philadelphia, admired and beloved by all who knew her, when one of our most fashionable *roues* was favored with an introduction to her. No sooner did he become acquainted with her than she was marked as another victim to add to the list of those who had fallen beneath his cursed arts. He sought her society whenever occasion offered, and by his winning address and protestations of affection soon won the unsuspecting confidence of the unsophisticated, young and inexperienced girl.

Ellen Roberts, alias Margaret Yagers

"After an intimacy of some time, which it was calculated would end in marriage, he persuaded her to elope with him to the city. She was deluded into this act of folly by his promise of marriage, the moment they arrived at his residence in New York, where he assured her she would find his friends and connections pleased to receive her. But imagine her horror when upon arriving here, she discovered her lover to be a mere adventurer, without home or friends, and that she was forced to submit to a fate worse than death.

"For some time they occupied apartments in one of our fashionable hotels, where they passed as man and wife, until the *honorable* gentleman's cash ran low, when he was obliged to shift his quarters to some less expensive situation. Without informing his wretched victim of her destination or of his intentions, he engaged a room for her in the notorious house where she met her untimely fate. Engaging board for himself in a different part of the city, where he passed as a merchant of Pearl Street, he visited the unhappy girl, and endeavored to reconcile her to her situation, but without avail. She had now become aware of her degredation, and her grief was inconsolable.

"She had lived in this situation but five days when the flames of the burning National drove the degraded inmates of the house into the street, with barely sufficient warning to escape to safety. Ellen was among the latest to leave the building, not having been sufficiently hardened in guilt to meet unblushing in open day, the gaze of the assembled multitude. Hardly had she reached the street when she remembered a little locket which contained a miniature of her deserted mother, and which she had left behind. With an air of desperation, notwithstanding the representations of the danger, by the crowd, she rushed back to her room to save the keepsake, when the high walls of the theatre, overhanging the roof of the building in which her room was situated, fell with an awful crash, burying the frail but unfortunate Ellen Roberts in the fiery ruins. An exclamation of horror, from the crowd, sounded above the din of crashing engines, and two or three daring firemen rushed into the house; but it was too late to save! The once beautiful, gay and happy belle of society was borne from the ruins of a vile brothel, a charred and disfigured corpse!—B"

How flash those eyes! How curls the wavy hair!
How sweet that smile!—that brow how rich and fair;
With soul as lofty as her form of grace,—
Her mind the image of her lovely face.

One first, false step to ruin paved the way—
As false the man first led that step astray;
Had he been true thy life had calmly past,
All pure and uncorrupted to the last.

How fondly lovers hung upon those lips,
Where now the grave-worm in his luxury sips;
How lightly rung thy gay and merry laugh,
Where bawds th' envenom'd cup with panders quaff.

Young, lovely, charming, and yet full of sin—
All beauteous outward and all crime within;
Thy life, of late one scene of daily woe,
Corrupt and poisoned in its early flow.

St. John's Park

NAIAD Hose Company No. 53, the second of that name*, was organized January 21, 1852, and occupied a three-story house at 179 Church Street which became one of the most completely equipped firehouses in New York. It was the first to outfit a kitchen and a library for the use of its bunkers.

Among the earliest members of No. 53 were John Garcia, William Thompson and Siro Delmonico, one of the brothers of restaurant fame. (Brother Lorenzo was a member of Engine Company No. 42 and Charles was active with North River Engine No. 30, later Metamora No. 29.) Samuel Ward's poem eulogizing Siro Delmonico, which is reproduced here, is taken from "The Book of Verse of New York."

He lieth low whose constant art
For years the daily feasts purveyed
Of wayfarers from every mart,
The Paladins of every trade.
That soon for us shall mark the tread
Of mourning friends and chanting priests.
Ah! there are other banquets spread
Than Siro's memorable feasts.

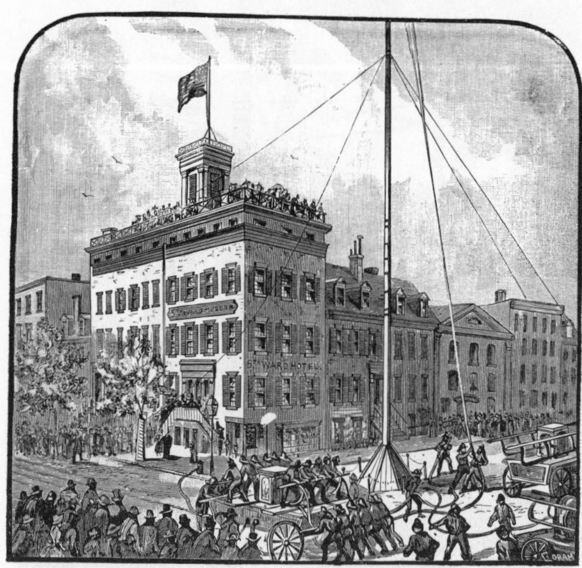

The Liberty Pole at Tom Reilly's Fifth Ward Hotel. Here, pumping contests or musters were held by the city's firemen, the most colorful waged annually on Thanksgiving Day.

The house of Naiad Hose Company No. 53 at 179 Church Street was one of the most beautiful in the city.

The Naiad Company distinguished itself at the William T. Jennings and Company clothing store fire at 231 Broadway on April 25, 1854. Upon the collapse of the walls three of her boys, Michael Flynn, Dan MacKay and his brother Alexander, who tried to save him, were killed. A number of firemen from other companies also were lost, including Andy Schenck, of Truck No. 1, who had promised his fiancée that this would be his last fire.

A huge crowd turned out on January 24, 1855, to watch the water-playing trials at Tom Riley's Hotel at Franklin Street and West Broadway. Companies No. 53 and 54 supplied the hose to Engine No. 42, the "Mankiller," most powerful engine in the city. The famous liberty pole, which had been marked off to indicate the height of the competing streams, showed that the Mankiller, winner of the muster, played to a height of 137 feet. The first pole, erected on Washington's birthday, 1834, was struck by lightning and removed the following year. It was soon replaced and finally removed in 1858.

**The first Naiad Hose Company, No. 16, was organized in 1827 and was located at 24 Beaver Street. Its firehouse was destroyed in the great fire of 1835 and also the famous old bell from the Provost Gaol which had tolled out the alarms in the days of Engineers John Lamb and Tommy Franklin.*

rare view of the St. John's churchyard taken in 1858, by Nathaniel Fish ... ore, President of Columbia University, who was also an amateur photog- ... her. The steeple of old St. John's Chapel can be seen in the background.

Of an evening the boys of Naiad liked nothing better than to tip their chairs back against the firehouse wall and gab about races and washings and sometimes to dwell on the morbid tale of Juliana Sands*, the unfortunate victim of New York's first noted murder mystery. Poor Juliana disappeared from her home a few days before Christmas, 1799. At the climax of a ten-days' search, her body was discovered in the well of the Manhattan Water Works near the present juncture of Spring and Greene Streets.

Suspicion immediately fell on Juliana's fiancé, Levi Weeks. In one of the greatest criminal trials in New York history, Weeks was acquitted. His defense had been conducted by one of the ablest groups of barristers ever assembled, including Alexander Hamilton, Aaron Burr and Brockholst Livingston.

The mystery was never solved, and Levi Weeks lived out his life under the burden of gossip and suspicion that never entirely died away. The well in which Juliana's body was found can still be seen in an alley in Spring Street near Greene.

What was once among the most beautiful spots in New York City is now a parking lot for motor trucks near the eastern end of the Holland Tunnel.

**Also referred to as "Alma."*

In 1753 this site was a camping ground for the Indians who came to New York in a great body, to hold a "treaty" or conference with the governor, the last ever held in the city. Their appearance was grand and imposing, and made a powerful impression on all beholders.

Some manuscript memoranda left by David Grim, who was then a youngster, described the gaudily painted braves and the scene in great detail: "They were Oneidas and Mohawks; they came from Albany, crowding the North River with their canoes; bringing with them their squaws and papooses; they encamped on the site now Hudson's Square, before St. John's church; from there they marched in solemn train, single file, down Broadway to Fort George, then the residence of the British Governor, George Clinton. As they marched, they displayed numerous scalps, lifted on poles, by way of flags or trophies taken from their French and Indian enemies."

On the following day, the residents of this area witnessed a spectacle equally unusual. To accord with the Indian concept of the honor and dignity due them,

St. John's Chapel after its renovation in 1856. The structure was demolished in 1908.

the governor and officers of the colonial government, accompanied by a large number of the citizenry, marched in a long procession to the Indian encampment and ceremoniously presented the chieftains with gifts of amity and peace.

In early times, before it was made into a well by the Manhattan Company, Bayard's Spring was an "effulgent and copious flow" of water, set on a knoll in the midst of an abundance of hickory and chestnut trees. Whole families liked to go there for picnics and to gather nuts in autumn. Altogether, it was a popular meeting place. On the meadow which surrounded the spring, the Vestry of Trinity Church began construction of St. John's Chapel in 1803.

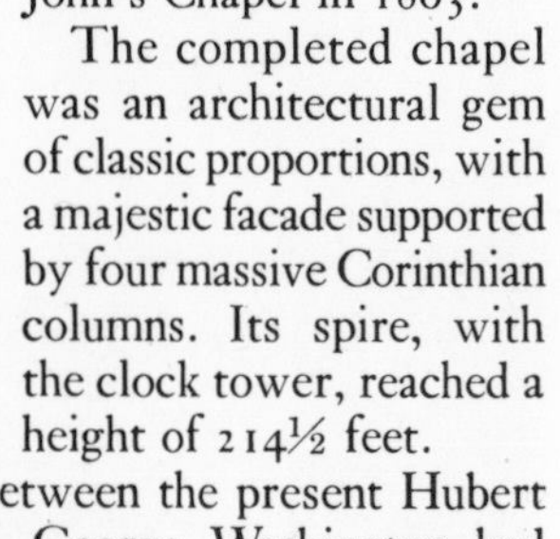

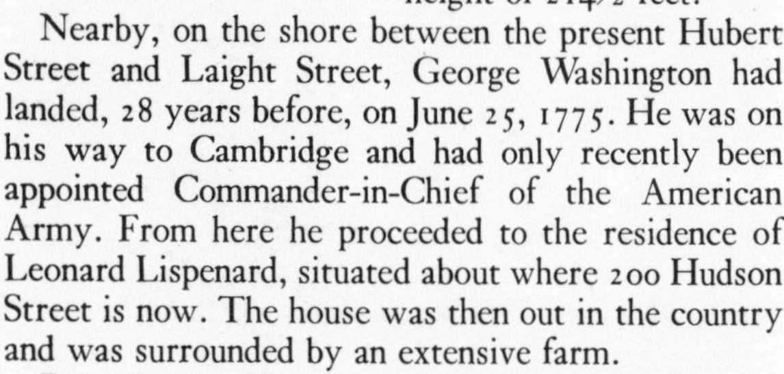

The strawberry girl

The completed chapel was an architectural gem of classic proportions, with a majestic facade supported by four massive Corinthian columns. Its spire, with the clock tower, reached a height of 214½ feet.

Nearby, on the shore between the present Hubert Street and Laight Street, George Washington had landed, 28 years before, on June 25, 1775. He was on his way to Cambridge and had only recently been appointed Commander-in-Chief of the American Army. From here he proceeded to the residence of Leonard Lispenard, situated about where 200 Hudson Street is now. The house was then out in the country and was surrounded by an extensive farm.

By 1807, the Vestry was able to turn its attention to the improvement of the neighborhood. A park was laid out, bounded by Varick, Beach, Hudson and Laight Streets. It came to be called Hudson Square or St. John's Park. The wooden picket fence which first surrounded it was replaced by a handsome iron one in 1830. The park was graded and planted, leading citizens built substantial brick homes there, wells were dug and pumps erected at the corners, and it was not long before the value of the property had greatly increased.

St. John's Chapel a short time before its demolition. The old houses flanking it were destroyed at the same time.

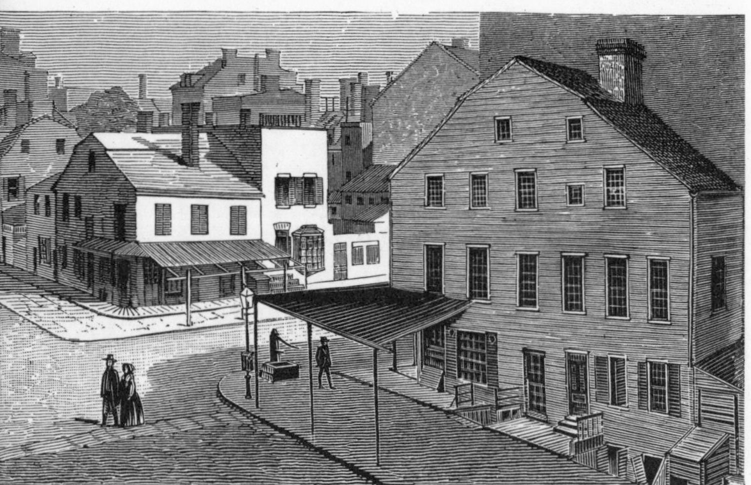

The home of Juliana Elmore Sands, on the southwest corner of Greenwich and Franklin Streets, 1799.

The entire area had been transformed. The park and churchyard became noted for the variety and beauty of their trees. The beauty of the homes and the abundance of graceful shade, made it a place of imposing grandeur. And the continuous line of balustrades framing the square and the houses and running up the stairs to the doorways added a peculiar aspect of European style and magnificence. Many of the homes had window panes that, in the course of time, acquired a violet tint, similar to those in old houses on Beacon Hill in Boston.

The families of Alexander Hamilton, General Schuyler, General Morton, the Aymars, the Drakes,

A view of the Hudson from St. John's Park, showing the neighborhood about Collister and Hubert Streets. The photograph is dated 1860 but it was probably made earlier.

the Coits, the Delafields, and others of equal distinction lived in this residential oasis. During its best years, the square was tenderly cared for by "Old Cisco," a former slave who was a great favorite with all the folks in the neighborhood.

Commodore Vanderbilt purchased the park for the New York Central and Hudson River Railroad in 1866. By the time it had been replaced by four acres of freight stations and storehouses, the only remaining inhabitant was John Erickson, the builder of the Monitor, who refused to leave and stayed until his death.

Today, St. John's Lane, Vestry Street and Erickson Place are all that remain to remind us of St. John's former glory.

Commodore Vanderbilt's Hudson River Railroad Depot on Hudson Street and detail over the main arch showing the Commodore and his varied transportation activities.

Skinner Road

IN 1827 Christopher Street extended northeast of Sixth Avenue under its old name, Skinner Road, until it reached a point between the present 11th and 12th Streets at which point it turned at a right angle in a northwesterly direction parallel with Greenwich Lane. The latter segment became known as Union Road. Both Skinner Road and its extension, Union Road, were abandoned when the new city street plan came into effect, but a relic of them can be seen today in the small triangular plot which is all that remains of a large cemetery on the south side of 11th Street just east of Sixth Avenue. This Jewish cemetery is said to have been established there in 1804-5.

Beyond the Union Road in the 1827 view is the willow-lined Minetta Brook, which formed the dividing line between the country places fronting on the Greenwich Lane and those facing the Bloomingdale Road (the present Broadway) which can be seen in the far distance.

The house in the immediate foreground was the residence of Mrs. Nicholson, widow of Commodore James Nicholson, Commander-in-Chief of the American Navy in 1777. The property was managed for her by her son-in-law, Colonel William Few, a signer of the Constitution and United States Senator from Georgia. In later life Few gave up his residence in that state and came to live in New York. In the last year of his life—1828—he had become president of the Bank for Savings. His city residence at that time was at No. 10 Park Place—the site today covered by the Woolworth Building. In 1813 he had followed Samuel Osgood in the presidency of the City Bank.

Another son-in-law of Mrs. Nicholson was Albert Gallatin, Minister to France and Secretary of the Treasury. Letters record that Gallatin stayed at this house on occasional visits to New York from his residence in western Pennsylvania. A friend of the family was Thomas Paine who lived nearby in Greenwich Village. Although Sixth Avenue had been opened through the farms in 1825 as far north as 21st Street the rural nature of the community had not been greatly affected, but in 1828, Ninth Street was cut through and as the house stood directly in its line it had to be moved to a new foundation on the south side of the street. Shortly after this the property was sold, and as the other nearby side streets were cut through, the neighborhood rapidly built up with city residences and its rural character was permanently gone.

In November 1832, the Jefferson Market at the intersection of Sixth Avenue and Greenwich Lane was completed, and a fire alarm bell tower attached to it.

Northeast from 9th Street and Sixth Avenue, about 1946

This market became the headquarters of one of the first police courts in the city when the Police Department was organized in 1845. The police "Watch," consisting of one hundred marshals, twelve hundred night watchmen, and other officers, had long been a scandal and a source of public dissatisfaction. The marshals received a fee for each arrest and were too often inclined to encourage crime rather than to prevent it.

When the new department was first organized, the police were expected to light lamps and ring fire alarms in addition to their other duties. Fortunately this was soon found "impracticable" and their burdens were lightened.

Mayor Harper had hoped to put the members of the force in uniform but they objected violently, maintaining that it was the livery of servants, and that as American citizens they were born free and equal, and would not wear it. Not until 1853 did they agree to any designation other than a star-shaped badge worn on the left breast.

On November 20, 1851, a false cry of fire in Ward School No. 26, in Greenwich Avenue, north of Jefferson Market, created a panic among the children and caused the death of about fifty of them, by their falling from the upper stories down a stairway shaft. The outer doors, swinging inward, could not be opened. This shocking occurrence led to the passage of an act, on January 2, 1852, compelling doors on public buildings to be made to swing outward.

Jefferson Market and the famous bell tower

NEW YORK FIRE ENGINE NO. 34

AQUATINT, COLORED ARTIST: J. W. HILL

ENGRAVER: JOHN HILL CIRCA 1830 COURTESY OF HAROLD V. SMITH, ESQ.

Howard Engine Company No. 34, affectionately known as "Red Rover," was organized in 1807. It was located in Amos (West 10th) Street in 1813, Gouverneur Street in 1820, at the northwest corner of Hudson and Christopher Streets at the time the above print depicts, and in 1864 at 78 Morton Street, where it remained until the Volunteer Fire Department was disbanded in 1865.

The engine shown in this scene was a side-stroke "goose-neck" or "New York style" machine, manufactured by James Smith of New York. The company was named for Harry Howard, later Chief Engineer of the New York Fire Department.

The exact location of the fire shown has not been determined but the surrounding neighborhood has features suggestive of the Sheridan Square vicinity. The direction taken by the engine, if proceeding from their house, most likely is southeasterly.

This company was the victor in the famous Mackerelville washing, in which rival Engine 14 went down in glorious defeat.

GANSEVOORT ST.
HORATIO ST.
JANE ST.
TROY ST.
ABINGDON SQUARE
BETHUNE ST.
BANK ST.
HAMMOND ST.
PERRY ST.
CHARLES ST.
AMOS ST.
CHRISTOPHER ST.
GROVE ST.
BLEECKER ST.
BARROW ST.
COMMERCE ST.
MORTON ST.
BEDFORD ST.
LEROY ST.
CARMINE ST.
DOWNING ST.
CLARKSON ST.
HAMMERSLEY ST.
KING ST.
CHARLTON ST.
VAN DAM ST.
SPRING ST.
WEST FOURTEENTH ST.
W. THIRTE
W. TWELFTH
W. ELEVENTH
W. TENTH ST.
W. NINTH ST.
W. EIGHTH ST.
WAVERLEY PLACE
WEST WASHINGTON PL.
WASHINGTON SQUARE
FOURTH ST.
JONES ST.
CORNELIA ST.
MINETTO LANE
MINETTO ST.
AMITY ST.
BLEECKER ST.
HANCOCK ST.
McDOUGALL ST.
SULLIVAN ST.
THOMPSON ST.
LAURENS ST.
WOOSTER ST.
GREENE ST.
PRINCE
SIXTH AVENUE
GREENWICH AVENUE
FOURTH ST.
FACTORY ST.
GAY ST.
WASHINGTON ST.
GREENWICH ST.
HUDSON ST.
WEST ST.
WEEHAWKEN
Hudson River Railroad
Trinity Church Cemetery
R. P. Getty's Packing House
Coal Yd.
Lumber Yd.
Stone Yard
N.Y. Sugar Refy.
Stewart Greer & Co's Sug. Refy.
St. Clement's Place
9
8
49
47
46
45
44

Greenwich Village

The original Greenwich Village was in the neighborhood of what is now Christopher Street, southwest of Washington Square. It was near the remains of an Indian fishing village called Sappanikan, the site of which was contained in the farm of Governor Wouter Van Twiller and later of the Mandevilles, facing the Hudson River. Its present name was given to it by the English.

Van Twiller had a tobacco farm in Greenwich in 1638. On August 19th of that year, it was recorded that because of "the high character it (the leaf) had obtained in foreign countries," any adulterations would be punished with heavy penalties.

North of the Village lay meadowland known as Clapboard Meadows. Although some authorities maintain that the name originated from the clapboard fences which surrounded certain property at the edge of the meadow, it is the opinion of the author that the name derived from the old Dutch word "Clopboes" (gooseberry), and that the meadow was so named for the profusion of berries that grew there.

Richmond Hill was originally the property of Abraham Mortier, from whose hands it passed to Admiral Peter Warren. The Warren family was a famous and prolific one, and many of the Village streets were named for its members. Of these, only Abingdon Square and Warren Street remain today, however. Southampton and Fitzroy Roads have disappeared and Skinner Road has long since become part of Christopher Street.

The house built by Mortier at Richmond Hill played host to many of the greatest people of the time. George and Martha Washington stayed there and Martha Washington took great delight in traveling up the Fitzroy Road to Chelsea and Bloomingdale. In the days when New York became the seat of the

MacDougal Alley in 1908

The Washington Mews, from University Place to Fifth Avenue

Contrasting views of a frame house on Weehawken Street, believed to be one of the City's oldest.

Grace Godwin's Garret, once a famous Village coffee house

The house Thomas Paine lived in, 309 Bleecker Street

Aaron Burr's residence at Richmond Hill

Old buildings on West Street between 11th and Perry Streets

Old building still standing at Greenwich Avenue and 10th Street

"Cast iron lace" on 11th Street between Sixth and Seventh Avenues

national government, Richmond Hill was the home of Vice-President Aaron Burr. Burr spent a great deal of time in beautifying it and in laying out gardens around it. There was originally a brook on the property, which Mortier diverted and dammed into a pond.

Eventually, as the city developed, Mortier House, which had been one of New York's most beautiful homes, ran down and became derelict. Before it was finally abandoned it was used as a theatre and when Varick Street was widened some years ago and excavations made in this section a portion of its old stage was found by the workmen. Just northeast of the Mandeville homestead on the north line of 14th Street between Eighth and Ninth Avenues stood the great Obelisk monument built in honor of General Wolfe. The road which led to the monument from 8th Street and 6th Avenue was called Monument Lane.

Greenwich Village grew spontaneously and irregularly. When the city plan took in the Village in 1803, it was not ruthlessly changed like other rural communities which became part of the city. Greenwich Village had already developed such an individual character, had so many buildings, and was so hotly defended by its inhabitants that it was allowed to remain largely unchanged.

The yellow fever plague of 1822 had a great deal to do with the growth of the Village. It precipitated a rush of people from the city to this much healthier region. A business quarter was quickly established and

A typical old Greenwich Village studio

Country home of General Jacob Morton on Morton Street (named for the General)

Paisley Place, once occupied by a colony of Scotch weavers

Grove Street to Hudson Street with St. Luke's Chapel in the background

St. John's Lutheran Church

St. Clement's Church on West 3rd Street, circa 1860

eman's daguerreotype album

Bank Street still retains its name from that period, when a row of banks were established there. A good-sized Scottish colony settled in the Village during the same period, composed of weavers who came from Paisley, Scotland. They built a row of frame dwellings and named their street Paisley Place after their home town.

Though the Village grew with intense rapidity, it has never lost its individuality. It is exceptionally rich in historical associations, and many famous men made their homes there. The house of Edgar Allan Poe was on Carmine Street. The home of Thomas Paine, the author-hero of the Revolution, was at 309 Bleecker Street and a tablet can still be seen on the house which stands at 59 Grove Street, marking the place where he died. Barrow Street was originally called Reason Street, after his book, "The Age of Reason."

Greenwich Village was for a time quite staid and sober. Perhaps it would simply have remained a quiet little community within a community, had it not been for the generation of writers who made it famous, including O. Henry, Alan Seeger, Willa Cather, Edna St. Vincent Millay and Frank Norris.

No part of Greenwich Village is more beautiful or less architecturally typical of machine-age civilization than Washington Square—and none differs so much from the rest of the Village in character.

Whereas the Village has a very mixed population, both in national origins and cultural background, the Square, particularly its north side, has been extremely reserved, patrician and aristocratic. The houses have been occupied by many of New York's oldest and wealthiest families and by its most famous literary and artistic figures.

One of the Rhinelander Houses, built in 1839 and now to be torn down, was the scene of Henry James' novel, "Washington Square." William Dean Howells and Edith Wharton lived on the northeast corner of Fifth Avenue and the park, and Rockwell Kent and John Dos Passos lived at the corner of University Place. John Barrymore and Eleanor Roosevelt had apartments on the west side of the Square. Walter Lippmann, sculptor Gertrude Vanderbilt Whitney and Grover Whalen were all residents of Washington Mews.

Narrowest hous in the Villag

The fittingly named "House of Genius," now torn down, was on the south side of the Park, east of Thompson Street, and among its tenants were Willa Cather, O. Henry and Theodore Dreiser. A famous poet, a noted painter, and one of the world's great explorers, Edwin Arlington Robinson, John Sloan and Viljhalmur Stefansson lived a few doors away, just west of Thompson Street. Another explorer and artist, F. W. Stokes, who accompanied Peary to Greenland, lives in a stove-heated apartment on the

Hose carriage and house of Amity Hose Company No. 38 at 130 Amity Street. At the time this carriage was the most beautiful in the City.

Fireman's button

Square, in the same building as artist Edward Hopper. Lincoln Steffens lived on the south side of the Park, at MacDougal Street, and Edgar Allan Poe lived on old Amity, now West 3rd Street, between Thompson and Sullivan, when "The Raven" was published.

Few people are aware that Samuel F. B. Morse was ever a professor of art, but it was while he was teaching this subject at New York University that he invented the telegraph. Nathaniel Currier, the founder of Currier and Ives, lived on the west side of MacDougal below 4th Street. A few doors away, at the Provincetown Playhouse, Eugene O'Neill's first plays were presented. The first head of the Metropolitan Museum, John Taylor Johnson, lived at the corner of Fifth Avenue and 8th Street, where he maintained the Museum's first art gallery in his home in 1856.

Washington Square was originally a potter's field and later a parade ground. The beautiful arch, which leads into the Park from Fifth Avenue, was erected in 1889 to commemorate the Centennial of Washington's inauguration.

Although, to most tourists, Greenwich Village is synonymous with Bohemia, it is loved by the people who live within its boundaries, not because it is an American "left bank" but because, having retained its small-town air, it has an atmosphere more relaxed, more friendly and more tranquil than can be found almost anywhere else in our metropolitan colossus.

Commodore Vanderbilt, a resident of the Square, had a horse named Mountain Boy, who did 2:16¼ on the Historic track at Goshen. So attached was he to the splendid animal that he had the famous trotter buried in his own sideyard.

Engine panel which once adorned the goose-neck engine of Howard 34

Hat fronts worn by members of Engine 34
The foreman's hat belonged to David Broderick.

The Village has been the home of many fire companies. Among the best known and most popular were Columbia (Wide Awake) Hook and Ladder Company No. 14, New York (Phoenix) Hook and Ladder No. 5, and Howard (Red Rover) Engine 34.

Phoenix Company was organized in 1804 and at various times in its sixty-one years of service was located on Greenwich Street near Barrow, at Hudson and Christopher, at Horatio near Hudson, and at Amity Street near Sixth Avenue. Edgar E. Holley, her first foreman in Amity Street, introduced an improved hook and ladder truck in the New York Fire Department. Because no one in New York could be found to carry out his designs, the truck was built in Newark, New Jersey. The boys voted it a success as soon as it hit the New York pavements, and very soon other trucks were built on the same model. Holley was the first man to use eliptic springs in hook and ladder trucks and his inventiveness was also displayed in making designs and improving the construction of hose carriages.

Columbian No. 14 was, during its brief existence, one of the most envied companies in the volunteer department. Organized in 1854 with only 14 members, they began duty in a temporary house, built at their own expense in Greenwich Street near Amos, now 10th Street. In 1857, they moved their new Pine and Hartshorn truck into one of the finest and most beautifully furnished houses in the city, at 96 Charles Street. If the life of a fireman had difficulties, it also had charms on an evening spent in such a resort as this new firehouse. It contained a splendid meeting room and parlor, a neat and well-appointed bunk room and truck room. There was a large and excellent library contributed by friends and in the back was a beautiful little garden where the boys could spend their leisure hours. Hugh Curry (the champion pipe holder) was a member of the company, and Charles O. Shay, later chief of the Paid Department, was an assistant foreman. During the Civil War five members of No. 14 enlisted with the First Regiment Fire Zouaves and nine more served with various other volunteer regiments.

Howard 34's side-stroke, piano-style engine, photographed in front of the equestrian statue of George Washington in Union Square

Red Rover started with 26 men on its rolls in 1807 and probably took her nickname from the James Fenimore Cooper novel.

One of the 34's runners was the possessor of such magnificent, shining golden hair that he was known as "Goldilocks," and it was said that 34 never needed a signal lamp when he was at the head of the rope. The company's most famous member was the adventurous Dave Broderick, who joined No. 34 on May 13, 1844, became a foreman of the company, and resigned on April 17, 1849 to go to California.

First known in New York as a good stone-cutter, Broderick was lifted to the heights of fame by his qualities of intense personal magnetism, his marvelous knowledge of men and his absolute integrity. "I tell you," he used to say to the man who hesitated in which course to follow, "you can make more reputation by being an honest man instead of a rascal."

Broderick was elected Senator to the first California Legislature and in 1856, was elected to the United States Senate. He was the trusted friend of Stephen A. Douglas, and not even Douglas was more scathing in his denunciations of any kind of political infamy than Broderick. An instinctive foe of slavery, he was idolized by the liberty-loving Californians. He became such a danger to the pro-slavery bloc that his enemies reputedly held a caucus in San Francisco in 1859 to decide how he could be removed from the political scene. They decided to challenge him to a duel, the equivalent of a death sentence, for Broderick knew nothing of duels and was a poor shot. David S. Terry, a southerner, was chosen for the grisly job and carried it out efficiently.

Broderick was mortally wounded and his dying words were, "They have killed me for opposing the extension of slavery." One year later these words were inscribed on the banners that bore his image during the Lincoln campaign. Edward D. Baker, who died a general during the Civil War, delivered one of the finest funeral orations of modern times over his bier. William S. Seward stated in the U. S. Senate that impartial history would rank the dead man with Winthrop, Villiers, Raleigh, Penn, Baltimore and Oglethorpe, the organizers of our states.

Side lamp from engine run by the "Red Rovers"

Merchant Prince

"DIED SENNIGHT . . . Col. Benjamin Tredwell, a gentleman who . . . was remarkable for his hospitality . . ." This sentiment, taken from the obituary of the grandfather of the owner of what is now known as the Old Merchant's House, aptly expresses the aura that even today permeates the gracious rooms of No. 29 East 4th Street, New York City. This home is maintained as a public museum by The Historic Landmark Society, Inc., of which George Chapman, a member of the original family, is president.

In 1835 Seabury Tredwell, Esq., took possession of this stately Federal residence which had been built about 1830. Joseph Brewster, a descendant of Elder William Brewster of the Plymouth Colony, built it on what was originally part of the old Elbert Herring farm.

Seabury Tredwell, Esq., reactionary, Tory and nabob. Notice the queue.

Mr. Tredwell, Senior, seems to have been a domineering reactionary of the old Tory school and is said to have been the last man in New York to wear a queue. As a hardware merchant and importer, he belonged to that group of men who helped build America into a thriving industrial nation.

Secret chambers, in the best mystery story tradition, were built in on both sides of the two parlors and are so cleverly camouflaged that even a careful observer would never know they existed. These secret rooms are large enough to accommodate several people but why they were built or how they were used is not known.

Six daughters and two sons were born to the Tredwells and the most interesting of these seems to have been the youngest, Gertrude Ellsworth. She made her appearance in 1840 in the luxuriously canopied four-poster mahogany bed in the second floor front bedroom and died ninety-three years later in the same bed, in the same room.

Close by the Tredwell home was the house of Clinton Hose Company No. 17. It goes without saying that on many an occasion "boys" of this company cast flirtatious eyes at the proud Tredwell girls in the windows or on the stoop, as they swung their shining black and gold machine past No. 29 East 4th Street. Clinton Hose was founded in 1838 and was located at 5th Street near Second Avenue. Later the company moved to First Avenue and 5th Street.

Life at No. 29 was not without its touches of romance. As a girl, Gertrude fell in love with a wealthy young doctor who, for some reason, did not meet with her father's approval as a suitor. Nevertheless, according to one report, the doctor used to climb through a window of the back parlor after the family had retired

Stately entrance to No. 29 East Fourth Street.

View of front parlor showing dining-room. Secret chambers are well concealed by sliding doors.

Entrance to secret chambers.

Secret chamber for the storing of finery and wealth.

and meet Gertrude who had crept stealthily downstairs. The lovers would then sit for hours on one of the red silk brocaded benches before the front windows. Unfortunately, these trysts were discovered by Mr. Tredwell who ordered bars attached to the rear window which were fastened across the closed shutters when the house was locked for the night. Many years later the doctor died suddenly in London, still unmarried. Perhaps in the packets of old letters that were burned when the house was on the verge of being put up for auction was the full story of this ill-fated love affair.

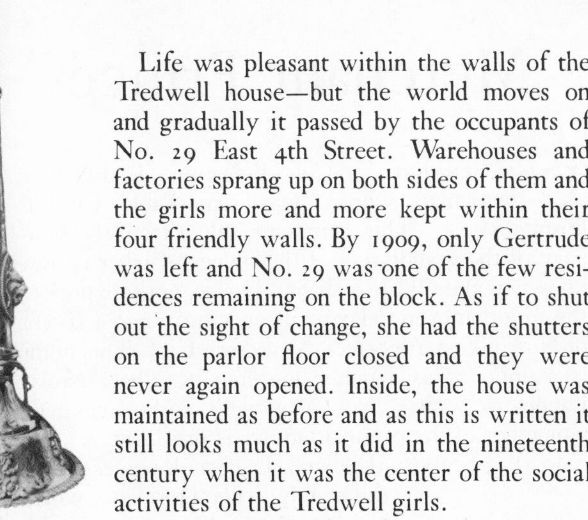

Presentation trumpet.

In 1885 Mrs. Tredwell died and was followed two years later by her unmarried son. The three daughters were then left alone in the house. Fortunately, they had their faithful Negro servant, George Tredwell, to look after and protect them. George was a great comfort to them, for by the closing years of the nineteenth century fashionable New York had moved north, and street rowdies and other undesirables had taken possession of this locality.

It was probably the piquant Gertrude who assumed the duties as head of the family upon the death of her brother. This position she retained until a favorite

Bed in which Gertrude Tredwell was born and in which she died ninety-three years later.

nephew, the dashing Tredwell Richards, took up his residence in the large front bedroom on the third floor. Life at No. 29 continued to drift leisurely on. The family sat up late at night reading *The Churchman* or for variety the sensational novels of the period. Gertrude who was an ardent chemistry enthusiast studied all the latest scientific discoveries. Many evenings they entertained their clairvoyant for they all believed in and were deeply interested in preternaturalism.

Coupling wrench and alarm keys.

Life was pleasant within the walls of the Tredwell house—but the world moves on and gradually it passed by the occupants of No. 29 East 4th Street. Warehouses and factories sprang up on both sides of them and the girls more and more kept within their four friendly walls. By 1909, only Gertrude was left and No. 29 was one of the few residences remaining on the block. As if to shut out the sight of change, she had the shutters on the parlor floor closed and they were never again opened. Inside, the house was maintained as before and as this is written it still looks much as it did in the nineteenth century when it was the center of the social activities of the Tredwell girls.

At sixty-nine this sole survivor was a dainty little lady in immaculate black silk with her hair worn in a French twist and an Italian lace shawl drawn tightly around her pretty shoulders who never grew too old or too tired to exclaim over the latest Parisian fashions of her callers.

Gertrude's suppressed love for the theatre lasted throughout her life for even at the age of eighty-six she dressed up and acted in a private production. Unquestionably it was this fondness for things dramatic that taught her the importance of a good "entrance," for she always kept her guests waiting at least half an hour before joining them in her drawing room. As she was a lavish spender, it was a good thing that her nephew managed her estate. His restraining influence must have ultimately waned for as she grew older she maintained additional establishments at Saratoga and Westport where she occasionally spent her summers. At the latter, she recklessly kept a companion throughout the year and put a car and charge accounts at her disposal. Although her last years were not spent in abject poverty, her income had depreciated perceptibly. All in all, however, she had had a full life.

Carriage and house of Clinton Hose Company No. 17 at 5th Street and Second Avenue.

E. P. CHRYSTIE

Astor Place

LAFAYETTE Street, along with Centre Street, now forms one of the lower city's most important thoroughfares. A section of intense commercial activity, it is less impeded and more fluid as a traffic route than either Broadway or the Bowery. Originally, Lafayette Street was composed of three streets, now extended, widened and joined together. Until 1902, Lafayette Place extended south from Art Street (now Astor Place) only as far as Great Jones Street. The section of Lafayette Street from Prince to Broome Streets was known as Marion Street (see map of vicinity) and the narrow road from Broome to Chambers Streets was known as Elm Street. All of this route, from Astor Place to Worth Street, was included in Lafayette Street when revised, though the street was later made to bear east at Worth. For some odd reason, a section of Elm, still in its original width, between Worth and Chambers Streets, has been renamed Elk Street. The open area formed by the intersection of the Bowery Road and Art Street was known as the "Crossroads." Art Street later became Astor Place in honor of John Jacob Astor who was the leading developer of the neighborhood.

This area, a century ago, was one of the city's most fashionable neighborhoods. Here were located the imposing Astor Place Opera House* and Saint Ann's Roman Catholic Church, originally Dr. Mason Murray's Church of Murray Street, which, in 1841, had been transplanted stone by stone to Astor Place. The building later became the Aberle Theatre and, in 1902, was demolished to make room for Wanamaker's. The Opera House was also replaced by Clinton Hall and the Mercantile Library in 1857, which later became the largest book market in the city. Fourth Avenue, from Astor Place to 14th Street, is still the principal second-hand book center of the city.

Vauxhall Garden, which occupied a considerable space between the Bowery and Broadway, 4th and Art Streets (Astor Place), from 1804 on was a place of resort for residents of the lower portion of the city. It was surrounded by a board fence, with the main entrance near the center on the Bowery. Along the inside wall of the fence were a series of boxes or dining booths with tables, at which light refreshments were served. The garden was embellished with walks, trees, shrubs, flowers, etc., and in the center was a large building in which theatrical performances were given, with interludes of songs and dances. An equestrian statue of George Washington, which may have been the wooden statue originally erected in Bowling Green stood in the Garden.

In 1849 Catherine Havens wrote in her diary: "These gardens were in Lafayette Place near our house, and there was a gate on the Lafayette Place side and another on the Bowery side . . . I remember going with my nurse to the Vauxhall Gardens and riding in a merry-go-round."

The scene of the so-called Macready-Forrest riots, May 10, 1849, in which thirty-four persons were killed.

Lafayette Place Reformed Dutch Church, Lafayette Place and 4th Street, 1863. This church was dedicated May 9, 1839 and closed February 27, 1887. Note hand-pump engine (piano-style) at right foreground.

Part of the Vauxhall Garden was later occupied by the Astor Library, which Washington Irving was largely instrumental in persuading John Jacob Astor to establish. The building is now occupied by the Hebrew Sheltering & Immigrant Aid Society.

Near the Astor home, nine marble dwellings known as LaGrange Terrace were built in 1831 by Seth Geer. Four of the dwellings are still standing. The stone cutting and fitting were done with prison labor at Sing Sing, which caused a great deal of disturbance among the local stone cutters. Built before any other important buildings in the vicinity, the colonnaded edifice was referred to as "Geer's White Elephant,"** but it was an extremely successful enter-

***Seth Geer was somewhat of an elephant himself and it was said he was so fond of oysters that although it was easy for him to pass through the narrow basement doorway of the Astor House Oyster Bar on Broadway and Vesey Street, after repleting himself he could barely get back through the same entry.*

The house and engine of Marion Engine Company No. 9 at 47 Marion Street, now Lafayette Street.

prise and was soon occupied by a number of the city's leading families. When five sections were taken down to make way for Wanamaker's warehouse it was necessary to blast some of the walls apart.

Marion Engine No. 9 occupied the house of Engine No. 3 at 47 Marion Street before that street became a part of Lafayette Street. Engine 9 was later quartered at 52 Marion Street, the offices of Joseph Pine, the fire engine manufacturer, while a new house was being built for them at No. 47. The photograph of the company was reportedly snapped by an assistant of Matthew B. Brady from the upper floor window of Pine's. The occasion was probably the company's procurement of their brand new James Smith steam engine and tender in 1862. It seems almost certain that the figure at the extreme right is the great Brady himself.

In the view which looks south on Elm Street at the right and Marion Street at the left, toward the Centre Market, the dome of the Odd Fellows Hall is visible at Grand and Centre Streets. The lofty brownstone building, one of the most famous old halls in New York, still exists today minus the dome and with extra stories added. The Grand Rendezvous of the Independent Order of Odd Fellows, it was built at a cost of $125,000. It was magnificently furnished and its high, spacious rooms were designed and ornamented in a striking variety of styles, including Grecian, Elizabethan and Egyptian. The Odd Fellows Order at this period numbered many thousand members, organized in about 90 lodges and 12 encampments and its receipts were estimated at about $75,000 annually.

The Chinese Assembly Rooms on Broadway above Prince Street were built by John Jacob Astor in 1830. P. T. Barnum moved here after the disastrous Barnum's Museum Fire in 1865 but this building was also totally destroyed by fire in 1868.

St. Patrick's Cathedral, the prelatical church of the Catholic Archbishop of New York, was built during 1811-12 at the corner of Mott Street and Prince, through the efforts of Bishop Du Bois. James Gulick, who was a hero to every New York fireman, was credited with having saved the cathedral when its existence was endangered by fire in the 1830's.

On January 8, 1854, two of the brightest spots in the fashionable life of New York, Tripler Hall and the LaFarge House on Broadway, were burned to the ground. T. F. Goodwin, foreman of Hose Company No. 35, persisted in holding his pipe to the flames until his boots were burned to a crisp on his feet. The conspicuous heroism of the firemen was lavishly praised by the newspapers. "Where," asked the *Herald* of the next day, "can man go and find deeds of greater heroism than in the history of the New York firemen?"

The Tripler family, builders of Tripler Hall, were prominent members of the old department. Built specially for Jenny Lind, the hall was first known as the Jenny Lind Concert Hall and was opened to the public October 17, 1851. It was the scene of the triumphs of Madame Anna Bishop, Catherine Hayes, Alboni, and Madame Sontag. After the fire it was rebuilt as the New York Theatre and became successively the Metropolitan Opera House,* Laura Keene's Varieties, Burton's New London Theatre, and finally

**The Academy of Music, 14th Street and Irving Place, was built in 1854. Primarily devoted to Grand Opera, it was destroyed by fire, May 22, 1866, and rebuilt and opened in February, 1868. This was New York City's home of Grand Opera until the present Metropolitan Opera House, 1887.*

A rare stereographic view of Lafayette Place, circa 1855. Courtesy of the New York Historical Society.

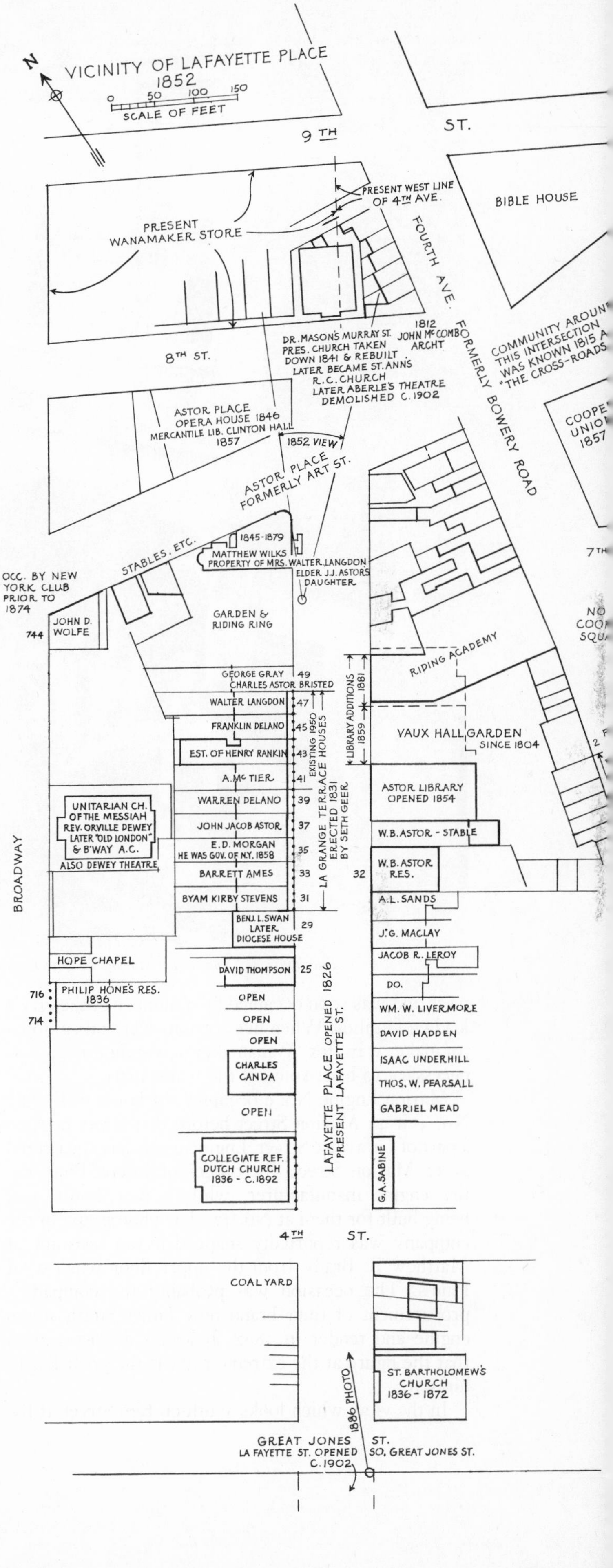

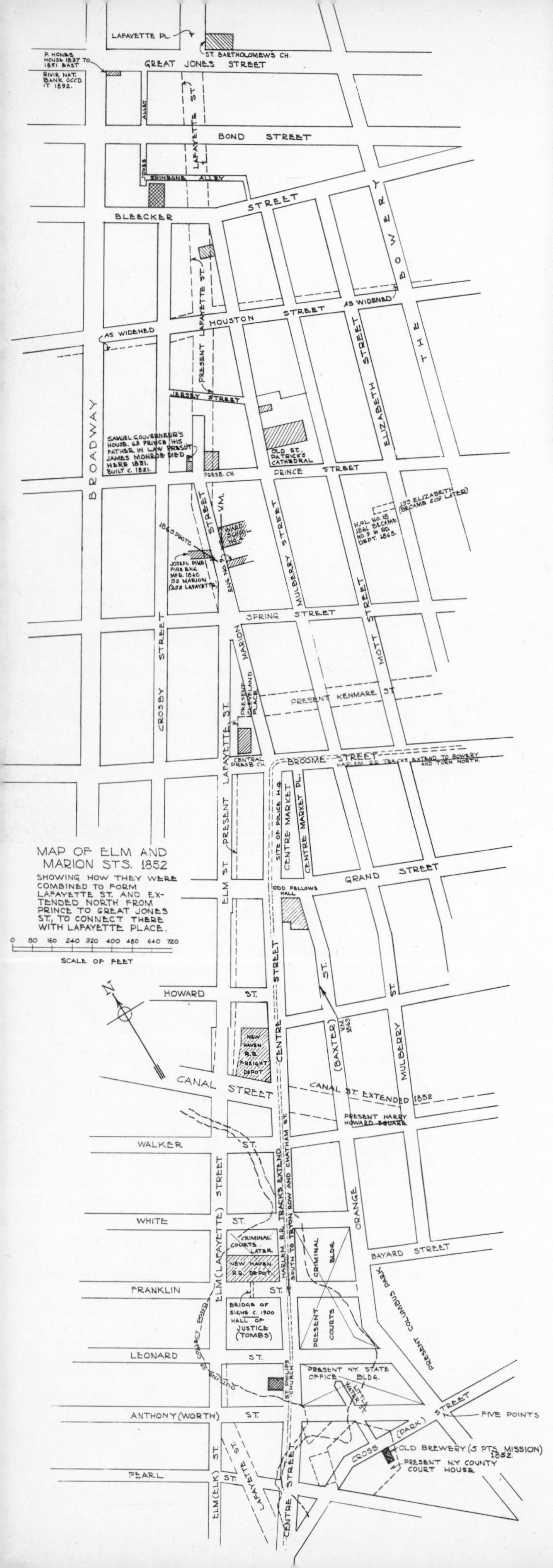

A section of the LaGrange Terrace, built in 1831, remains on Lafayette Place. Part of it is now occupied by Conte's Restaurant.

MAP OF ELM AND MARION STS. 1852
SHOWING HOW THEY WERE COMBINED TO FORM LAFAYETTE ST. AND EXTENDED NORTH FROM PRINCE TO GREAT JONES ST., TO CONNECT THERE WITH LAFAYETTE PLACE.

the great Winter Garden, where Edwin Booth played Hamlet for 100 consecutive nights.

The LaFarge House, which opened in 1856, was named for its builder, a sagacious French investor. After his death and the burning of the Winter Garden in 1867, the property was acquired by E. S. Higgins, the carpet manufacturer, who on August 25, 1870, opened the most palatial hotel in New York on the site once occupied by both Tripler Hall and the LaFarge House. The beautiful dining hall of the hotel, known first as the Southern Hotel and later as Grand Central Hotel, was on the same spot as the stage on which Adelina Patti had made her first local appearance, and on which the great French tragic actress Rachel met her first American audience. The hotel made the fortunes of a number of proprietors, and was finally acquired by Tilly Haynes, who refurnished and modernized it, renaming it the Broadway Central Hotel.*

A structure that aroused more than an ordinary share of curiosity a century ago was the Lord & Taylor store at the corner of Grand** and Chrystie. It was unusually light and spacious for that era, with a large central rotunda and cathedral-like windows.

The Lord & Taylor firm won fame as business pioneers. They were the first store to move to Broadway when the uptown trek began and after their recovery from the panic of 1873, became the first firm to occupy a store on Fifth Avenue. They began in a typically humble fashion as a drygoods store on Catherine Street, established in 1826 by a young English immigrant named Samuel Lord and his wife's cousin, George Washington Taylor. They won popularity by discarding the custom of "pullers-in," who forcibly helped customers decide their "choice."

In 1850 Arnold Constable occupied a building on the northeast corner of Canal and Mercer, which is still standing. In 1827 they had occupied a small building east of Broadway near Centre Street.

Niblo's Garden was in Grand Street on the site of

**The stairway where Edward Stokes shot John Fisk, Jr. in 1872 can still be seen.*

***Grand Street had its own forerunner of the "dime" store which hung out a display sign reading "3-9-19¢ store" and called itself "The Surprise."*

the old horse training ground and many wonderful exhibits and plays were given there. Among the special features was the famous trotting mare Flora Temple, exhibited on the stage, January 19, 1861.

During a fire in a furniture store on Grand Street,* opposite the Essex Market, on the first of July, 1854, the heroic Johnny Garside (Dandy Gig) rescued two women and a boy of fourteen years before the building collapsed. The brave and reckless Garside climbed up the gutter of the building helped by one of his "associates" and hanging on to the window succeeded in passing them to safety in an adjoining building. At a dinner in his honor over which the mayor presided, given at Odd Fellows Hall, he was presented with a jewelled gold watch and chain and $250 in gold in recognition of his courage.

Garside was a member of Columbian Hose Company No. 9, known as the "Quills" and the "Silk Stocking Company." Their carriage, one of the most beautiful in the city, was called "Silver Nine" from the elaborate plating of the metal parts. Young Havens was among its members, as Catherine Havens mentioned in her diary on October 1, 1849. "When he (my brother) was twenty-one years old," she wrote, "he joined a fire company and it was called 'The Silk Stocking Hose Company' because so many young men of our best families were in it. But they didn't wear their silk stockings when they ran with the engine, for I remember seeing my brother one night when he came home from a fire and he had on a red flannel shirt and a black hat that looked like pictures of helmets the soldiers wear. He took cold and had pain in his leg and Dr. Washington came and he asked my mother for a paper of pins and he tore off a row and scratched my brother's leg with the pins and then painted it with some dark stuff to make it smart and it cured him."

The biggest fire suffered by New York since the conflagration of 1835 started in a two-story dwelling and grocery on the southeast corner of Chrystie and Delancey Streets on March 31, 1842. It raged all afternoon and evening along Eldridge, Delancey, Broome, Chrystie and Forsythe Streets, destroying more than 100 buildings.

George Templeton Strong, who raced uptown in the evening to watch the fire, confided to his diary that "it was a field day for those superstitious beings who believe in bad luck coming in threes. On the same day, while the firemen valiantly attempted to keep this fire within some sort of bounds, one of their favorite taverns, the 'Brown Jug' at Pearl and Elm Streets and several other buildings at Pearl and Elm Streets and a building at 151 Washington Street were destroyed."

**Firemen dreaded a call to Grand Street because it was two miles across from river to river. Father Knickerbocker was fattest at that point. The engines, working in line, formed nearly a mile of hose and it required twenty companies pumping from one to the other to get a stream of water from the river to the fire.*

An earlier fire in Broome Street near Sheriff, on April 23, 1829, was responsible for the injuries that incapacitated the famous Conk Titus and ended his days of glory in the Volunteer Department. A front wall fell out, breaking a ladder on which Conk stood with two companions. Strangely, he was the only one hurt of the three. When he recovered from his injuries he became bellringer in the City Hall Tower.

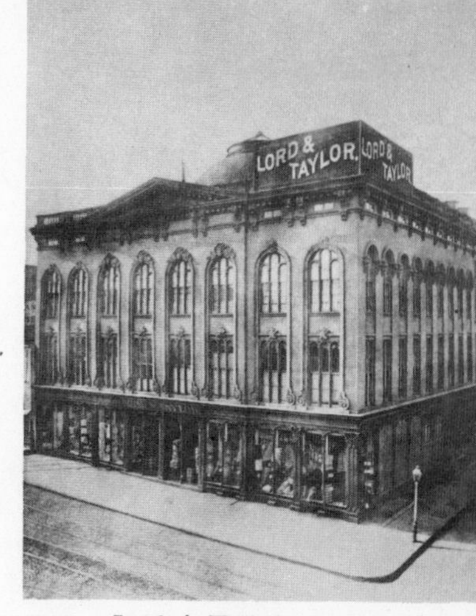

Lord & Taylor's "cathedral" at Grand and Chrystie Streets.

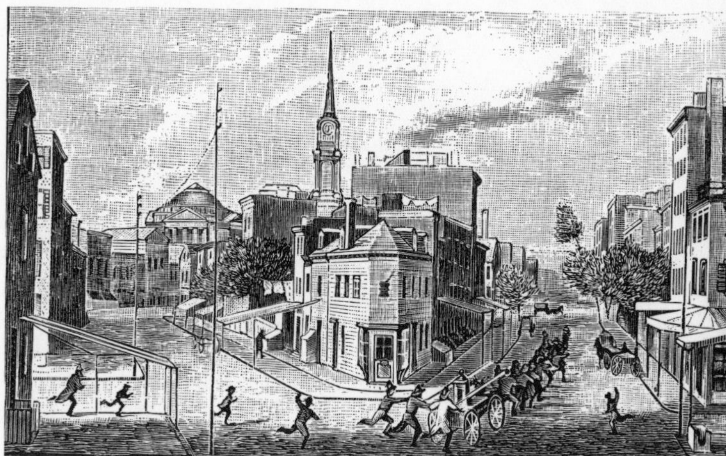
Looking south on Elm Street toward the Centre Market. The dome of Odd Fellows Hall can be seen in the background at the left.

The LaFarge House, Broadway and Bond Street. This structure was destroyed by fire in 1854.

Union Square

UNION Square, once Union Park, has often been thought of as a geographical accident. Originally a rugged and hilly area, it was the junction of the Bowery Road with the extension of Broadway. In 1811 it was referred to as the Union Place and then extended from 10th to 17th Streets. It was reduced to its present size in 1815.

The area was graded by 1834 and opened for public use five years later. By 1849 it was a beautiful residential neighborhood having much of the character of St. John's and Gramercy Parks. Samuel B. Ruggles, the originator of the Gramercy Park idea of real estate selling who also had much to do with the development of this section, lived at No. 24 Union Place, now a part of 17th Street. There was a time when it was seriously proposed to move the City Hall to Union Square but the idea, fortunately, was not long-lived.

The heroic, equestrian statue of George Washington*, now within the limits of the park facing down Broadway, used to stand at the intersection of Fourth Avenue and 14th Street, in the triangular area occupied by the World War II Memorial. This, incidentally, is said to have been the exact spot where George Washington and his victorious troops were greeted and first acclaimed by the New York folks on Evacuation Day, November 25, 1783.**

Little Catherine Havens wrote in her diary in 1849, "I roll my hoop and jump the rope in the afternoon, and sometimes in the parade ground on Washington Square, and sometimes in Union Square. Union Square has a high railing around it, and a fountain in the middle. My brother says he remembers when it was a pond and the farmers used to water their horses in it."

Daniel Drew of Erie Railroad fame, when at the height of his career, purchased property fronting Union Square on 17th Street, where he built a four-story mansion of brownstone. Drew moved uptown from Bleecker Street because, as he described it†, "it gave me standing among the boys on the Street" and he could "live in the company of the money kings."

**After the speeches at the statue's unveiling, the string was pulled—and nothing happened. Brave hook-and-ladder laddies finally loosed the shrouds and won the cheers of both the viewers and the red-faced officials.*

***Stokes says it was near the present junction of the Bowery and Third Avenue.*

†The book of Daniel Drew, Doubleday-Doran.

R. H. Macy & Co., the birthplace of a new idea. 204 East 14th Street

As Drew had "a hankerin' for cattle" (having built the foundation of his wealth as a cattle dealer) and was "bent on keeping at least a milch cow" he built a barn in his spacious lot with a cow-shed and fine stable attached. The former cattle-drover and owner of the Bull's Head Tavern and sales yards swore that the smell of the cow-barn and stable helped to ease the weight of his worries.

Drew was not the only one to keep a cow in Manhattan†† at this period. A somewhat bewildered cow could be seen grazing amidst the "handsome but unhappy peacocks" on the grounds of Peter Goelet's mansion at the northeast corner of Broadway and 19th Street.

††Within the present limits of the City of New York, on a 400 acre farm at Willet's Point, Bayside, in northern Queens, the late Charles G. Meyer of the Cord-Meyer Corporation and a director of The Home Insurance Company, bred and maintained a prize herd of Jersey milk cows until 1943. These, including several champions, were consistent top producers of high-test butterfat milk. Fort Totten is built on land once a part of this farm.

Orbach's and Klein's are now a part of this scene. View of 1872.

The "crack" Metamora Hose Company, 1863. Foreman John R. Platt (trumpet in hand).

Metamora Hose

A NUMBER of crack fire companies were located in this vicinity, at the time 23rd Street was considered "uptown"—among them, Metamora Hose 29, which was organized in 1854. Composed principally of clerks and merchants, the company was located at 21st Street and Broadway—and later moved to 21st Street and Fifth Avenue. They ran a beautiful four-wheeled carriage, the reel of which was made of richly gilded rosewood of a dark plum color.

They adopted the name "Metamora," from the successful play, written by John August Stone, based on the life of King Phillip, Mohican Indian Chief. Edwin Forrest, the tragedian, enacted the role for the first time at the Park Theatre in 1829 and gave a magnificent performance as the heroic Indian.

Charles Delmonico, the famous caterer and restaurateur, was a member of Metamora as was John R. Platt (previously of Hose 36), ex-president of the Fire Department who was a foreman of the company and "one of the quickest firemen in the city."

As a boy, Frederick Van Wyck* was a runner with the "Metties." Young Fred had a fine looking black and white dog named Rollo who used to run along with him to fires. Rollo was a great fighter and often upheld the honor of Metamora Hose against the dogs of other companies. He became so close to the hearts of company members that when they received a night alarm the firemen often signalled the lad by tapping his window with a slate pencil tick-tack, in the hope that he and Rollo could get out and run with them.

The company distinguished itself in the Draft Riots of 1863 when they saved Mayor Opdyke's house from destruction after the rioters had set fire to it, winning a citation from the press of the city the following day.

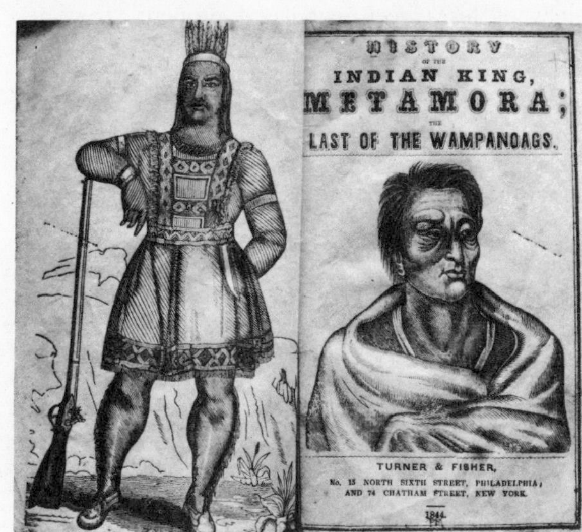

History of Metamora, a best-seller of its day.

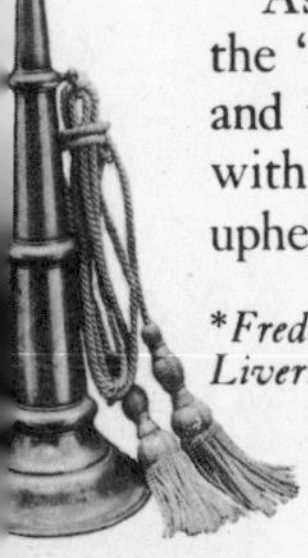

Frederick Van Wyck—"Recollections of an Old New Yorker," Liveright, Inc.

rade Trumpet

Every. One. of You
(Stand. By. Your. Brakes. Men)
Put. in. the. Butt.
(Play. Away. Men)
Now. Will. You. Work.
(Mackerelville. Night Fire.)
14
34
Work. Her. Lively. Lads. (Up. To. The. Bend.) You. Dont. Half Work.
Now. You've. Got. Her. (Old. 14. Washed. By. 34) Stave. Her. Sides. In.
(Say. You. Will. Now.)

UP TO THE BEND

Primitive painting on linoleum

ARTIST UNKNOWN   CIRCA 1850   COURTESY OF HAROLD V. SMITH, ESQ.

One of the worst catastrophes that could befall an old volunteer fireman was to have his engine "washed" by a rival company. In order to play water on a fire, the engines had to line up from the source of water to the fire, one engine pumping water into another, and so on down the line. If, in the process of transferring the water, one engine could not handle the water pumped into it by the preceding machine, she was washed and in the eyes of her own men and the members of other companies, temporarily disgraced.*

One of the most talked-of contests of this nature was the one between Engine Company No. 14 and their new Philadelphia style double-deck engine which was one of the most beautiful in the department, and Howard Engine Company No. 34, the celebrated "Red Rovers" from Greenwich Village who ran a "New York" goose-neck style engine.

The engagement took place at Mackerelville and was one of the most exciting contests of its day. After a battle that lasted for hours, the Red Rover bested 14's engine and when the cry of "She's up to the rabbits" and "She's over" came ringing through the cheers of 34's men, the members of No. 14 were stunned, some of them crying as if they had lost a friend or a relative. While the lusty cheers of Howard 34 echoed throughout the shanties of Mackerelville, Harry Venn, the foreman of 14, who had coaxed, cursed and coddled his men for hours made preparations to let the water out by the tail screw while his men lay on the ground exhausted by the evening's ordeal.

Mackerelville was located in the neighborhood of 14th Street and Avenues A and B. One of the roughest, toughest districts in the city, it was the common battleground for the "fightingest Irish" on Manhattan Island. It was a place of squatters' sovereignty and many of the tincan mansions were, as in the other "suburbs," built on the public domain.

When this Hibernian heaven was cleaned out, many of the "citizens" moved to Odelville and Seneca Village in the area now occupied by 85th Street and Eighth Avenue, and to Hell's Kitchen and to Dutch Hill, where Tudor City now stands.

A notorious character of this neighborhood was the famous "Mackerelville Dwarf" who achieved notoriety by virtue of his ugliness and ferocity. It is said that this dwarf was one of the most savage street-fighters in the city.

Once a part of the museum collection of the New York Volunteer Firemen's Association this primitive contemporary painting was lost for many years. It was recently found in an old trunk, among other New York fire relics, in Philadelphia.

* A term used by the volunteers to indicate the flooding of their own or a rival machine.

Sunfish Pond

Where, in a twilit eddy of my dream,
Thine image, Isaak, pored upon a bream.

As one stands at the corner of Fourth Avenue and 31st Street today it may be difficult to imagine this pleasant scene as the place where the stage horses, having passed the second milestone from the city below, were stopped to nuzzle the cool waters of Sunfish Pond and to drink, while the passengers stretched their legs before continuing the journey to Harlem and on beyond to Boston.

This was a favorite spot for the young farm lads of the vicinity, and their dads, to do a little fishing or muskratting. This lovely little body of water provided in winter a natural ice rink on which to cut a fancy figure eight or, on new thin ice, to risk the exciting game of "tickelly bender" over its heaving bosom.

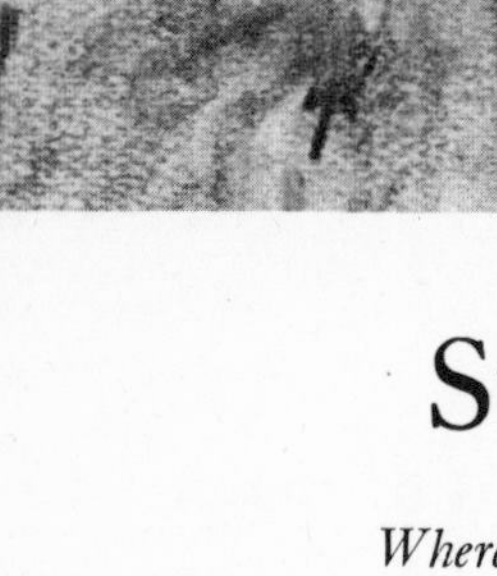

A carved wooden decoration from the old Murray Hill Hotel, owned by Mary Muller

The pond was fed both by springs and by a brook which also carried its overflow down to the East River at Kip's Bay. This brook, in olden times called "t'Oude Wrack," by tradition got its name from an early Dutch ship which was wrecked in the Bay.

British scouts and advance patrols came through here upon the landing of their main force in Kip's Bay, following the defeat of the American forces on the heights of Brooklyn. A considerable number of this amphibious army undoubtedly made good use of the pond's refreshing waters during the securing of their beachhead and while their commander, General Howe, with his staff, was being entertained at Inclenberg, the farm of wealthy Quaker merchant Robert Murray, on September 15, 1776.*

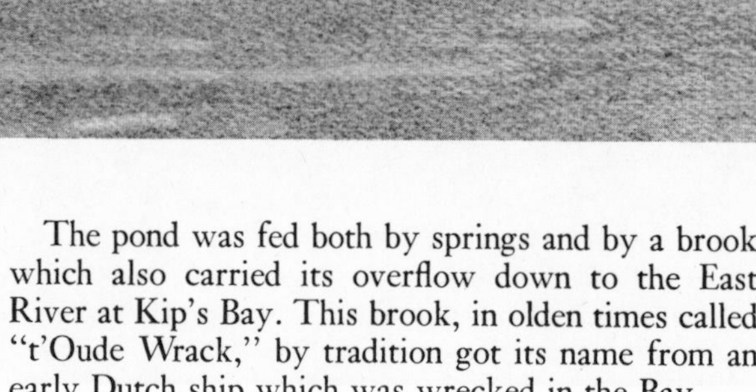

Sunfish Pond was a favorite spot for fishing and muskratting

The story of the charm and seductive hospitality of Mrs. Murray, the former lovely Mary Lindley of Philadelphia, has often been told. The Madeira wine with which she and her handsome daughters regaled the British gentlemen was the favorite vintage of George Washington who but a few days before had been a guest at the Murrays'. At the moment of the British toasting Washington was in a very precarious frame of mind, due to the wavering of his opposing patrols in the fields near Bryant Park.

The Murray farmhouse was on the high ground of what is now known as Murray Hill, beyond the Quarry Hill, which abutted the north side of Sunfish Pond. Murray built this house prior to 1764 and it remained one of the show-place farms until destroyed by fire in 1834. It was a most productive farm and Murray's favorite cornfields lay just north of his home beyond his orchards and included the entire area which now extends nearly to Grand Central Station.

Lindley, the eldest of the twelve Murray children, in 1793 agreed to sell "Bellevue," the old Keltatas mansion facing the East River at 26th Street, later used as a tavern and club, to Brockholst Livingston**, treasurer of the University of the State of New York. Livingston in turn conveyed the tract to the city in 1798. Bellevue Hospital, named for this estate, now stands on this land. Incidentally, Lindley Murray, the famous grammarian, was a cripple, reportedly because of a youthful attempt to jump across Peck Slip, a distance of some twenty-one feet, on a dare.

The Murray house faced the East River, overlooking Kip's Bay and the quaint little settlement of Kipsborough. The original Jacob Kip mansion built in 1655 was the oldest house in Manhattan at the time it was removed in 1851. It stood approximately at the present corner of 35th Street and Second Avenue. As the Kip family increased, the streets of the town, all of which have now disappeared, were named for the various members of the family.

When the House of Refuge on the old "Parade" (Madison Square) caught fire in 1838, Knickerbocker Engine Company No. 12 ran up from William and Duane Streets and took position with their suction dropped in Sunfish Pond, one of the nearest sources of water.

Several machines worked in line in order to carry a stream as far as the fire. "Old Nick's Boys" played into Black Joke Engine No. 33 and gave her a good washing on that occasion. The latter's boys loudly protested, to the amusement of the whole department, that mud from the pond choked their engine.

**Mr. Brower who saw the British force land in Kip's Bay, as he stood on the Long Island heights, says it was the most imposing sight his eyes ever beheld. The army crossed the East River, in open flat boats, filled with soldiers standing erect; their arms all glittering in the sun beams. They approached the British fleet in Kip's Bay, in the form of a crescent, caused by the force of the tide breaking the intended line, of boat after boat. They all closed up in the rear of the fleet, when all the vessels opened a heavy cannonade.—Watson, 1828.*

***Born Henry Livingston, brother of James R., he abandoned his given name for "Brockholst."*

Charles Haswell shown working on his delightful reminiscences

Charles Haswell, that delightful octogenarian and diarist, recalled Sunfish Pond as a place of cherished memories, but that eventually the glue factory which Peter Cooper built nearby so contaminated the waters that the pond had to be drained and filled in 1839.

Part of the bed of Sunfish Pond later became the site of the old Harlem Railroad stables and in turn, the 33rd Street trolley car barns. It is now occupied by the handsome 100 Park Avenue Building.

VICINITY OF SUNFISH POND 1819

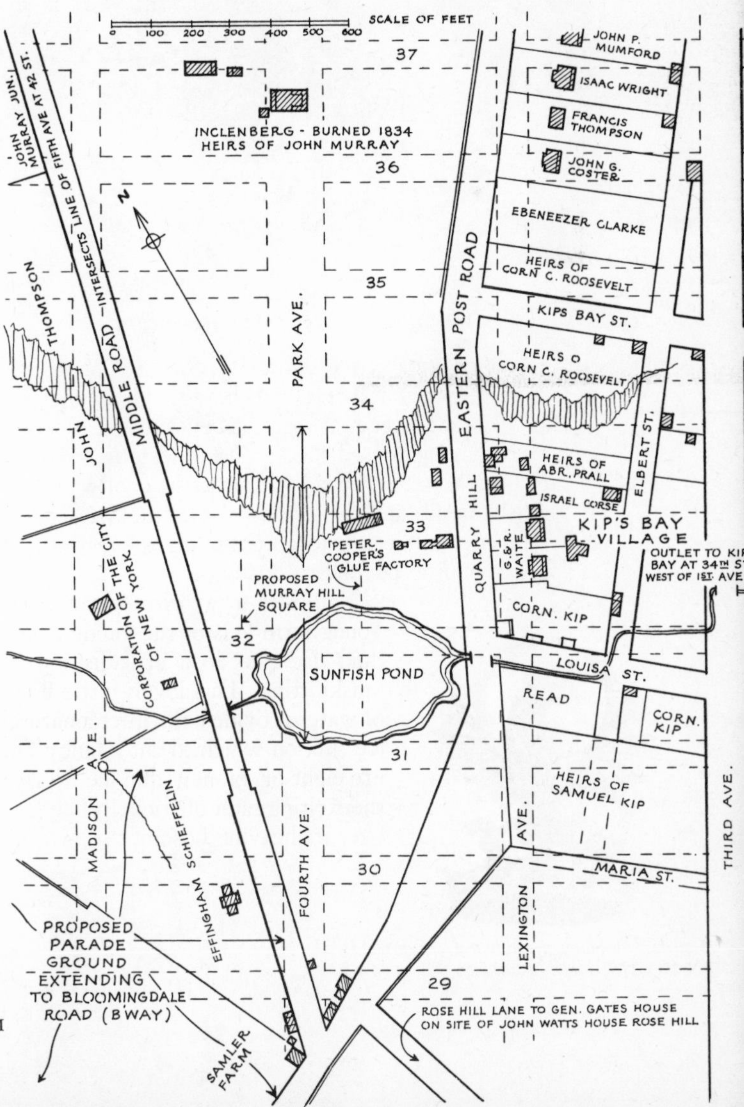

The Fork in the Road

TODAY as you pass by the intersection of the former Abingdon and Bloomingdale Roads (21st Street and Broadway), if you look north toward Madison Square, you will see a gradual widening of Broadway as far as 23rd Street. This widening, now extended as a throughway to Fifth Avenue, was originally the fork of the Bloomingdale and the Eastern Post Roads and existed long before Fifth Avenue was projected south beyond the old Middle Road.

Madison Square Park was originally planned and laid out in 1803 as "an area sufficient to drill and maneuver the entire militia of the State of New York." It was bounded by 23rd and 26th Streets, Third and Eighth Avenues. The old arsenal, built in 1807 as a part of this military project, was abandoned for this use in 1824. This swampy area had been a potter's field (1794) which was discontinued when a new potter's field was opened in what is now Washington Square. Within the arsenal enclosure the House of Refuge occupied parts of old buildings with new ones added, some of which remained even after Fifth Avenue was opened through the grounds in 1837.

After the War of 1812 this area was gradually pared down to the approximate limits of the present park and was opened as a public park in 1847.

The Madison Cottage in 1852.
This site was later occupied by the Fifth Avenue Hotel.

The east branch of the Minetta Brook had its source practically in the center of Fifth Avenue just north of 21st Street. South of 21st Street the neighborhood east of Broadway was known as Gramercy or Bowery Hill. The brook was known as the Crommessie Fly and as Cedar Creek just below 26th Street, Park and Madison Avenues.

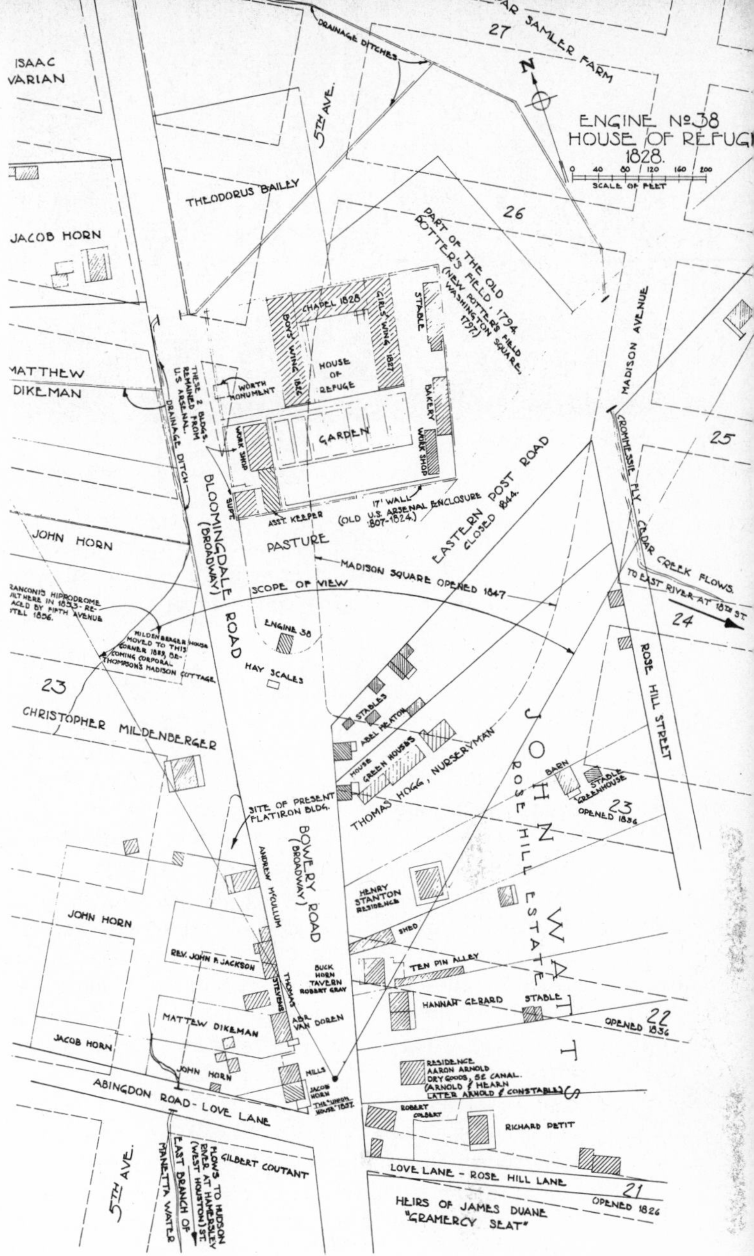

As early as 1812 the original Buckhorn Tavern, which stood in the bed of the present 22nd Street*, was a noted resort with its ten-pin alley and horse sheds. For many years the pair of antlers which hung over the entrance were familiar to sportsmen and travelers. When 22nd Street was cut through in 1826, this tavern was removed and the antlers were transferred to the old Mildenberger house. Opened as a place of refreshment in 1839, this was known to several generations of New Yorkers, including the trotting horse enthusiasts, as the Madison Cottage.

On this site Franconi's Hippodrome was opened in 1853. It was a huge two-story brick structure with an open arena where circus performances were given. As its shows were too daring and sensational for the

**The residence of Aaron Arnold stood two doors below the original Buckhorn Tavern. Arnold was a partner in the firm of Arnold and Hearn, drygoods dealers at 52 Canal Street, who were the founders of the great department store of Arnold Constable.*

Franconi's Hippodrome, Fifth Avenue and 23rd Street, in 1853.

The bird house built in Madison Square to house the sparrows imported from England, 1859.

period, it was not a financial success and within two years was abandoned, after which the location was occupied by the Fifth Avenue Hotel, constructed in 1856.

Anthony Tieman's paint factory was located in the bed of the present 23rd Street, between Rose Hill Street and Fourth Avenue. This was removed in 1836 at the time 23rd Street was cut through. Daniel F. Tieman, his son who became Mayor of New York, lived in that house as a boy in 1788.

The Common Council directed that the firehouse of Engine No. 38 be removed from its location on Love Lane at 21st Street to the ground adjoining the House of Refuge.* The reason for the move was that the grading in preparation for the opening of 21st Street left the house standing considerably below the level of the street. No. 38 remained at this location until 1849 when it was stationed in Nassau Street near Cedar and in later years at 28 Ann Street.

In 1876, the arm of the Statue of Liberty, which had been sent from France for display at the centennial exposition in Philadelphia, was brought to Madison Square Park and erected on the Fifth Avenue side. It remained in this prominent place until 1884 while funds were being raised by the school children of America to pay for the base of the statue. The arm was then returned to France for completion of the entire work, which subsequently arrived at Bedloe's Island later in that year and was placed in the position in which it is seen today.

Lexington Engine No. 7, organized in 1849, originally in the house occupied by Engine No. 46 in Third Avenue between 26th and 27th Streets, was transferred to the house in 25th Street which it occupied throughout the rest of its career. Many members of Lexington were distinguished in the affairs of the city.

**The year following the House of Refuge fire, Cataract Engine No. 25 was replaced by Engine 38 in the little frame building with its small tower and bell on the House of Refuge grounds. This company was composed mostly of mechanics.*

The arm of "Miss Liberty" in Madison Park, 1876.

The Varian Tree, Broadway and 26th Street, 1864.

The ten horse stage at Madison Square in 1876. This stage was one of the most beautiful ever seen in old New York. Photograph courtesy of the New York Historical Society.

The first Madison Square Garden (formerly called Gilmore's Concert Garden) in 1870. This building was demolished in 1889 to make way for the new Garden on the same site, designed by McKim, Mead and White.

This company did excellent service during the draft riots of 1863. After extinguishing a fire in the Provost Marshal's office at Third Avenue and 46th Street, they raced to the third Bull's Head Hotel, then at the corner of Fifth Avenue and 44th Street. They next were called to the colored orphan asylum at Lexington Avenue and 43rd Street, where a row of dwellings was in flames. After helping to extinguish this fire, they were immediately sent to the Armory at Second Avenue and 21st Street and worked there until the flames were overcome. The Lexington Engine house became headquarters for what little law and order there was in the district at that time, and a great many conscientious citizens rallied with the members of Engine No. 7 and in the ensuing struggle, the rioters were severely punished and were held in check until the arrival of the militia.

There were a number of serious fires in this locality including one terrifying night when three large stables in the city were destroyed within a few hours, perhaps the work of a vicious group of incendiaries. Of this George Templeton Strong wrote, November 20, 1848: "I took a walk in the morning to look for the locality of a fire that had made somewhat of a light and a good deal of an uproar during the night. Found that Murphy's Stables, corner of Third Avenue and 25th Street, had been burnt, together with several adjoining buildings, some little houses, a church, a meeting house, and a public school . . . About 120 horses had been burnt—the stable was fired by an incendiary, who came near being caught. I wish he might be and hanged."

Crommessie Hill was earlier described as an eminence near Governor Stuyvesant's farm. The general locality was identified as Crommessie Fly or Valley.

Originally part of a farm "north of the Crommessie," which Judith Stuyvesant sold to one of her freed Negroes, Gramercy Park has been one of New York's most conservative and delightful residential areas since 1830. It has been preserved against the encroachment of commercial progress by the inviolate right of the owners and occupants of the surrounding

The Eden Musee, a famous waxworks, at 55 West 23rd Street, 1886. It was demolished in 1916.

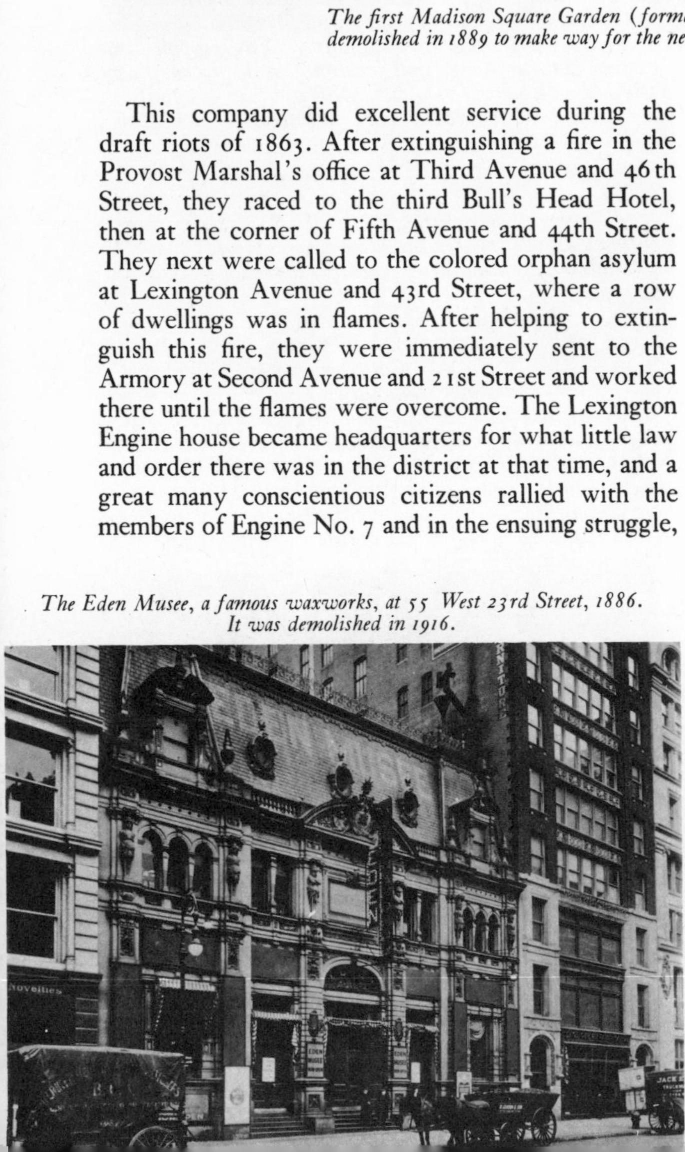

St. Gaudens' Diana atop the second Madison Square Garden (see Appendix).

lots of land. The property was purchased and laid out by Samuel B. Ruggles who, among other things, was one of the most energetic real estate developers in New York during the 1830's and '40's. He acquired this parcel of land from the old Duane Farm and named the park for Gramercy Seat which was James Duane's title for his country estate.

Much has been written about Gramercy Park but fresh information has recently come to light in the diary of George Templeton Strong, who married the "wonderful and Beautiful" Ellie, daughter of Samuel Ruggles. Together, George's father, George Washington Strong, and Sam Ruggles built a house on the north side of the Park for the young folks. The story of George's downfall or Ellie's conquest, with incidental comments on their home at the Park, is abridged somewhat here.

APRIL 9, (1848), SUNDAY—4 P. M. . . . *this last month of my life has been so utterly different . . .*

Perfect—entire happiness—so new and strange . . . Happiness that . . . bewilders me . . . the happiness of loving and being generously loved by a beautiful high-principled noble-hearted frank affectionate good girl—possessed of everything that refinement and cultivation and taste and intelligence can adorn womanhood withal . . .

APRIL 20, THURSDAY—12 M. "*Noon of night*" *that is. Just in from Union Square . . . Out this morning with my glorious little E. making calls on divers people. Mercy on us. I'm terrified at that* "my" *in the last line . . .*

APRIL 24, MONDAY NIGHT . . . *Whatever E. does is right, ipso facto. . .*

Wish I had the man here that invented the Polka —I'd scrape him to death with oyster shells . . .

APRIL 28, FRIDAY—12 . . . *Orders given to commence excavating in Twenty-first Street Wednesday . . .*

Gramercy Park at the turn of the century. The fountain was installed in July, 1853.

What the object may be of putting us into a 40 ft. house—and how soon such an establishment is going to reduce us to an insolvent state . . .

The wedding is to be noisy to a degree—perfectly vociferous . . . If any one had told me six months ago that I should be utterly indifferent to such a prospect I should have looked at him with serene incredulity—and if he'd repeated the statement offensively, should have kicked *him with violence for his impudent mendacity.*

APRIL 30, SUNDAY AFTERNOON . . . *Just striking five and I'm in a fidget to be back in Union Square—and here's a whole fortnight . . . to be got rid of . . . before the 15th of May . . . It strikes me that I'm in love—a little. And tomorrow I've got to do some work . . . and I'd rather take a dose of physic . . .*

MAY 8, MONDAY . . . *Poor dear good innocent little E.—thinking so much of me and so grateful for every little attention I'm able to show her. It really seems incredible that I should have gained such an unprecedented combination of all sorts of excellence as she is . . . making her happiness the one great object of my life.*

Enter conscience and common sense with a bucket of cold water and a knout. "Mr. G. T. S., don't you know what a miserable selfish thoughtless good-for-nothing vagabond you really are? Don't you know that five years hence or ten years hence your Wife will be an every-day affair and not the lovely novelty that she is now—and that there will be cares and anxieties and worriments and vexations and temptations to bad humor . . . Won't you be lazy and tempted to neglect her . . . Now if you forget your feeling of this time . . . and cease to keep it all fresh and a living spring of action in your heart . . . then look out for yourself . . . so we shall come down with a knout of retributive vengeance . . ."

MAY 13, SATURDAY . . . *So one era of my life is ending . . .*

Preliminary mass meeting of bridesmaids and groomsmen came off Thursday night. Monday at 12½ all hands reassemble at No. 24 Union Place (Ruggles' residence).

(On Monday, the 15th inst., married
by Rev. Dr. Taylor,
GEORGE TEMPLETON STRONG, to ELLEN, daughter of
Samuel B. Ruggles, Esq.)

MAY 15th . . . *The house building plans have undergone a series of mutations. First there were to have been 3 houses on the 4 lots. Then Aunt Olivia Templeton concluded that she would not live in anything so big, and insisted on a single lot. Then the remaining 3 lots were to have been divided between the two other houses, but when plans and estimates came in, my father became refractory and struck for a single lot too. Now our architectural arrangements are ordered . . .*

Nos. 1 and 2 have got the start and are going on fast. No. 3 advances more slowly. Mr. R. very kindly gives us a stone front and a kind of architectural bay window for Ellen's boudoir or snuggery on the west side. The house will cost a clean $25,000—of which fact I don't think my father has yet a full realizing sense. As to furnishing, I've called in a little $2000 investment, which will do something, and for the balance I trust to economy of income . . .

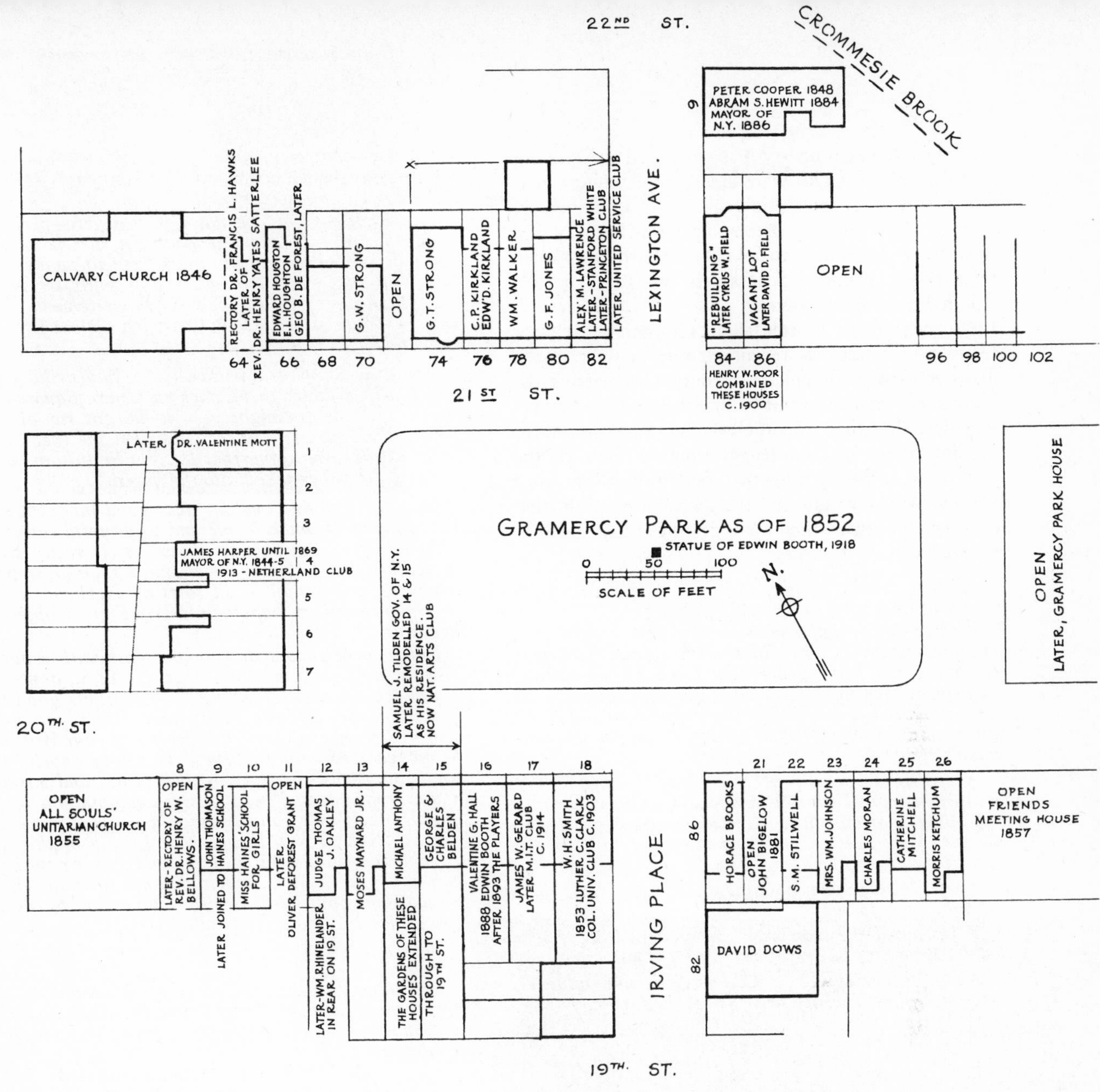

September 10, Sunday . . . *Gramercy Park houses prospering, . . . How I shall furnish the Schlossam Square when it's finished, without borrowing, is an inscrutable problem . . .*

September 20, Wednesday . . . *Bought a pew in Calvary Church, Saturday, of Isaac S. Hone—$550 . . .*

September 25, Monday afternoon . . . *Poor little E. is decidedly ill: no worse thing the matter than influenza, I hope . . . She wasn't well on Saturday, but the indomitable little woman would be up and busy with her little household arrangements at No. 54 Union Place . . .*

September 30, Saturday . . . *XXIst Street Palazzo coming on fast. In the name of the Sphinx,* what *will I do with it when it's finished and possession delivered???? Carpets and mirrors and Louis Quatorze chairs and buhl tables and ormolu gimcracks cost money and unless I happen to pick up a large roll of $100 bank notes in the street and fail in discovering the loser . . . or am somehow favored with some sort of unprecedented good luck, I don't see but that I'm likely to find the question of Ways and Means complicated and embarrassing.*

October 7, Saturday afternoon . . . *E. is better . . . On Thursday afternoon, the long expected emigration took place, and we became Housekeepers . . . at No. 54 . . .*

October 17, Tuesday afternoon . . . *Domestic life comes on famously. It's cheaper than I expected . . . If it were not for the housefull of rosewood and red satin that's got to be bought and paid for in a year or so, I should feel quite pecuniary and comfortable . . .*

October 30, Monday . . . *Looked into the* Schlossam Square *before I came downtown. "Scratchcoated" throughout and coming on reasonably fast. It looks spacious, scrumptuous and imposing . . .*

Chelsea

Clement C. Moore,
author of "A Visit From St. Nicholas"

UPON his return from the French and Indian Wars, Captain Thomas Clarke, an officer in the Colonial provincials, bought a farm from Jacob and Tunis Somerindyck which he named Chelsea after an institution for retired soldiers in London. The farm lay roughly between the present-day 19th and 24th Streets, Eighth Avenue and the shore of the Hudson River which then extended along the line of the present Tenth Avenue. It was reached by Love Lane (Abingdon Road, approximately the present 21st Street) and the Fitzroy Road running slightly to the west of the present Eighth Avenue and roughly parallel with it.

The original house appears to have stood north of the line of 23rd Street and slightly west of Fitzroy Road. This house burned down shortly before Captain Clarke's death and when Mrs. Clarke rebuilt in 1777, she chose a site only a short distance away on a bank overlooking the Hudson. After her death in 1802 the house passed to her daughter Charity and son-in-law, Benjamin Moore, who was Episcopal Bishop of New York. Bishop Moore and his wife in turn conveyed the house in 1813 to their son, Clement C. Moore.

Professor of Greek and Oriental literature at the nearby General Theological Seminary in Chelsea Square, Clement Moore won lasting fame for the beloved verses, "A Visit from St. Nicholas" which were written in Chelsea. It is said that the inspiration for the poem came to Moore in 1822 as he was driving from the lower city to Chelsea, his sleigh filled with toys for the children. The jingle of the bells on old Dobbin gave him the idea for verses which he recited that night before the great fireplace in his home. During the autumn of the following year Miss Harriet Butler, one of Moore's visitors, saw the poem and

The home of Clement C. Moore in Chelsea, circa 1822

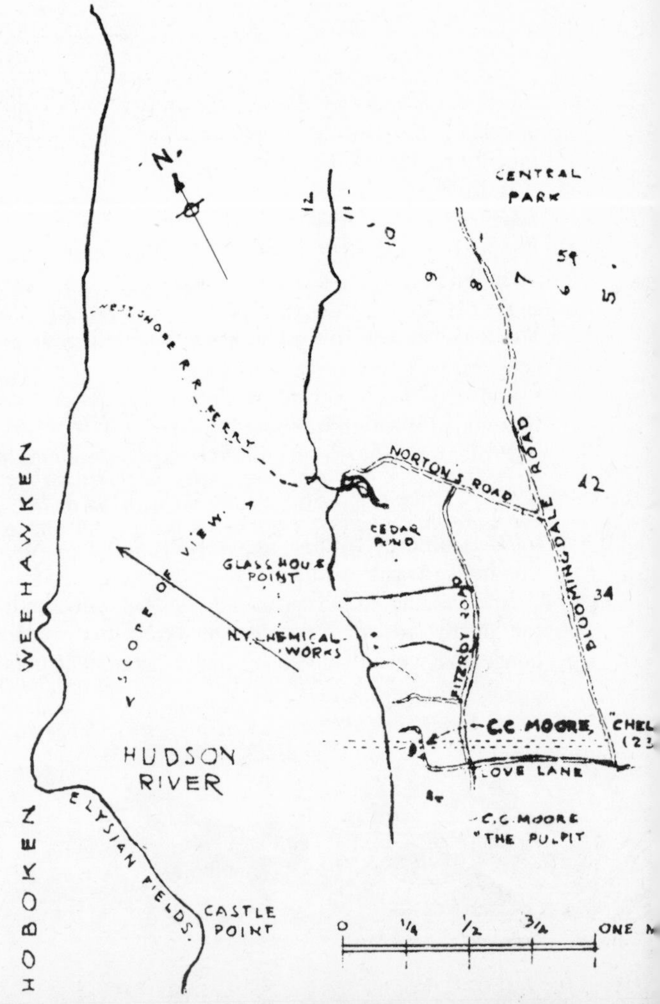

The Moore residence from a daguerreotype by Nathaniel Fish Moore

asked if she might make a copy. She took it to Troy, New York, and sent the transcript to Orville Holley, editor of the Troy Sentinel, who published it anonymously on December 23, 1823. It was not until 1838 that its authorship became generally known when it appeared on December 25 of that year in the Troy Budget over Dr. Moore's name. In 1844, the poem was included in a volume of verse written by Dr. Moore.

At the time the famous poem was written the house was two stories in height with a pedimented gable roof. However, a short time later, due to the increasing size of Professor Moore's family, a third story was added with a hip roof above it.

Professor Moore occupied the old house until about 1850 when Corporation improvements along the waterfront made its abandonment necessary. The city had by that time built up all around the blocks in which this house was the sole occupant.

In 1845 the London Terrace houses were built on the north side of 23rd Street opposite the Moore house and were tenanted by many prominent persons including Edwin Forrest and Lily Langtry. These houses were of uniform design, three stories high, in the Greek Revival style and gave the effect of a continuous colonnade from one end of the block to the other. The houses were set back from the street with gardens in front, about thirty-five feet in width. At the same time similar houses, two stories in height, and known as the Chelsea Cottages, were built along the 24th Street side of the block. Most of the houses remained until 1929 when they were demolished to make way for the present London Terrace Apartments.

Wood engraving of Santa Claus, used to illustrate Moore's poem when originally published and first four lines of the poem in the author's own handwriting.

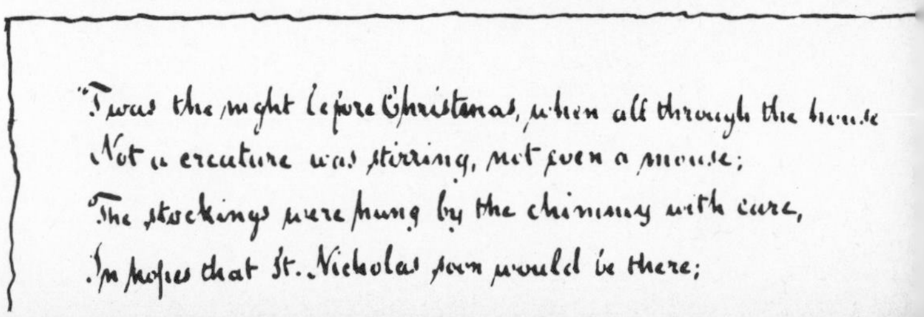

'Twas the night before Christmas, when all through the house
Not a creature was stirring, not even a mouse;
The stockings were hung by the chimney with care,
In hopes that St. Nicholas soon would be there;

Members of the Moore family on the porch of their residence in Chelsea. Also taken by the family's amateur photographer cousin, Nathaniel.

Nathaniel Fish Moore, cousin of the poet and one-time President of Columbia University

When Clement Moore died in 1863 his house had been torn down but its site was to be remembered forever afterward as the birthplace of the famous lines:

"*'Twas the night before Christmas,*
when all through the house,
Not a creature was stirring, not even
a mouse;
The stockings were hung by the chimney with care
In hopes that St. Nicholas soon
would be there . . ."

Professor Moore's cousin, Nathaniel Fish Moore, was President of Columbia University. During a visit to the Crystal Palace in London in 1851 he became highly enthusiastic about photography. Many of his excellent photographs still survive, providing interesting sidelights on the life of the Moore family.

The name of Moore's estate is perpetuated today by the Chelsea Hotel on West 23rd Street, one of the few old residential hotels still surviving. Here Thomas Wolfe is said to have written some of his later novels. An earlier writer, O. Henry, lived a few blocks away on 24th Street.

Chelsea Square* today retains a quiet, peaceful character in sharp contrast to the changes that time has brought to the surrounding neighborhood. The old square should not be confused with modern Chelsea (Alexander Hamilton) Park, bounded by 27th and 28th Streets, Ninth and Tenth Avenues, which was opened in 1907.

**Chelsea Square was bordered by the present Ninth and Tenth Avenues between 20th and 21st Streets. Clement C. Moore had deeded the property to the General Theological Seminary.*

The Moore residence in 1845 with the first London Terrace in the background

Yellow Bird

Miniature wooden hose carriage and display case

AMONG the number of fire companies in the Chelsea district, one of the most prominent was Washington Hose Co. No. 12, organized in 1837 at 244 West 17th Street. They were a very lively company and received their nickname, the "Yellow Birds," from the decorations and color of their beautiful yellow, white and silver four-wheel carriage.

In 1847 they were transferred to Horatio Street near Hudson and were disbanded in 1858. Washington Hose Company was later organized again and was then located in 43rd Street near Tenth Avenue. William P. Daniels of the Washington Association was a member of this company. The photograph shows some of the members assembled in front of their firehouse on 43rd Street. It is the only photograph of this company known to exist. The last house used by them was within the general area known as "Hell's Kitchen." Damon Runyon's friend, Lieutenant Michael "Micky" McNamara of the 13th Precinct, has an interesting story of how "Hell's Kitchen" got its name.

In 1917, when "Micky" lived in "Hell's Kitchen," there was an old, disfigured character by the name of "Buttermilk John" who ran a small grocery at the corner of Tenth Avenue and 34th Street. The man, who then must have been over eighty years of age, liked to reminisce. He said his father had often told him that in the early days when that section of New York was largely rural, there was a hermit who lived in a shack under a bluff at the northeast corner of 39th Street and 11th Avenue. Such independent, enviable souls, living out of the normal pattern of social intercourse, have always been bait for the intolerance of youth. Probably because of rumors and wild tales of the youngsters, the old man became known locally as "that ole Devil livin' under the hill." At any rate, the hermit's fire at night lit up the bluff above his shack and the "farm" kids of the vicinity referred to the eerie light as the fire in the devil's kitchen.

There have been many versions of how Hell's Kitchen got its name but Lieut. McNamara vows this is the true one.

gine lantern

The boys of Washington Hose Company No. 28, their mascot and their firehouse at 244 West 17th Street

Signal lamp of Mazeppa 48.

Mazeppa

THEIR title inspired by Byron's tale of Jan Mazeppa, hetman of the Cossacks, Mazeppa Engine Company No. 48 enjoyed a long and active career in the old Volunteer Department. Organized in 1828, the company served in five different locations until their disbandment in 1865. They were the first to use the name Mazeppa, although Hose Company No. 42 also adopted that title in 1848. The latter was located on Thirty-third Street near Ninth Avenue, where their carriage was often exhibited to visitors from all over the city. The reel was an ornate affair and deserving of the attentions it received from admirers at the Castle Garden Fair in 1851.

Fire scene from a Firemen's Ball ticket

Theatrical poster advertising Adah Menken's sensational portrayal of Mazeppa.

Perhaps its most daring accessories were the paintings on its panels which dramatized the exploits of Mazeppa. One showed him fleeing from a pack of wolves in the forest and the other illustrated the scene in which the hero is partially pinned beneath the wild horse to which he was lashed.

The house of this company was set on fire in September, 1862, by an incendiary. While the flames rushed up the stairway cutting off possible escape in that quarter, John Bothe, one of the runners of the company, improvised a fire escape by detaching a rope used for hoisting hose to dry and persuading another member, Billy Timms, to hold one end of it while he slid to safety. It wasn't until after Bothe had made the yard that Timms realized there was no one left to hold the rope for him and he had to take his chances with the stairway.

There is no record of their meeting after the fire but one can easily visualize the tenor of Billy's greeting to his more imaginative compatriot.

Old poster of firemen's burlesque minstrel show.

Name plate of Mazeppa Engine 48.

Longacre Square

WAYFARERS of long ago on the dusty Bloomingdale Road used to pause and quench their thirst with the cooling waters of a little spring which bubbled forth on the site now occupied by the Bond building at Broadway and 45th Street. These early travelers would have been dumbfounded could they have foreseen that one day an artificial waterfall* would flow high above Times Square's throngs.

Just a short distance west of this spectacular advertising display, near Father Duffy island, is the site of the meeting between Washington and Putnam** during the Revolutionary War. Putnam and his garrison were

Longacre Square, looking south, circa 1900

retreating from lower Manhattan after the defeat by the British of the American forces in Brooklyn and Long Island, while Washington, his staff and a few troops were rushing down from his headquarters at the Apthorp House to meet him and render what assistance they could. The exact site of their meeting was marked by a tablet now in the lobby of the Paramount Building, 1501 Broadway.

The intersection at 42nd Street, Seventh Avenue and Broadway was settled sparsely in 1850 and the

Looking west along West 42nd Street from Broadway. The site of the present Times Building

locality was slow in developing. Probably the most stimulating factor in its progress was the daring of showmen like Oscar Hammerstein who moved "uptown" and purchased the east side of Broadway between 44th and 45th Streets for the purpose of erecting his Olympia Music Hall.† Hammerstein's venture was successful and his initiative soon inspired others to erect restaurants like Rector's, Shanley's and Reisenweber's; the new Astor Hotel; and other theatres and business establishments, some of which are still in existence.

For many years this locality was known as Longacre Square. In 1904, shortly after *The New York Times* moved uptown and erected their Times Building, the title "Times Square," gradually came into use.

**The Bond Store's advertising display.*

***They probably met at "Frogg Hall," the house of Daniel Horsmanden, which lay in the bed of 44th Street, west of Broadway.*

†Hammerstein once kept a cow in a stall at his Victoria Theatre. He claimed her milk was the best in New York.

An old New York street scene

The "Great White Way" about 1900. Hammerstein's Victoria is at the left and Shanley's Restaurant at the right. Excavations in the foreground are for the construction of the New York Times Building.

In Bryant Park

Looking north on Sixth Avenue, 1857.

BRYANT Park (Sixth Avenue and 42nd Street, at the rear of the New York Public Library) was once part of a field where an American force—dispatched from Harlem to prevent the British from cutting off Putnam and his New York garrison—wavered and fled before the advancing British columns. As most of these American troops were militia, never before tested in battle, the sight of the formidable, martial line of the sturdy red columns disorganized them and nearly turned their retreat into a terrified rout. The wild disorder of the American troops seemed so ludicrous to the advancing British that the trumpeters of the Light Infantry and the 33rd Regiment took up the fox-hunting call soon followed by the taunt from the line of "Hoicks, Hoicks!"—the traditional cry of encouragement to the fox hounds at a quarry so quickly flushed. The steady skirl of the pipers of the Black Watch added their ominous taunt as the unswerving tartaned columns rolled westward.

It was at this time that Washington despaired of his leadership and, but for the quick thinking of an aide who seized the bridle of the General's horse and drew him from the field, he might have been captured by the British. The troops, noticing the General's evident despair re-formed and took a stand that temporarily halted the British, giving Putnam and his men the time needed to escape.

A potter's field for many years, Bryant Park became the scene of a great disaster October 5, 1858, when the famed Crystal Palace was destroyed by fire. Opened July 14, 1853, for the "Exhibition of the Industry of All Nations," this glass and iron "fairy palace" vanished in smoke in the short space of half an hour, burying a tremendous collection of art and manufactured products from all over the world.

The Latting Tower and Crystal Palace shortly after opening in 1853.

Crystal Palace fire.

xth Avenue and 40th Street, looking north; Dime Savings Bank on the left

Armor from the Tower of London, Sevres china, Gobelin tapestries, jewelry, Marochetti's statue of Washington and innumerable other works of art perished in a molten mass of ruins.

Firemen repeatedly rushed into the building in an attempt to save the fire apparatus that was on exhibition but their efforts were unsuccessful except for the hose carriages of Empire Hose Company No. 40 and Oceana Hose Company No. 36, which they salvaged from the flames. However, the hose carriages of Eagle No. 1 and Croton No. 6 were destroyed along with the engines of Gotham Company No. 16 and Pacific No. 28, the hook and ladder truck of Mutual No. 1, and several steam fire engines.

There were well over 2,000 persons in the building when the fire started but not a life was lost, though one man was rescued just a few seconds before the dome fell. The work of incendiaries who ignited papers in the lumber room, the fire resulted in a total loss of $2,000,000.

Engine on the ·h Avenue "El."

·king south on Fifth Avenue from 42nd Street, Croton Reservoir at the right.

Destruction of the Latting Tower.

North of the Crystal Palace on the south side of 43rd Street (the site now occupied by Stern Brothers) was the Latting Observatory, which was also destroyed by fire August 30, 1856. Over 350 feet high, the tower, together with the pattern formed by the dome of the Crystal Palace, was similar to the design used in the New York World's Fair of 1939.

Many living residents of the city can still recall the Croton Reservoir which was built on the southwest corner of Fifth Avenue and 42nd Street, on the site now occupied by the New York Public Library, as well as the world-famous Hippodrome, on Sixth Avenue between 43rd and 44th Streets, the greatest and largest place of amusement of its time. Opened April 12, 1905, by Thompson and Dundy, the Hippodrome cost $1,750,000 to build and boasted one of the world's largest stages. The auditorium was decorated in ivory, gold and silver on a Roman red background while the lobbies were of marble and Caen stone. The whole theatre was illuminated by 40,000 electric lights. The Hippodrome was the original home of the spectacle and extravaganza but it also had its popular favorites one of whom was that incomparable master of pantomime, Marcelene.

The building was torn down in 1939 and a new structure is now being built on its site.

Marcelene.

The old Hippodrome.

Looking north along Fifth Avenue in 1893

Along Fifth Avenue

Beginning at Washington Square and ending above Harlem, Fifth Avenue has been thought of since its opening in 1824 as Manhattan's "Queen of the Avenues." It was, even before invaded by the colorful shops and fashion stores, one of the handsomest and most exclusive streets in the city. The avenue's long line of brownstone palaces and imitation Versailles chateaux had housed the wealthiest and most prominent families in New York, driven northward from lower Manhattan and Washington Square by the endless waves of immigrants. Its name today is synonymous with quality and elegance.

Empire State Building

The view looking north along Fifth Avenue in 1893, affords a brief glimpse of a few of the mansions on the avenue. Here were the homes of John Jacob Astor on the 33rd Street corner (left), and of William W. Astor on the 34th Street corner (center), as they appeared before the former was demolished for the construction of the old Waldorf Hotel. A few years later the 34th Street house made way for the addition to the Waldorf, which then became the Waldorf-Astoria in 1898. The latter building was designed by H. J. Hardenbergh who also designed the Ritz-Carlton Hotel in 1907. The site is now occupied by the Empire State Building, the tallest structure in the world, which soars 1472 feet above the city.

Nearby, on the south side of 34th Street between Madison and Park Avenues, Commodore Vanderbilt kept a stable which later became the Tally-Ho restaurant. Stalls once occupied by some of the greatest trotting horses in America, such as Aldine, Early Rose, Maude S. and Jay Eye See, were converted into attractive dining booths and the entire restaurant was replete with name plates, rigs, original harness, blankets and other equipment used in the old stable.

Farther north, at Fifth Avenue and 42nd Street, was, in early times, Burr's Corner, a favorite meeting place for farmers. The farm of Isaac Burr stood at the northeast corner of Loew's Lane or Steuben Street and the old Middle Road, a diagonal road running approximately from 44th Street and Broadway to 41st Street near Third Avenue. The intersection,

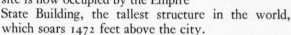

Sixth Avenue between 32nd and 33rd Streets in 1864. This site now occupied by Gimbel's.

Vanderbilt stable; later Tally-Ho Restaurant

The famous bell in Herald Square, formerly mounted on Herald Building

Looking north from Fifth Avenue and 56th Street in 1865.

however, came on the exact spot where 42nd and Fifth Avenue now meet.

The first traffic light in New York was installed at the intersection on the corner of 42nd Street and Fifth Avenue in 1920, a pioneer of the 12,800 traffic lights in New York City which are automatically controlled from central switchboards. The first light stood in the middle of the street and was mounted on a tower occupied by a policeman at all times.

The Chrystie view above, looking north along Fifth Avenue, shows the row of three frame houses which stood on the corner of 56th Street and Fifth Avenue in 1865, the site now occupied by Bonwit Teller. For years after the houses were demolished the trees remained standing. At the extreme left of the view (on the southwest corner of 58th Street) can be seen the brownstone houses demolished to make room for the enlargement of the Cornelius Vanderbilt home.

Looking north along Fifth Avenue and 56th Street in 1948. Bonwit Teller at the right.

Fifth Avenue parade in the 90's. Note solitary motorcar

Northeast corner of Lexington Avenue and 42nd Street in 1879, site now occupied by the Chrysler Building. The elevated spur ran from Third Avenue to Grand Central Depot.

Roar of the Rails

THROUGH the first quarter of the nineteenth century Manhattan had no city-wide public transportation facilities. In order to get to businesses, churches and places of amusement, those who were not fortunate enough to own a carriage either hired one or walked. In 1830, however, Abraham Brower put into service a local coach called the Accommodation, which plied Broadway from Bowling Green to Bleecker Street. This service proved so popular that two years later he added a second coach appropriately named Sociable because all the passengers sat in a single compartment.

In 1831 the first of the so-called omnibuses began to lumber along Broadway and soon afterwards the double-deck coaches made their appearance. The coaches during that era were known as the Broadway Beauties.

In 1831 also, John Stephenson, who later operated a successful car manufactory in Harlem, secured a patent for a "horse-drawn carriage running on rails" which soon afterwards was used to initiate the first streetcar system in the world, the New York and Harlem Railroad. The road was formally opened on

Chrysler Building, 1951.

Tenth Avenue "cowboy" who rode ahead of train clearing way of traffic.

November 14, 1832, with sixty distinguished guests riding the full length of the line—from Prince Street to 15th Street. These coaches resembled the omnibuses of the period except that they were equipped with iron wheels and were drawn by horses over iron rails laid in the middle of the street.

The New York & Harlem subsequently ran down

Tryon Row Station, first terminal of New York & Harlem Railroad, in 1839. Now occupied by Municipal Building.

the Bowery to a barn in Tryon Row, opposite the Hall of Records in City Hall Park. In 1833, four years later, it was extended to 32nd Street. An open cut construction was then made at Murray Hill to accommodate that route. The New York and Harlem was ordered to cover it, August 8, 1850 and by 1852, it had assumed its present shape as a tunnel. This tunnel, later used as a trolley route to Grand Central Station, is used today as parking space for nearby air terminal busses.

Hudson River Railroad depot at West Broadway and Chambers Street, 1868. Cars drawn by horses to Madison Square Depot.

nd Central Depot in 1875. Note early street cars, hacks, gas lamps and cobbled streets.

The idea of an underground railway was broached as early as 1857. In October, 1904, the first subway—the I.R.T.—was officially opened by the city's Mayor, George B. McClellan, son of the famous Union general. This line ran from City Hall to 145th Street and Broadway and marked the inception of New York City's present rapid transit system.

The first Grand Central Depot was constructed in 1869-1871, housing the New York & Harlem Railroad, the New York Central & Hudson River and the New York, New Haven and Hartford lines under the control of the "Commodore," Cornelius Vanderbilt. The depot was remodelled in 1899 and torn down in 1910 to make way for the present terminal, one of the most magnificent transportation structures in the world.

A fairly general idea of the city's early railroad system, which grew along with the need for rapid and more adequate internal communications, may be gained from a glimpse of the accompanying sketch.

Millions of New Yorkers owe a deep debt of gratitude to George McAneny who spent a number of years of his life in planning and pressing the development of the rapid transit system which New York enjoys today. New York's golden era of development followed this great achievement.

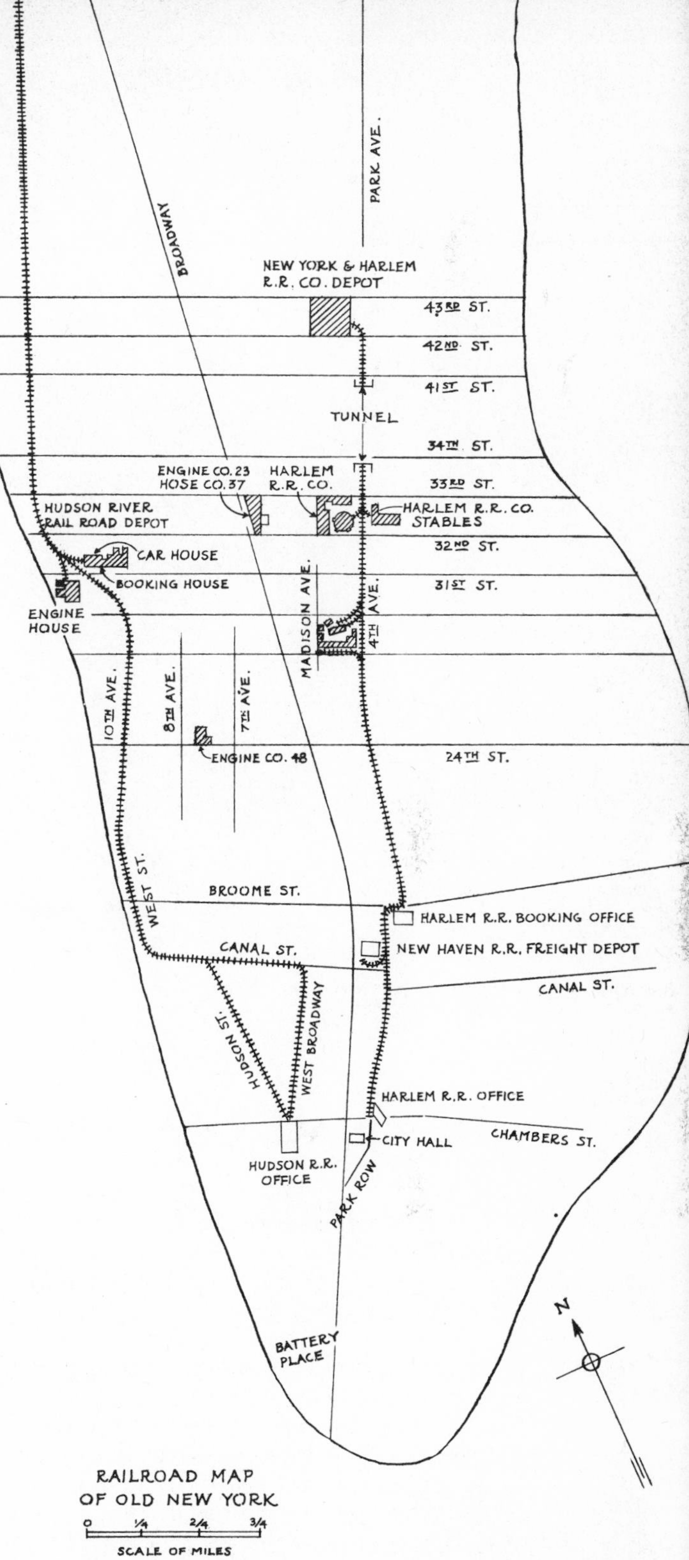

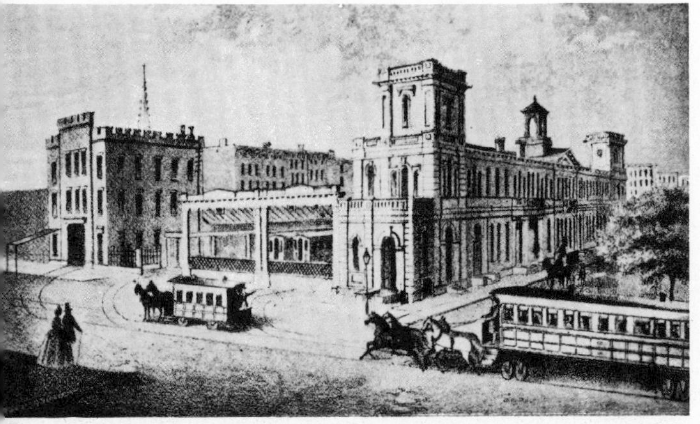

Old Madison Square Garden when used as a terminal.

The "Commodore Vanderbilt" which ran on the Sixth Avenue "El" from Rector Street to Central Park in 1886. Note portrait of the Commodore above front wheels.

Graceful Living

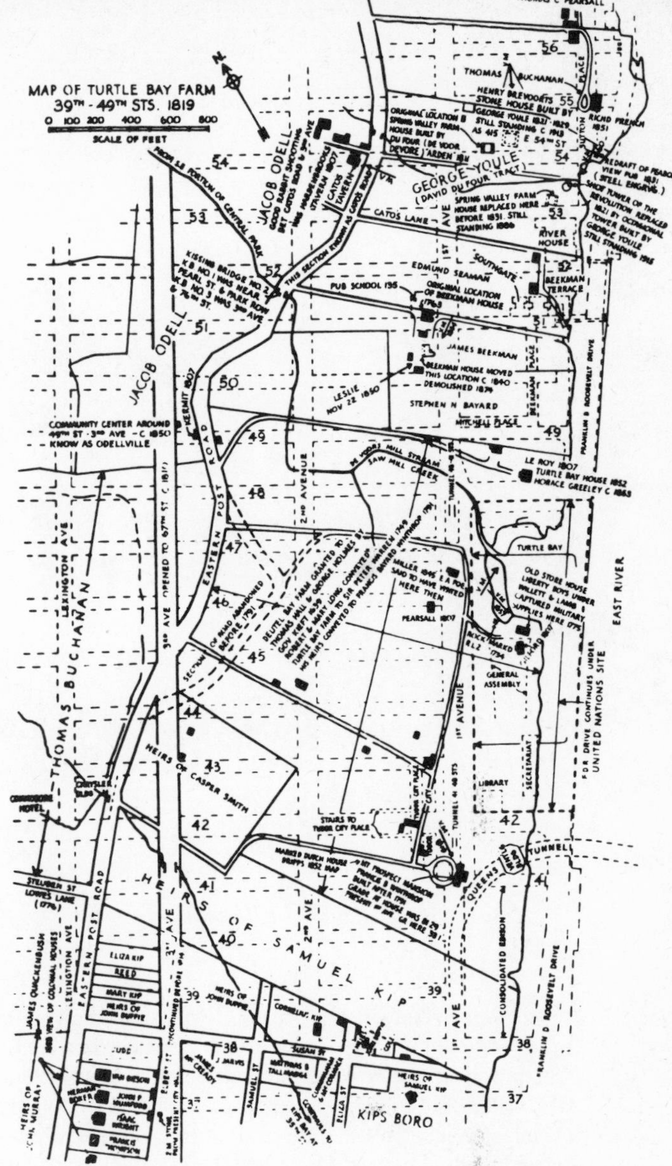

IN 1791, Francis Bayard Winthrop, a merchant with a city house in Wall Street, bought what was then known as the Turtle Bay Farm for about $3,800. This farm lay along the East River, approximately between 39th and 49th Streets, and extended back in a northwesterly direction to the Eastern Post Road, which wove to and fro across the present line of Third Avenue. The farm included Turtle Bay on the north, where it adjoined the Beekman farm. To the south the land belonged to the Kips, whose house stood near 35th Street and Second Avenue.

The present Turtle Bay is undoubtedly an English phonetic derived from the old Dutch word "deutel," which was usually applied to a slightly bent blade. The term was, in all probability, applied originally to the rather shallow bay on the East River.

Winthrop built a large frame house with lower wings at either end, selecting as his site the highest point of the property, which happened to be at the intersection of the present 41st Street and First Avenue. He called the house Prospect Hill, which perhaps explains the later name of Prospect Place, only a few yards distant from the site of the house. Just to the south of the farm several creeks or brooks ran into the East River. In the early records references are made to "the Kill of Schepmoes where the Beech tree lies over the water" which ran into the East River at approximately 39th Street. Below, at about 35th Street, was "the Kill where the water ripples over the stones" and at 34th Street, the Old Wreck Brook ('t-Oude Wrack) which flowed down through the fields from Sunfish Pond.

A water color drawing of Kip's Bay in 1830. The view was taken from a pier at the foot of 37th Street. Courtesy of the Stokes Collection, N. Y. Public Library.

The old storehouse at Turtle Bay where Sears and Willett led a daring raid on British military stores in 1775.

Sir Peter Warren purchased Turtle Bay Farm in 1749, only three years prior to his death in England. The purchase was made from Robert and Mary Long, who had come into its possession after several stages of inheritance, from George Holmes. Holmes, together with Thomas Hall, had received a grant in 1639 from Governor Kieft for what was then designated as the Deutel Bay Farm. Hall, a heroic Englishman who was adopted by the Dutch and became an extremely important citizen of old Manhattan, undertook the development of a tobacco plantation on the Turtle Bay Farm with Holmes. Hall soon tired of the arrangement and sold back his share of the enterprise to Holmes in return for a dog, a gun and a boat and the right to stop off if he should be overtaken by bad weather in the vicinity without other shelter. He then returned to the old Governor Van Twiller farm in Sappanikan.*

It seems unlikely that Warren ever occupied the Turtle Bay Farm as he already owned a large estate on the west side near Greenwich Village. After his death little was done about the division of the property until 1787. In that year, five heirs entered into a strange partition agreement whereby each was to nominate a representative to cast dice for the three parcels into which the real estate had been divided. A representative of Susannah Skinner, granddaughter

**Sometimes spelled Sapokanican.*

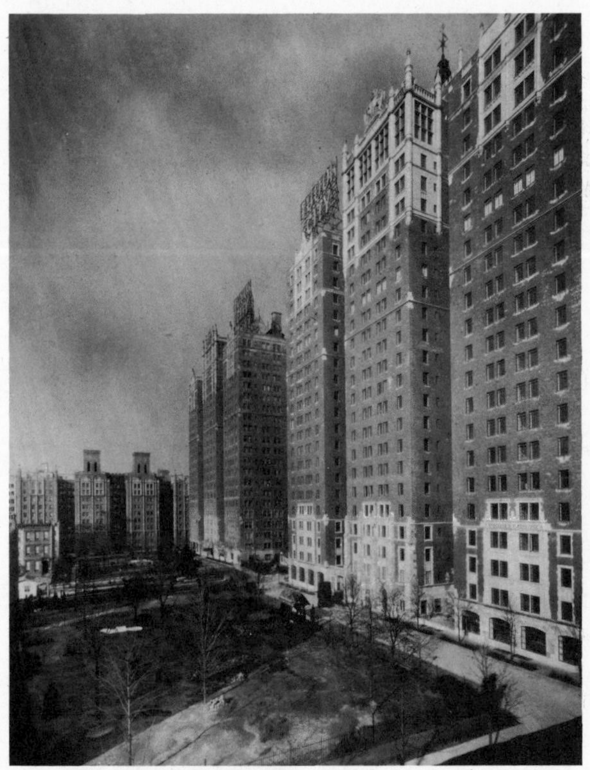

Tudor City, west of First Avenue between 41st and 43rd Streets, covers the site once occupied by the south end of the Turtle Bay Farm.

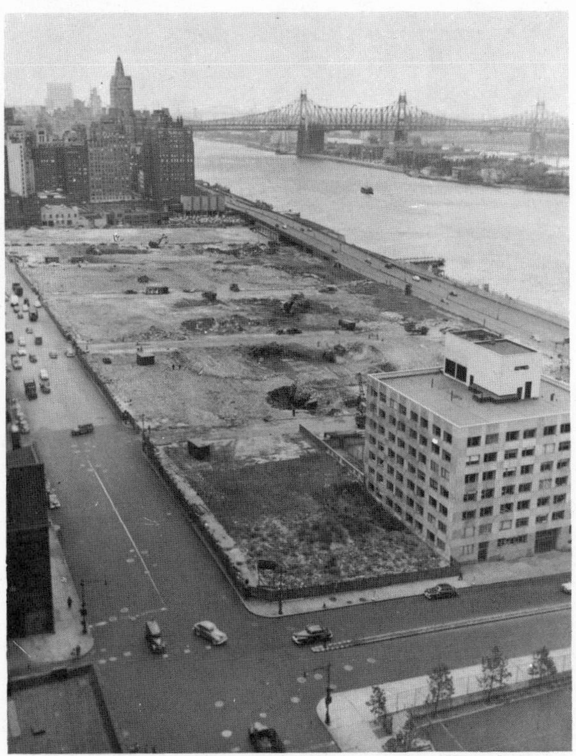

Site of the present United Nations Building before construction. This site falls on a section of the old Deutel Bay Farm.

Just northeast of this site today, overlooking the East River, stands the new United Nations headquarters. It is significant to note that approximately a block away from the present General Assembly building, a stroke for liberty was ventured in 1775 by the Liberty Boys who seized a British magazine of military stores in a daring amphibious raid. The American party "under the direction of Lamb, Sears, Willett and McDougal, procured a sloop at Greenwich (in Connecticut), came stealthily through the dangerous vortex of Hell Gate at twilight, and at midnight surprised and captured the guards. They secured the stores, a part of which was sent to the grand army at Boston, and a part to the troops then collecting on Lake Champlain to invade Canada."

The leader of this raid was a merchant named Isaac Sears, who was a persistent thorn in the British side before and during the Revolution. Sears' daring and audacity as leader of the Sons of Liberty and his active and open resistance to the Crown resulted in his being classified as a marked man. He survived the war, however, and later became vice-president of the Chamber of Commerce.

of Sir Peter Warren, who was one of the heirs, rolled the highest number and was awarded parcel "A" which included, besides the Turtle Bay Farm, a good sized plot on Broadway, just north of Exchange Alley, with a garden in the rear extending down the hill to the Hudson.

There were several houses on Susannah's estate. One of them, facing Turtle Bay, between 46th and 47th Streets, was visited by Edgar Allan Poe in the summer of 1845. At nearly the same period, Horace Greeley occupied another of the houses as a country residence. This stood on the rocky bluff just north of Turtle Bay facing the East River on the line of 48th Street.

The two houses and Prospect Hill itself were approached by separate lanes which led in from the Eastern Post Road. The course of the Post Road at this point is marked by the slanting east wall of the Chrysler Building, which follows the property line facing the old road.

In 1852 the old Winthrop home bore the designation of "Dutch House," suggesting its use as a tavern and the origin of the name Dutch Hill.* It probably did not stand much longer, as shortly afterwards First Avenue was cut through the knoll on which it stood.

**Present Tudor City.*

The United Nations Building in 1950.

Rural Manhattan

The Shot Tower in 1831. Located on the Old Post Road, on the banks of the East River, the tower was erected in 1823 by George Youle.

WHILE a history of upper Manhattan Island would require a volume in itself, perhaps the illustrations shown in the following pages will convey some idea of its provincial atmosphere in early times. Originally an area of meadows, deep valleys, projecting hills and precipitous, rocky cliffs, many of the island's finest and most prosperous farms were once located uptown, as were several of the city's most sumptuous mansions and summer retreats.

A score of towns and villages were located uptown, the most important of which were Yorkville, Bloomingdale, Harsenville, Manhattanville, Carmansville and Harlem, and it was a hunter's paradise—the entire area abounding in woodcock, English snipe and rabbits. Historian Charles Haswell reported that he himself shot a grouse at Breakneck Hill near the Jumel Mansion (161st Street between Ninth and Tenth Avenues) in 1831. Haswell believed his quarry to be the last of its type to suffer such a fate in Manhattan.

The Shot Tower in 1864.

As the nearest trotting course was the Union Course in Long Island, the Bloomingdale, Middle and Eastern Post Roads were often used as racing lanes, vying in popularity with Harlem Lane for Sunday drives. Flora Temple, the famous trotter originally purchased for $80, first gave evidence of her brilliance on a short trotting track laid out on the site of the "Red House" on Second Avenue between 110th and 113th Streets.

The mansions and estates in the uptown area numbered among the city's finest—the Apthorp Mansion on the Bloomingdale Road, Hamilton Grange, built in 1802 by Alexander Hamilton (with its 13 gum trees each named for a state in the Union) on 143rd Street, the Claremont Inn on Riverside Drive, which was only recently torn down, and the historic Beekman House facing the East River, which was built in 1763 and was the headquarters of Generals Howe and Clinton during the occupation of the city. It was in this house that Major Andre was imprisoned and it was there also that Nathan Hale was held and questioned prior to his death. The Gracie Mansion was the home of Archibald Gracie, one of New York's outstanding merchants, who bought the site in 1798. Here Gracie played host to Washington Irving, Josiah Quincy and other celebrities of the day, probably

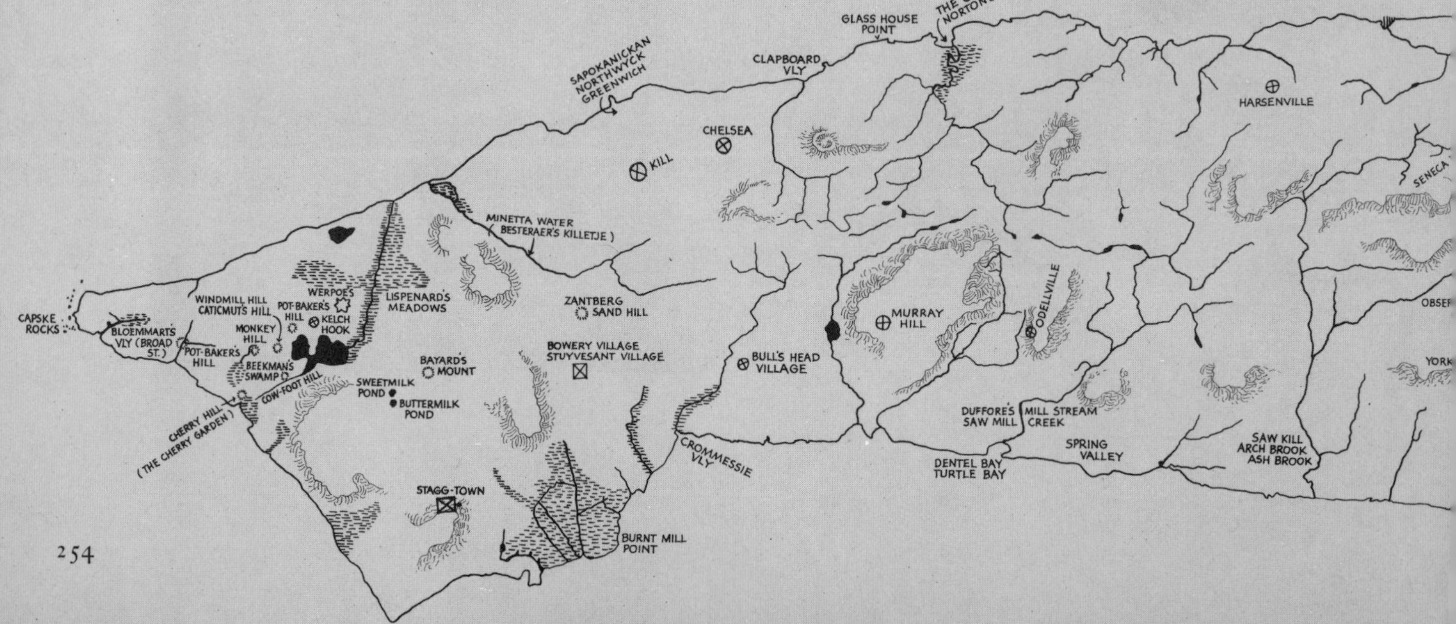

Beekman House, 50th Street and the East River.

North from 58th Street between Tenth and Eleventh Avenues, circa 1862, Hudson River in background.

with no inkling that one day his home would be used as a residence for the Mayors of the city.

One of the oddities of upper New York was the "jet" in the Harlem River which disposed of the overflow from the Croton Aqueduct. In periods of high water or excessive pressure this jet was opened and it was a source of awe to all who saw the seemingly mysterious spout or fountain arising from the center of the Harlem River. One witness, Catherine Havens, recorded in 1850: "On one Fourth of July my father got a carriage from Hathorn's stable and took my sister and my brother and me out to see

Riverside Drive and the Hudson River from the Claremont Inn.

A "Shanty-town" at 98th Street between Madison and Fifth Avenues in 1893.

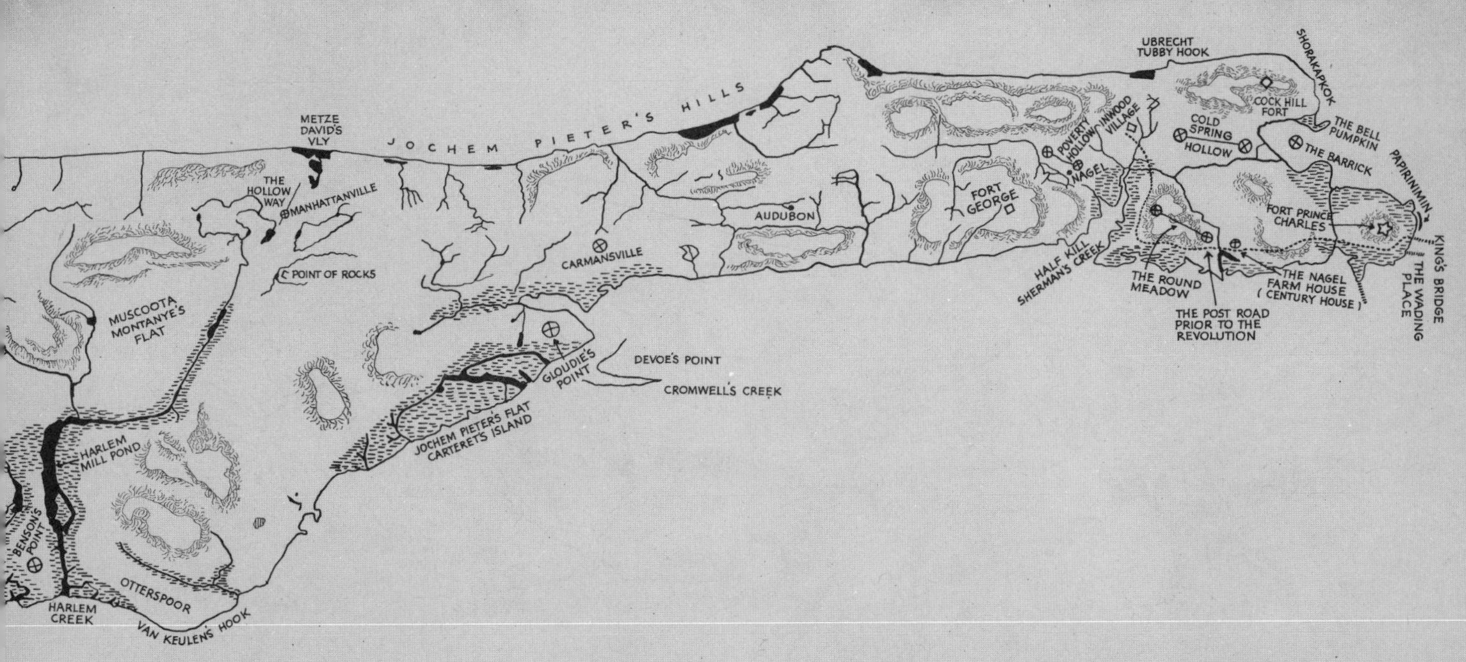

The mansion of Archibald Gracie (now the home of New York City's Mayors) in 1869 and 1949. During the Revolution a redoubt of 9 guns, called the Thompson Battery, was built on this property to protect Manhattan from the English fleet approaching from Long Island.

the High Bridge. It is built with beautiful arches and brings the Croton water to New York. My brother says he remembers riding to the place where the Croton Aqueduct crossed Harlem River by a siphon before the bridge was built, and the man who took charge of it opened a jet at the lowest point and sent a two-inch stream up a hundred feet."

The last of the trotting races on the Speedway, taken by Arthur Brown in 1898.

A skating scene in Central Park in 1866. Fifth Avenue in the background.

Cycling was a favorite sport along Riverside Drive in the 90's.

The Firemen's Exempt Company and their goose-neck engine at a muster in the Bronx, circa 1891.

A view of "Niew Harlaem" in 1765.

View of the jet on the Harlem River. The Jumel mansion is at the left.

The town of Bloomingdale in 1867.

Harlem Town in 1798.

Gouverneur Morris said to be the champion scythe-swinger in Manhattan.

The trotting race-course on the Harlem Lane.

The Hurlgate Ferry Hotel, foot of 86th Street and the East River, a fashionable summer retreat in old New York.

Sleighing on the Bloomingdale Road, circa 1850.

The Riker and Lawrence Houses by the Arch Brook (East River and 75th Street) in Yorkville, July, 1869.

The Last Stand

SINCE the brent mersters of '86 (that is, the fire laddies of 1686) hung their new hickory ladders on the Stadt Huys fence and called it home, firehouses have been placed as pawns against the demon fire where it has seemed most prudent or politic to the city fathers in order to provide protection in each section of the growing city. The first sheds built behind the old City Hall in Wall Street in 1731, like other early firehouses, were of the simplest construction, usually resembling small barns. They were placed in the handiest spots available and, in their simplicity, often dominated the surrounding scene. As the city expanded and the number of engine companies increased, more and better firehouses were built. Eventually these were constructed of brick and after a time were two and three stories high.

Former house of Engine Company No. 20—believed to be the oldest firehouse in the city—at 126 Cedar Street, between Greenwich and Washington Streets.

The earliest firehouse still standing in Manhattan, now 126 Cedar Street, was probably built in 1819 and was at one time the home of Washington Engine No. 20, among other companies. Half of the building is low, long and narrow and was the habitat of several early hook and ladder trucks of a bygone day. It is at present occupied as a place of refreshment known as "The Fire House Tavern." The old fireplace in this tavern served as a model for the fireplace in the old firehouse recreated for the H. V. Smith Museum in The Home Insurance Company Building.

Built in 1859, the building at 146 John Street shown above (formerly 18 Burling Slip) was originally the home of Edwin Forrest Hose Company No. 5. The membership of this company, named after the great tragedian, was composed almost entirely of oyster dealers and boatmen, who ran a hose carriage built by William Williams. The building still carries the shields of Engine Company No. 32 which occupied it after 1865. This company, and its steamer, may be seen above in the photograph of 146 John Street, taken in 1867.

The old firehouse at 246 West Broadway was built in 1823 to house Hope Engine No. 31. West Broadway was then known as Chapel Street. This building is probably the second oldest firehouse in existence in New York City today. It is now being used for commercial purposes.

Firehouse on the south side of Charles Street near Hudson Street which was formerly occupied by the volunteer Hook and Ladder Company No. 14. It has been occupied by Hook and Ladder No. 5 since the earliest days of the Paid Fire Department (see cut).

he City purchased the site of 195 Elizabeth Street, September 26, 1862, to erect a house for ibernian Hook and Ladder Company No. 18. The new house was completed in 1864 and later rved as headquarters for Hook and Ladder No. 9, which recently gave up the building to Police mergency Squad No. 2. The two-story houses adjoining it were built before 1830.

An old volunteer house at 70 Barrow Street which has been converted into a modern apartment building. The structure was built for Empire Hose Company in 1843 and rebuilt in 1851 for the same company.

The Firemen's Hall at 155 Mercer Street (see color plate) still stands, although stripped of its exterior ornamentation (the figures of the fireman and other carvings are removed). Constructed in 1854, this building was the third Firemen's Hall of New York's Volunteer Fire Department. The first, situated in a court yard on the north side of Fair Street, now Fulton, just east of Gold Street, housed four or five machines. It was abandoned and sold in 1829 and the second Firemen's Hall, a two-story brick house, was

An old print of an exceedingly handsome volunteer firehouse thought to have stood on the Bowery, below Chatham Square.

use of Oceana Engine Company No. 11, at 99 Wooster Street, which is warehouse. The first sliding pole used in New York City is said to have troduced here by Captain Daniel Lawlor. The house was occupied by Company No. 13 at that time.

Now a wholesaler's warehouse, this old firehouse located at 91 Ludlow Street once housed Manhattan Engine No. 8 in the volunteer days and, after 1865, Metropolitan Engine No. 17. The latter company, seen above, ran a Stedman 2nd class engine, built in 1861.

Former Firemen's Hall at 155 Mercer Street, now occupied by Engine 13 and Hook and Ladder 20. In the color plate at the right the figure of the fireman atop the building is of Harry Howard, Chief Engineer of the Volunteer Fire Department from 1857 to 1860. Since the days of the volunteers, Firemen's Hall has been used as the regular battalion headquarters and later as the telegraph bureau for the department. The inside of the building has been altered a number of times.

erected in the same year at 127 Mercer Street. Twenty-five years later it was replaced by the third hall on the same site.

In the following pages may be seen other survivors of the Volunteer period, most of which are still used by the Paid Fire Department. However, a survey undertaken in recent years by the city's Fire Commissioner in a drive for more efficiency, may doom the old houses to the glamorous past of which they were so much a part. Conceived to house the city's hand, and later, horse-drawn machines the old houses are inadequate for modern use and thus may well be making their last stand in the life of the metropolis.

Another volunteer firehouse still standing on the south side of 14th Street be First and Second Avenue, probably built in 1864. This building hous Metro Steam Engine Co. No. 5. This company ran the third class steam e built by Joseph Smith in 1863.

An old firehouse located at 223 East 25th Street which once housed Lexi Engine No. 7 serving in the first and second districts during the volunteer Their machine was a Philadelphia style second class engine with a nine-inch s It was operated by over 40 men and manufactured by James Smith in 1855.

The house of Hook and Ladder Company No. 13, at 159 East 87th Street is s use. As may be noted in the contrasting views the lower part of the buildin been remodelled.

The house of Engine No. 15 on the north side of Henry Street west of Gouverneur Street; formerly occupied by the Volunteers of Engine No. 6.

The house of Engine Company No. 8 at 153 East 51st Street. It formerly housed the volunteer company, Engine Company No. 8.

A photograph of the two firehouses now located on the southeast corner o Street and Eighth Avenue. Both of these houses were built before 1855, pr in 1851. In 1855, the Empire Hook and Ladder Co. No. 8 was located her firehouse in the rear on 48th Street was the home of volunteer Index Hose Co No. 32. The room which once held their carriage now houses the car battalion chief.

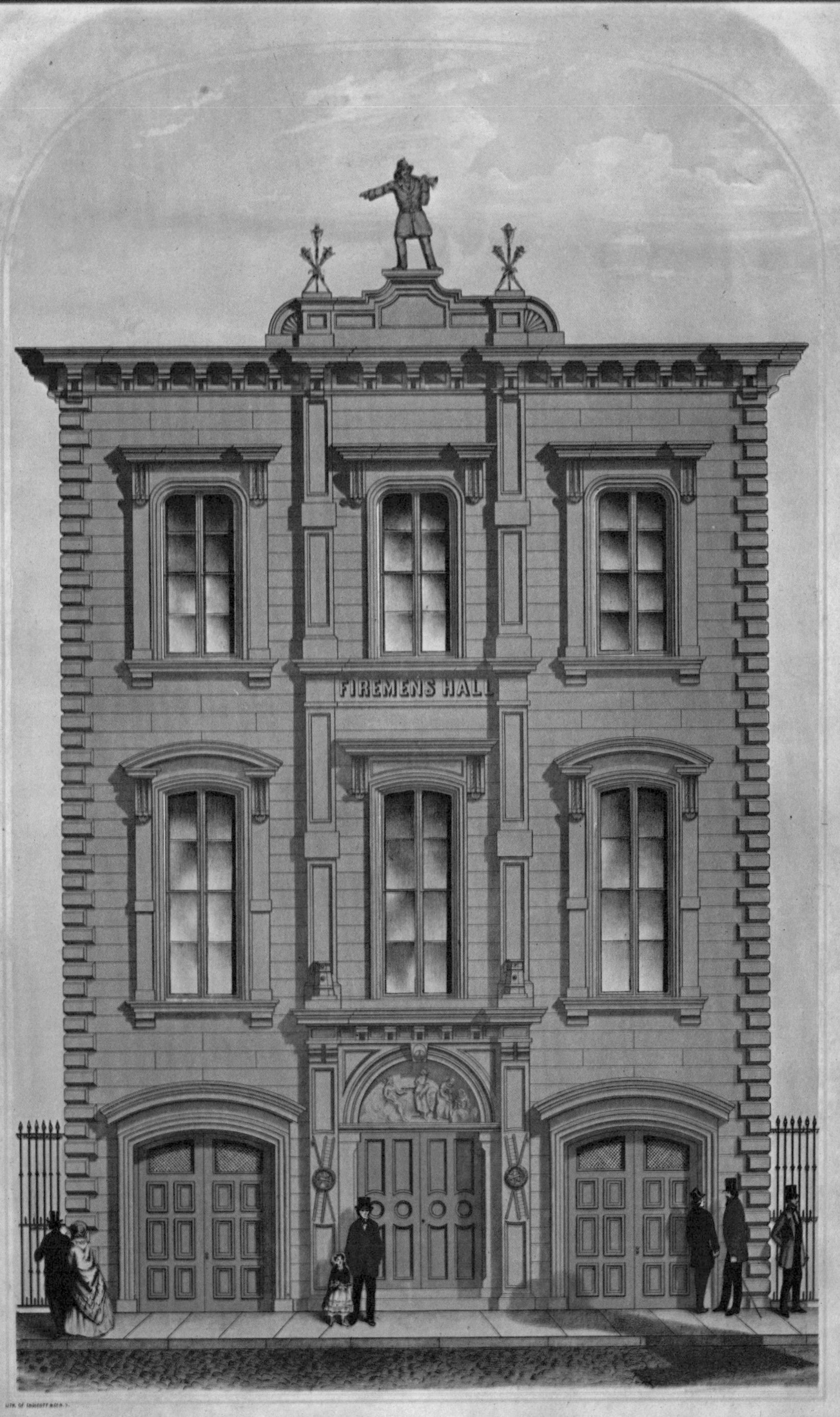
FIREMENS HALL
FIREMEN'S HALL.
NEW YORK.
ERECTED AD 1854
Elijah H. Purdy
COMMITTEE ON FIRE DEPARTMENT.
Wesley Smith
James M. Bard
S. L. H. Ward
Board of Aldermen
Charles H. Ring
Saml R. Mabbat
Edwd Anderson
Board of Assistants
Field & Correja Architects
James L. Miller & Co. Carpenters
Platt & Fisher Masons
James L. Miller
James F. Wenman
William D. Wade
John S. Belcher
James Donohue

The Upshot

ONE of the benefits gained from the creation of any historical work is the gratifying opportunity of communing with historians and scholars of the past whose love of their subject (in this case, the Island of Manhattan) enriched their lives, sharpened their perspective and gave them the necessary inspiration to carry out their projects as they saw them. To I. N. Phelps Stokes, Charles Haswell, Felix Oldboy, Martha Lamb, James Grant Wilson, Henry Collins Brown, "Florry" Kernan, Augustine Costello, A. E. Sheldon, Philip Hone, George Templeton Strong, David T. Valentine, John Fanning Watson, and many others goes my profound admiration. Their efforts in recording and interpreting the political, social, industrial and artistic development of the city have been truly monumental and will live as long as men are interested in the past and in each other.

The compiling of a history is an arduous task, for very often the writer is overwhelmed by the mass of material confronting him. This, I think, is a most trying time for any author.

The writer has been particularly fortunate in having the loyalty and support of his friends and associates throughout the preparation of this book. To Harold V. Smith, in particular, the author again wishes to express appreciation for his vision and support.

To my friends and professional associates listed below goes my deepest gratitude: to Mr. George McAneny, a true lover of New York whose energy and idealism are responsible for so much of the city's progress and who so courteously consented to write the introduction; to my friend, Ed Chrystie, not only for his contribution to the book via his remarkable charcoal restorations, but also for his guidance and cooperation no matter how many times called upon; to Halsey Thomas and Columbia University for the use of the remarkable material from the diary of George Templeton Strong; to Grace Mayer, curator of the print department at the Museum of the City of New York, Arthur B. Carlson, curator of maps and prints at the New York Historical Society, and Harry Shaw Newman of the Old Print Shop. To King Rich for artistic assistance and to Albert Frank-Guenther Law, Inc. To my good friend, Richard Doyle, for his untiring efforts in the reading and checking of proofs and in the efficient handling of numerous details; to my secretary, Eleanor Gilbert, for her encouragement and patience; to the late Sol Sayles for his appreciation and sincere interest in my efforts. To each member of the staffs of Morrell & McDermott, Inc., Typography, and Newman-Rudolph Lithographing Company who contributed that extra effort in craftsmanship to this volume. My thanks go also to David and Jane Dunshee who have been so steadfast and understanding and to three little boys, Ted, Pete and Tom, who some day may understand better than now why they so often had to do without their regular second baseman during the past three years.

This book is set in Janson monotype, chapter headings in Caslon No. 471.

INDEX

INDEX

APPENDIX

TO OCEANA HOSE COMPANY NO. 36 (*Air: "Life on the Ocean Wave"*)

Huzza for brave Thirty-six, Ever prompt at the fire-bell's call,
Three cheers for brave Thirty-six, Huzza for the Firemen all,
When the red flames wildest flash On the startled midnight air,
Where the crackling embers crash, Oceana's men are there.
Huzza, etc.

Secure may the mother sleep, With her babe upon her breast,
Brave hearts her vigils keep, To guard them while they rest.
Though sudden flames alarm them In the dwelling where they lie,
The fires shall never harm them, Oceana's men are nigh.
Huzza, etc.

(*See Page 171n*)

"DIANA," The Goddess of Virtue—*Following facts from Art Digest, April, 1932*

New York has lost its slim bronze Diana, the Virgin Goddess who stood posed gracefully atop the old Madison Square Garden from 1895 to 1925. A conspicuous and loved figure in the City's Skyline. Designed by Augustus Saint Gaudens (1848-1907) out of friendship for Stanford White, the architect of the Garden. Diana has gone to Pennsylvania Museum as a gift. According to Eli Kirk Price (Pres.) the goddess will be placed in the court of the new $18,000,000 building at the head of Parkway.

Since the demolition of the Garden in 1925 Diana had reposed in a Brooklyn warehouse while the New York Life Insurance Company, its owner, tried to find her a permanent and appropriate home. In 1932 she was kept at the Roman Bronze Works in Corona, Long Island, undergoing a beauty treatment. It was to be given to New York University but they could not provide a suitable site.

The original Diana was 18 feet high and wore a flaring sash but Saint Gaudens felt that she was too tall so she was replaced by another 13 feet high. Her scanty drapery was swept away by a storm.

The elevation of this naked beauty shocked the tea-tables of the Victorian era into gasps of protest.

Age has not withered her. Storage has not dimmed her appeal to the eye which gave O. Henry the inspiration to write "The Lady Higher Up."

Elihu Root and Tex Rickard stood in silent admiration at her perfection the day the derrick lowered her slowly to the soil of Long Island.

The Rev. Mary Hubbard Ellis, pastor of the Primitive Methodist Church who was a crusader against pornography and indecent shows, tried to prevent her being on display in Philadelphia as not fit to be seen by Philadelphia children. Later Mrs. Ellis withdrew her objection.

(*See Page 235*)

THE MIRACLES OF ADVERTISING

CRAFTSMAN!—Oh, Lord! My God! Is there none to help the widow's son to some employment to prevent starving?—Rueff, linguist, bookkeeper.—"Herald," downtown.

THE young lady who rescued little girl from Broadway cable car on Thursday afternoon will receive substantial benefit and permanent gratitude by addressing Father, box 296 "Herald."

YOUNG HUSBAND—You need have no fears on account of your wife; we have a doctor in the building in constant attendance.—Siegel, Cooper Co.

A NOBLEMAN of the highest rank, 37 years of age, of distinguished appearance and mental attainments, would contract speedy marriage with lady of corresponding attributes, and wealth to sustain a European home; no agents.—Answer for one week, to R. N., "Herald" Bureau, Washington, D. C.

IS your nose red? Fould can bleach it. Call or address Fould, 214 Sixth Ave., New York.

LADY will teach whist, euchre and other games in ladies' own homes.—Accomplishments, 419 "Herald."

A GENTLEMAN would like to make the acquaintance of a young lady bicyclist matrimonially inclined.—Address Retired, 1,227 Broadway.

PROFITS from a scientifically conducted Frog Farm will excel any Gold Mine; $5,000 wanted to invest in establishing a frog farm, with duck ranch as a combine; success assured and strictest investigation solicited, and highest references given.—Answer, N.B., 138 Windsor Ave., Norfolk, Va.

PORPOISE FISHERY for sale; only completely equipped one in the world; send for information, circular.—Riggs & Co., 575 Philadelphia Bourse.

YOUNG lady, good figure, wants to pose for artists; references exchanged; positively no triflers.—Address E. L., 206 "Herald."

LARGE ears, pug noses, hump, flat ill-shaped noses, made to harmonize with the other features. Send stamp for book on Beauty. —J. H. Woodbury, 127 W. 42d St., N. Y.; inventor Woodbury's facial soap.

A.—MAGICAL BEAUTY. Instantaneous results from using Kosmeo Balm and Turkish Rose Leaves. A plain, ordinary woman, you are instantly a dainty, lovable creature, and the secret your own. —Thompson's, 947 Broadway and 177 Fifth Ave.

A HANDSOME young girl desires copying or some other profitable work.—M. P., box 335 "Herald," 23d St.

(*See Page 51*)

Since its organization in 1853 in the Directors' Room of the Continental Bank at No. 10 Wall Street, The Home Insurance Company has occupied a number of sites in lower Manhattan. In 1854 they temporarily occupied the basement of the St. Nicholas Bank on the corner of New and Wall Streets. The company moved into its own new building at No. 4 Wall Street in December, 1854. It soon outgrew these quarters and in May, 1858 moved to 112-114 Broadway. Beginning in 1863 the company occupied the northwest corner of Broadway and Cedar Street for sixteen years. In the spring of 1879 the main offices were moved to the Boreel Building at 119 Broadway. In 1902 the Mutual Life Insurance Company completed a building especially for The Home at 52-56 Cedar Street where they remained until their occupancy of the present site at 59 Maiden Lane in 1923.

(*See Page 20*)

A DIRECTORY OF FORGOTTEN STREETS

The writer would like to acknowledge here, his great indebtedness to I. N. Phelps Stokes, creator of the magnificent Iconography of New York City, George Henry Stegmann, Henry Collins Brown and other early New York historians who made his chore of compiling this directory an easier one.

Abattoir Place, now West 12th St. bet. 11th Ave. & the Hudson River.

Abingdon Place, now West 12th St. bet. Hudson & Greenwich Sts. Laid out c. 1807, it was known then as Cornelia St.; in 1817 as Scott St.

Abingdon Road, see Love Lane.

Academy Street, still in existence.

Achmuty Lane was in block bounded by Water, South, Pike, & Rutgers Sts.

Adams Place, now West Broadway bet. Spring & Prince Sts.

Albany Avenue once ran from 26th St. bet. 5th & Madison Aves. northwesterly, crossing 5th Ave. bet. 29th & 30th St. to corner of 6th Ave. & 42nd St. then northerly on the present line of 6th Ave. to 93rd St.

Albion Place, now East 4th St. bet. 2nd Ave. & the Bowery.

Alms House, south side of Chambers St. on the site of New Court House.

Amity Alley (*Amity Place*), formerly in rear of No. 216 Wooster St.

Amity Lane was a country lane (1752) which commenced at Broadway, 50 ft. north of Bleecker St. & ran northwesterly to 6th Ave. just south of 4th St.

Amity Street, now West 3rd St. bet. Broadway & 6th Ave.

Amos Street, now West 10th St. bet. Greenwich Ave. & the Hudson River.

**Amsterdam Street*.

Ann Street, now Grand St. bet. Broadway & the Bowery. Laid out in 1797; name changed in 1807.

Ann Street, now Elm St. bet. Reade & Franklin Sts. Name changed in 1807.

Anne Street. Horse & Cart St. in 1748.

Anthony Street, now Duane St.

Anthony Street, now Worth St. bet. Hudson & Baxter Sts. Laid out in 1795; known in 1797 as Catherine St.; in 1807 as Anthony St. from Little Water St.

**Antwerp Street*.

Arch Place was in rear of No. 109 Canal St. bet. Church St. & West Broadway.

Arden Street (*Ardens*) now Morton St. bet. Varick & Bleecker Sts.; changed in 1829. Also called Eden St.

Art Street, now Astor Place. Originally a lane leading from the Bowery to a part of the Stuyvesant Farm, it was known as Art St. in 1807.

Arundel Street, now Clinton St. from Division to Houston Sts. Laid out c. 1760; changed to Clinton in 1828.

Ashland Place, now Perry St. bet. Waverly Place & Greenwich Ave.

Asylum Street, now West 4th St. bet. 6th Ave. & 13th St. Originally Chester St. Changed to Asylum St. when N. Y. Orphan Asylum erected there.

Auchmuty Street, named after Rev. Dr. Samuel Auchmuty, Rector of Trinity Parish. Now Rector St.

Augustus Street, now City Hall Place. Laid out c. 1795 it was known as Augustus St. in 1797.

Bache Street, now Beach St.

Back of Jail, Chambers St. near Centre St.

Bailey Street was laid out through the New York Common Lands. It ran from Broadway to Albany Ave. bet. 25th & 26th Sts.

Bancker Street, now Duane St.

Bancker Street, now Madison St. bet. Catherine & Pearl Sts. It was projected c. 1750; known as Bancker in 1755; as Madison since.

Bannon Street, now Spring St.

Bar Street as laid out, ran from Grand St. to the East River bet. Scammel & Jackson Sts. It was also called Fir St.

Barley Street, now Duane St. from Rose to Hudson Sts. Laid out 1791; name changed 1807.

Barrack Street, former name of Tryon Row (now obsolete); known by this name in 1766.

Barrick Street, now Exchange Place.

Barrow Street, now West Washington Place bet. Macdougal & West 4th Sts.

Bartley Street, now Duane Street from Greenwich St. to Pearl St.

Batavia Lane, now Batavia St.

Battoe (Batan, Batteaux) *Street*, now Dey St.

Bayard Place, now Charles Lane; narrow street running from Washington to West St. bet. Charles & Perry Sts.

Bayard Street, now Stone St.

Bayard's Lane, now Broome Street.

Beaver Lane (Bever Lane) now Morris St. (Goelets St.).

Beaver's Path, now Battery Place. Old Indian road.

Bear Market, Greenwich St. between Fulton and Vesey Streets.

Bedlow Street, now Madison St. bet. Catherine & Montgomery Sts. Known by this name in 1797; as Bancker St. in 1817.

Belvedere Place, now West 10th St.

Bellevue Lane, a lane in the old Bellevue estate which led to the Old Post Road.

Benson's Lane, now Elm St.

Berkley Street, part of Barclay St.

Bethune Street, still standing. Opened 1827.

Beurs Straat was the Dutch name of Whitehall St.

Bever Graft, *Bever Straat*, *Bever Paatjie*, were the Dutch names of Beaver St. from Broadway to Broad St.

Birmingham Street, still standing.

Blindman's Alley, still standing—26 Cherry St.

Bloomfield Street formerly ran from No. 7 10th Ave. to the Hudson River.

Bloomingdale Road, once one of the city's main thoroughfares, started at 23rd St., being the continuation of Broadway at that point. It followed the present Broadway as far as 86th St. where it veered easterly, running bet. Broadway & Amsterdam Ave. At 104th St. it again followed the line of Broadway until reaching 107th St., where it turned slightly westerly until it met the present easterly roadway of Riverside Drive, following it to 116th St. Here it turned easterly, crossing Broadway at 126th St. & meeting Old Broadway at Manhattan St. (The present Old Broadway bet. Manhattan St. & 133rd St. is a part of the original road.) From 133rd St. it ran slightly east of the present Broadway into Hamilton Place at 138th St., following Hamilton Pl. to its termination at Amsterdam Ave. & 144th St., & from there running northeasterly & ending at the junction of Kingsbridge Road, just east of St. Nicholas Ave. & 147th St.

Bogart Street formerly ran from No. 539 West St. west to the Hudson River.

Boorman Place, former name of part of West 33rd St., between Eighth and Ninth Avenues.

Boorman Terrace, former name of part of West 32nd St. between Eighth and Ninth Avenues.

Borce (*Boree*) part of the old Bowery Road.

Boston Post Road. (See Eastern Post Road.)

Bott Street, now Elm St.

Boulevard, The, now Broadway from 59th to 155th Sts. It was opened in 1868 and changed to Broadway on Jan. 1, 1899.

Boulevard Place, now West 130th St. from 5th to Lenox Aves.

Bowery, The, part of an old Indian trail. Called Bowery Rd. or Bowery Lane, it dates from 1625.

Bowery Lane. The Bowery was so called in 1760; since 1807 known as The Bowery.

Bowery Place was in the rear of No. 49 Chrystie St. bet. Canal & Hester Sts.

Bowling Green, now Cherry St.

Brannon (*Branner*, *Brennan*) *Street*, now Spring St.

Breedweg (*Breedwegh*). Broadway bet. Bowling Green & Park Row was known by these names during the Dutch occupancy.

Brevoort Place, now West 10th St. bet. Broadway & University Pl.

Brewers Hill, now Gold St.

Brewen's Street (*Breurs*, *Brawer Straat*) now Stone St.

Bride Street, now Minetta St. from Bleecker St. to the bend in the street.

Bridge Street was one of the former names of Elm St.

Bridewell, Broadway opposite Murray St.

Broad Wagon Way, The, name of Broadway in 1670.

Broadway Alley runs from No. 153 East 26th St. north to 27th St.

Brook Street was the former name of Hancock St.

Brouwer Straat (Brewer's St.) was the name the Dutch first gave to the present Stone St. It was one of the earliest streets laid out by them and received this name on account of the Brewery of the West India Co. being located on it. Since 1797 has been known as Stone St., having been called High St. in 1674, and Duke St. in 1691.

Brugh Straat (*Bridge Street*) was one of the early Dutch streets and received this name on account of it being the street which led to the bridge over the canal in Broad St.; known as Bridge St. in 1674; as Hull St. in 1691; and as Bridge St. since 1728.

Brugh Steegh (*Bridge Lane*) was a narrow street, about twenty-two feet wide, which ran between Bridge and Stone Sts. It was closed about 1674.

Budd Street was the former name of Van Dam St.

Bullock Street was the former name of Broome St.; known by this name in 1766; since 1807 known as Broome St.

Bunker's Hill, Grand St., from Mott St. to Broadway.

Burger's Path was the Dutch name of a part of William St.

Burling Lane was a country road which commenced at the present Broadway, between 17th and 18th Sts., and ran southwesterly, meeting the Southampton Road at about the present 6th Ave. and 16th St.

Burnet Street was the former name of Water St. between Wall St. and Maiden Lane.

Burr Street was the former name of Charlton St.

Burrows Street was the former name of Grove St. In 1807 was known as Columbia St. and since 1817 as Burrows St.

Burton Street was the former name of LeRoy St. from Varick to Bleecker Sts.

Bushwick Street was the former name of Tompkins St.

Byvanck Street, formerly ran from Grand to Water Streets.

Camden Place was the former name of East 11th St. between Avenues B and C.

Caroline Street was at the head of Duane St. Slip.

Carroll Place was the former name of Bleecker St. between West Broadway and Thompson St.

Cartmans Arcade was an Alley which ran south at No. 171 Delancey St. now closed.

Catherine Place was the former name of Catherine Lane.

Catherine Street was the former name of Worth St.; known in 1797 as Catherine St.; in 1807 as Anthony St.

Catherine Street was the former name of Waverly Pl. between Christopher and West 12th Sts.; known by this name in 1807.

Catherine Street was the former name of Mulberry St. between Bayard and Bleecker Sts.; known by this name in 1797.

Catherine Street was the former name of Pearl St. between Broadway and Elm Street; was also called Magazine St.

Cato's Lane started at the Eastern Post Road, about the present 2nd Ave. between 52nd and 53rd Sts. and ran southeasterly to the East River at Ave. A between 50th and 51st Sts.

Centre Market Place—by Centre Market.

Centre Street—now Marion Street.

Chapel Street was the former name of West Broadway from Murray to Canal Sts.; known by this name in 1797; name changed to College Place in 1830.

Chappel Street was the former name of Beekman St.

Charles Alley was the former name of Charles Lane.

Charlotte Street was the former name of Pike St. between Cherry and Division Sts.; was known by this name in 1791.

Chatham Street was the former name of Park Row. This street was originally part of the Bowery; called Chatham St. in 1774, changed to Park Row in 1886.

Cheapside was the former name of Hamilton St. between Catherine and Market Sts.; was known by this name in 1797; name changed to Hamilton St. on Aug. 27, 1827.

Chester Street was the former name of West 4th St. between Bank and Christopher Sts.

Chestnut Street was the former name of Howard St. between Broadway and Mercer St.; known in 1807 as Hester St.

Chrystie Street was the former name of Cherry Street.

Church Lane was one of the first streets laid out in the village of Harlem; it ran from 117th St. between 3rd and 4th Aves. northerly to 120th St. then northeasterly, crossing 3rd Ave. at 121st St., 2nd Ave. at 123rd St. and ending at the Harlem River between 125th and 126th Sts.

Church Street was the former name of Exchange Place between Broadway and William St.

City Hall Lane, see Coenties Alley.

Cingle (*Single St.* or *The Cingel*), in Dutch times south side of Wall Street.

Clendening's Lane was a country road which started in Central Park about on line with 6th Ave. and 105th St. and ran westerly along the southerly side of 105th St. to the middle of the block between Columbus and Amsterdam Ave., then southwesterly to the Bloomingdale Road, at about a point fifty feet south of 103rd St.

Clermont Street was the former name of Mercer St.; known in 1797 as First St. and since 1807 as Mercer St.

Clermont Street was the former name of Hester St. between Centre St. and Broadway and of Howard St. between Broadway and Mercer St.

Clinton Place was the former name of West 8th St. from Broadway to 6th Ave.

Coenties Alley runs between Pearl and Stone Streets.

Col. Burr's, Richmond Hill, S.E. corner of Varick and Charlton Sts.

Colden Street was the former name of Duane St. from Lafayette to Rose St.; known by this name in 1803.

Collect Street was the former name of Centre St. from Hester to Pearl Sts.; known by this name in 1807 to 1817.

College Place was the former name of West Broadway from Barclay to Warren Sts.; known in 1755 as Chapel St.; name changed to College Pl. in 1830.

Columbia Place was the former name of a part of 8th St.

Columbia Street was the former name of Grove St.; was also called Burrows and Cozine Sts.

Columbia Street was the former name of Jersey St.

Columbian Alley, now Jersey St.

Commerce Street was the former name of Barrow St.

Commons Street. Name of Park Row at one time.

Concord Street was the former name of West Broadway from Canal to 4th Sts.

Congress Place was the former name of an Alley in the rear of No. 4 Congress St.

Coopers Street was the former name of Fletcher St.

Cop Street was the former name of State St.

Cornelia Street was the former name of West 12th St. between Greenwich Ave. and Hudson St.

Cottage Place was the former name of East 3rd St. between Avenues B and C.

Cottage Row was the former name of 4th Ave. between 18th and 19th Sts.

Cozine Street was the former name of Grove St.

Crabapple Street was the former name of Pike St.

Crolius', north side of Chatham St., between Pearl and Duane Sts.

Cropsie Street was the former name of State St.

Cross Street was the former name of Park St.

Crown Point Street was the former name of Grand St. from the Bowery to the East River.

Crown Point Street was the former name of Corlaers St.

Crown Point Street was the former name of Water St. between Montgomery St. and the East River.

Crown Street was the former name of Park St.; known in 1797 as Cross St.

Crown Street was the former name of Liberty St.; it was laid out about 1690; at one time called Tienhoven St.; name changed to Liberty St. in 1783.

Custom House St. was the former name of Pearl St. between Whitehall St. and Hanover Square.

Cuyler's Alley, formerly Mesier's Alley.

David Street was the former name of Bleecker St. between Broadway and Hancock St.; name changed in 1829.

David Street was the former name of Clarkson St. between Varick and Hudson Sts.

Decatur Place was the former name of 7th St. between 1st Ave. and Avenue A.

Depau Row was the former name of Bleecker St. between Thompson and Sullivan Sts.

DePeyster Street runs between Water and South Streets.

Desbrosses Street was the former name of Grand St. between Broadway and Varick St.

Dirty Lane was the former name of South William St. This street was opened about 1656 and was called by the Dutch Slyck Steegh, meaning Dirty Lane. In 1674 it was called Mill Street Lane; name changed to South William St. about 1832.

Division Street was the former name of Fulton St. between Broadway and West St.

Dixon's Row was the name given to a part of 110th St. between 8th and Columbus Aves.

Dock Street was the former name of Pearl St. from Whitehall Street and Hanover Square.

Dock Street was the former name of Water St. between Coenties Slip and Beekman St.

Dominic Street was the former name of Doaning St.

Donovan's Lane was near No. 474 Pearl St.

Duggan Street was the former name of Canal St. between Centre and West Sts.

Duke Street was the former name of Stone St. During the Dutch times a part was known as Brouwers Straat, and another part as Hoogh Straat; in 1674 it was known as High St. and a part as Stone St. In 1691 it was called Duke St. and since 1797 has been known as Stone St. This street was the first to be paved with stone in the City.

Duke Street was the former name of Vanderwater St.; it was known by this name in 1755.

Duncomb Place was the former name of East 128th St. between 2nd and 3rd Aves.

Dunham Place was an alley running south from West 33rd St. now closed.

Dunscombe Place was the former name of East 50th St. between 1st Ave. and Beekman Place.

Dutch Street runs from John to Fulton Streets between William and Nassau Streets.

Dwar's Street was the former name of Exchange Place between Broadway and Broad St.

Dyes Street. Dey Street was so called in 1767.

Eagle Street was the former name of Hester St.; it was laid out about 1750; known in 1755 as Hester St.; in 1766 as Eagle St., and since 1807 as Eagle St. bet. Bowery and Division Sts.

East Bank Street was an old road in Greenwich Village; it ran from 7th and Greenwich Aves. northeasterly to the Union Road in the block now bounded by 6th and 7th Aves., 13th and 14th Sts.

East Court was in West 22nd St. near 6th Ave., now closed.

East George Street was the former name of Market St.

East Place formerly ran in the rear of Nos. 184–186 East 3rd St. between Avenues B and C.

East Road was the former name of 4th Ave. between 37th and 90th Sts.

East Rutgers Street, now Rutgers Street.

East Street was the former name of Mangin St.

East Tompkins Place was the former name of East 11th St. between Avenues A and B.

Eastern Boulevard, connected Central Park with the East River.

Eastern Post Road started at the present Broadway and 23rd St. and ran northeasterly across Madison Square to about 30th St. just west of Lexington Ave.; it then ran northerly, parallel to Lexington Ave. to 36th St. there veering easterly, crossing 3rd Ave. at 45th St. and then running northerly, midway between 2nd and 3rd Aves. to 50th St. where it turned northeasterly. Crossing 2nd Ave. at 52nd St. from there it ran northerly, midway between 1st and 2nd Aves. at 62nd St., 3rd Ave. at 72nd, and Lexington Ave. at 76th St. It then ran northerly and northeasterly, recrossing Lexington Ave. at 77th St., then northeasterly, northerly and northwesterly, crossing 5th Ave. at 90th St., then northerly through Central Park, recrossing 5th Ave. at 109th St., 4th Ave. at 115th St., then northeasterly between Third and Fourth Aves. to the Harlem River at 130th St. and Third Ave. This road was also called the Boston Post Road. It was closed in 1839.

Eden Street was the former name of Morton St. between Bedford and Bleecker Sts. It was also called Arden St.

Eden's Alley; see Ryder's Alley.

Edgar Street was the former name of Morris St.

Edgars Alley was the former name of Exchange Alley.

Eighth Street was the former name of Hancock St.

Elbow Street (*Lane*) was the former name of Cliff St.

Eliza Street was the former name of Waverly Place.

Eliza Street was a country road on the Kips Bay Farm. It started in the block bounded by 2nd and 3rd Aves., 28th and 29th Sts. and ran northeasterly, crossing 2nd Ave. at 35th St. and ended at 39th St. between 1st and 2nd Aves. It ran at right angle to two other old roads, Kips Bay St. and Maria St.

Ellet's or Elliotts Alley was the name by which Mill Lane was known in about 1664.

Elm Street was the former name of Lafayette St. between Worth and Spring Sts.

Erie Place was the former name of Duane St. between Washington and West Sts.

Exchange Court was in the rear of No. 74 Exchange Place.

Exchange Street was the former name of Beaver St. between William and Pearl Sts.

Exchange Street was the former name of Whitehall St.

Exchange Street was the former name of Marketfield St. In 1791 was called *Petticoat Lane.*

Extra Place was an alley which ran north from 1st St. between the Bowery and 2nd Ave.

Factory Street was the former name of Waverly Place between Christopher and Bank Sts. It was also called Catherine St.

Fair Street was the former name of Fulton St. from Broadway to the Hudson River; east of Broadway it was called Partition St. It was laid out about 1720.

Farlow's Court was formerly in the rear of Nos. 153, 155, 157, 159 and 161 Worth St.

Fayette Street was the former name of Oliver St. It was known as Oliver St. since 1825. From Park Row to Madison Sts.

Federal Hall, was at the N.E. corner of Wall and Nassau Streets; now the Sub-Treasury bldg.

Feitner's Lane; see Verdant Lane.

Ferry Street was the former name of Bayard St.

Ferry Street was the former name of Peck Slip.

Ferry Street was the former name of Jackson St. between Division and Cherry Sts.; was known by this name in 1807; was also called Ferry Place.

Ferry Street was the former name of Scammel St.

Field Street, *Fieldmarket Street* were the former names of Marketfield St.

Fifth Street was the former name of Orchard St.

Fifth Street was the former name of Thompson St.

Fifth Street was the former name of Washington St.

Fir Street ran from Grand St. to the East River between Scammel and Jackson Sts. now closed; it was also called Bar St.

Fire Alley, believed to be behind the Old City Hall on Wall Street where the first two engine houses were built.

First Street was the former name of Chrystie St. from Division to Houston Sts. Known by this name in 1766.

First Street was the former name of Mercer St. Called Clermont St. in 1797; since 1807 known as Mercer St.

First Street was a former name of Greenwich St.

Fisher's Court was in the rear of Nos. 22, 24 and 26 Oak Street, between Roosevelt and James Sts.

Fisher Street was the former name of Bayard St. from the Bowery to Division St. Known by this name in 1755; since 1807 known as Bayard St.

Fitzroy Place was the former name of West 28th St. between 8th and 9th Aves.

Fitzroy Road; see Roy Road.

Flatten Barrack Street was one of the former names of Exchange Place, between Broadway and Broad St. It was known by this name in 1728.

Fly Market Street, foot of Maiden Lane.

Fourth Street was the former name of Allen St. between Division and Houston Sts.

Fourth Street was the former name of West Broadway between Canal and West 4th Streets.

Franklin Terrace was in the rear of No. 364 West 36th St.

French Church Street was the former name of Pine Street between Broadway and William St.

Front Street was the former name of Greenwich St.

Fulton Street was a former name of Nassau St.

Garden Lane was the former name of Exchange Alley; was also known as Tin Pot Alley.

Garden Row was the former name of Nos. 140 to 158 West 11th St.

Garden Street was one of the former names of Exchange Place. This street was laid out during the Dutch rule and was called by them Tuyn (Garden) Straat, in 1691 it was known as Church Street, in 1728 as Garden St. and a part as Flatten Barrack; in 1797 it was all called Garden St.

Garden Street was the former name of Cherry Street from Montgomery to Corlaers Sts.

Garden Street Alley, now Exchange Alley.

Gardiner Street was the former name of Tompkins St.

Garry Place was the former name of West 35th Street between 7th and 8th Avenues.

Gen. Greene Street was the former name of Gouverneur St.

George Street was the former name of Beekman St.

George Street was the former name of Bleecker St. between Hancock and Bank Sts.

George Street was the former name of Hudson St.

George Street was the former name of Market St. between Division and Cherry Sts. It was known by this name in 1791.

George Street was the former name of Park St.

George Street was the former name of Rose St.

George Street was the former name of Spruce St. It was laid out about 1725 as George St.; in 1817 it was known as Little George St.

Germain Street was the former name of Carmine St.

Gerritsen's Wagon Way, thought to be the early name of Astor Place.

Gibb's Alley ran from Madison St. between Oliver and James Sts. northwesterly about one-half a block.

Gilbert Street was the former name of Barrow St. between Bleecker and West 4th Sts.

Gilford Place was the former name of East 44th St. between 3rd and Lexington Avenues.

Glassmakers Street, *Glazier Street* was a former name of William St. between Pearl and Wall Sts.

Glover Place was one of the former names of Thompson St. between Spring and Prince Sts.

Golden Hill was the former name of John St. between William and Pearl Sts.

Gotham Court, name of street beginning at 19 Cherry Street.

Gould, now Gold St.

Gouverneur Alley, former name of Gouverneur Lane.

Grand Avenue was the former name of 125th St.

Great Dock Street was one of the former names of Pearl St. This street was known in 1657 as Pearl St.; in the same year was also known as Hoogh St. and the Waal; in 1691 as Great Dock and Great Queen Sts.; in 1728 as Queen St.; in 1797 it was known as Pearl St. as far north as Park Row, the rest being called Magazine St. Since 1807 the entire street has been known as Pearl St.

Great George Street was the name Broadway, north of the City Hall Park, was known by in 1791.

Great Kill Road; see Southampton Road.

Green Alley or Lane was the former name of Liberty Place.

Green Street was a former name of Liberty St.

Greenwich Lane was the former name of Gansevoort St. and Greenwich Ave.

Greenwich Street. Washington St. was called by this name at one time.

Hall Street, formerly Hall Place.

Hamilton Place was the former name of West 51st St. between Broadway and 8th Ave.

Hammersly Street was the former name of West Houston Street between Macdougal St. and the Hudson River.

Hammond Street was the former name of West 11th St. between Greenwich Ave. and the Hudson River.

Hanson Place was the former name of 2nd Ave. between 124th and 125th Sts.

**Hariot St.*

Harlem Lane. The present St. Nicholas Ave. from 110th to 123rd Sts. was called by this name; it was part of the Kingsbridge Road.

Harlem Road (The Old) was a country road leading to the Village of Harlem; it started at the junction of the Eastern Post Road in Central Park about on a line of 108th St. and between 5th and Lenox Aves. running northeasterly; crossing Madison Ave. between 113th and 114th Sts.; Park Ave. between 115th and 116th Sts.; Lexington Ave. between 117th and 118th Sts.; 2nd Ave. at 123rd St.; 1st Ave. at 125th St. and ending at the Harlem River at the foot of 126th St.

Harlem Road started at the Eastern Post Road about the present 95th St. between Madison and 5th Aves. and ran northeasterly, crossing Madison Ave. at 99th St.; Park Ave. at 108th St.; Lexington Ave. at 116th St. and ending at the Harlem River at 129th St.

Harman Street was the former name of East Broadway; it was originally a lane, known as Love Lane that led to the Rutgers' Farm.

Harsen's Lane was a country road which connected the Village of Harsenville (70th St. & Broadway) with the eastern part of the Island; it commenced at the Bloomingdale Road (the present Broadway) between 71st and 72nd Sts. and ran easterly about on line of the present 71st and ended at the Middle Road, the present 5th Ave. and 71st St.

Harsen's Road, formerly ran from the Eastern Post Road to Ninth Ave.; then northwesterly to Bloomingdale Road between 71st and 72nd Streets.

Hazard Street was the former name of King St.

Heer Graft (High Ditch) was the name given by the Dutch to the present Broad St. between Beaver and Pearl Sts. in 1657; it was one of the earliest streets laid out in the City; and received its name on account of the narrow canal which ran through the center. This canal was filled in about 1676 and street was called Broad St.; it was sometimes spelled Heeren Gracht.

Heere Dwars Straat, former name of Exchange Place between Broad and William Streets.

Heere Straat, Heere Wegh, Heere Waage Wegh, were the Dutch names for the present Broadway between Bowling Green and the City Hall Park.

Hell Gate Ferry Road was a country road which ran from the East River at the foot of 90th St. southwesterly, joining the Eastern Post Road at Madison Ave. and 82nd St.

Henry Street was the former name of Perry St.

Hereweg. The Dutch name of the present Park Row from Broadway to Chambers St.

Herman Place was in the rear of Nos. 194, 198 4th St. between Avenues A and B.

Herring Street was the former name of Bleecker St. between Carmine and Banks Sts. known by this name in 1817; name changed to Bleecker St. in 1829.

Herring Street was the former name of Mercer St.

Hester Court was formerly in the rear of No. 101 Hester St.

Hester Street was the former name of Howard St.

Hett Street, Hetty Street, were the former names of Charlton St.

Hevins Street was the former name of Broome St. between Broadway and Hudson Sts.; was also known as St. Hevins St.

High Street was the former name of Madison St. from Montgomery to Grand Sts.

High Street was the former name of Stone St.; known by this name in 1674.

Hoboken Street formerly ran from No. 474 Washington St. west to West St.; now a part of Canal St.

Hoogh Straat (High Street) was the name of Stone St. east of Broad St. prior to 1664.

Hopper's Lane was a country road which ran from the Bloomingdale Road (the present Broadway), just south of 51st St. westerly to the Hudson River at the foot of 53rd St.

Horse and Cart Lane was the name of part of William St.

Houston Street was the former name of Prince St. between Broadway and Hancock St.

Hubert Street was the former name of York St.

Hudson Place was the former name of West 24th St. between 9th & 10th Aves.

Hudson Street was the former name of West Houston Street between Broadway and Hancock St.

Hull Street was the former name of Bridge St. between Whitehall and Broad Sts.; known as Bridge St. in 1676; Hull St. in 1681; and Bridge St. since 1728.

Inwood Street, former name of Dyckman Street.

Jackson Avenue was the former name of University Place.

Jackson Place was an alley which ran north from No. 16 Downing St. now called Downing Place.

James Street, once known as St. James Street.

Jauncey Court was in the rear of Nos. 37, 39 and 41 Wall St.

Jauncey Lane was a country road which started between 93rd and 94th Sts. just west of West End Ave. and ran easterly crossing 8th Ave. at 94th St. and ending at the Eastern Post Road; about the present line of 96th St. between 5th and 6th Aves.

Jew's Alley was the former name of South William St. between Broad Street and Mill Lane.

Jew's Alley formerly ran from Madison St. between Oliver and James Sts.

Jones Court was in the rear of Nos. 48, 50 Wall St.

Jones Street was the former name of Great Jones St.

Judith Street was the former name of Grand St. between the Bowery and Centre St.

King Street was the former name of Pine St. It was laid out about 1691 and was known as Queen St.; known in 1728 as King St.; name changed to Pine St. in 1793.

King Street was the former name of William St. between Hanover Square and Wall St.

King George Street was the former name of William St. from Frankfort St. easterly to Pearl St.; known by this name in 1755.

Kingsbridge Road branched off from the Eastern Post Road a little north of McGowns Pass, about the present line of 108th St. between 5th and Lenox Aves. and ran northwesterly along the present St. Nicholas Ave. to 169th St.; from there it followed along the present Broadway to the Harlem River, crossing the river on the old Kings Bridge.

Kingsbridge Road. There was a second road known by this name which started in the Village of Harlem; about the present Sylvian Place, between 3rd and Lexington Aves., 120th and 121st Sts. and ran northwesterly to 124th St. and Park Ave., continuing northwesterly to 127th St. between Lenox and 7th Aves., then southwesterly to a point in the block bounded by Lenox and 7th Aves., 126th and 127th Sts., then northwesterly to St. Nicholas Ave. between 131st and 132nd Sts., where it joined the other Kingsbridge Road.

Kings Highway was one of the former names of Park Row and the Bowery.

Kings Road was the former name of Pearl St. between Franklin Square and Park Row.

Kip Street was the former name of Nassau St. between Maiden Lane and Spruce St.

Kips Bay Street was a country road which started at the Eastern Post Road, the present Madison Ave. and 35th St. and ran southeasterly, crossing 2nd Ave. at 34th St. and ended at the East River at the foot of 34th St.

Knapp Place was formerly in the rear of No. 412 East 10th St. between Avenue C and Dry Dock Street.

Koninck Street was the former name of Pine St.

Lafayette Place was the former name of Lafayette Street, between Great Jones St. and 8th St. It was opened July 4, 1826.

Lake Tour Road, formerly ran from 39th St. and Bloomingdale Road westerly along 39th St. to 7th Ave.; then northwesterly to 9th Ave. between 41st and 42nd Sts.

Lamartine Place was the former name of West 29th St. between 8th and 9th Aves.

Lambert Street was the former name of Church St. between Edgar and Liberty Sts.

Laurens Street was the former name of West Broadway, between Canal and 4th Sts.

Leandert's Place was formerly in the rear of No. 147 7th St. between Avenues A and B.

Leary Street was a former name of Cortlandt St.

Leather Street was the former name of Jacob St.

Lenox Place was the former name of 22nd St. between 8th and 9th Aves.

Leroy Place was the former name of Bleecker St. between Mercer and Greene Sts.

Leyden Place was the former name of Fourth Ave. between 11th and 13th Sts.

Liberty Court was formerly in the rear of Nos. 4 and 6 Liberty Place.

Lispenard's, west of Hudson St., between Desbrosses and Watts Sts.

Little Street was the former name of *Cedar St.* between Broadway and the Hudson River.

Little Ann Street was the former name of Elm St. now Lafayette St.) between Reade and Franklin Sts.

Little Chappel Street was the former name of College Place (now West Broadway) between Barclay and Warren Sts.

Little Division Street was the former name of Church St.

Little Division Street was the former name of Montgomery St.; known by this name in 1766-1767.

Little Dock Street was the former name of Water St. between Broad St. and Old Slip.

Little Dock Street was the former name of South St. between Whitehall St. and Old Slip.

Little George Street was the former name of Spruce St.; known by this name in 1725.

Little Green Street was the former name of Liberty Place.

Little Queen Street was the former name of Cedar St. It was laid out about 1690 and was known as Smith St.; known in 1728 as Little Queen St.; known since 1793 as Cedar St.

Little Stone Street was the former name of Thames St.; known by this name in 1766; known since 1791 as Thames St.

Little Water Street was the former name of Mission Place.

Locust Street was the former name of Sullivan St.

Lombard Street, Lombardy Street, were the former names of Monroe St.; known in 1791 as Rutgers St.; name changed to Monroe St. Jan. 10, 1831.

London Terrace was the former name of the north side of 23rd St. between 9th and 10th Aves.

Lorillard Place was the former name of Washington St. between Charles and Perry Sts.

Louisa Street, Kips Bay Farm was a country road which ran from the Eastern Post Road about the present Lexington Ave. and 32nd St. southeasterly, crossing 2nd Ave. at 31st St. and ending at the East River at the foot of 30th St.

Love Lane, also called the Abingdon Road, was a country road which commenced at the Roy Road; about the present 8th Ave. and 21st St. and ran easterly on about the line of the Eastern Post Road at the present 3rd Ave. and 23rd St.

Love Lane was a country road which ran from Chatham Square easterly to the Rutgers' Farm, about the line of the present West Broadway.

Lowe's Lane was a country road which commenced at the Eastern Post Road about the present 41st St. slightly east of Lexington Ave. and ran westerly crossing the Middle Road (5th Ave.) at 42nd St. and ending at the Bloomingdale Road (present Broadway), between 43rd and 44th Sts.

Low Water Street was the former name of Washington St. between Battery Place and West Houston St.

Low Water Street was the former name of Water Street between Broad and Wall Sts.

Lower Robinson Street, former name of Robinson Street.

Ludlow Place was the former name of West Houston St. between Sullivan and Macdougal Sts.

Lumber Street was the former name of Trinity Place between Morris and Liberty Sts.

Lumber Street was the former name of Monroe St.

Maagde Paetge (*Maidens Path*) was the name of Maiden Lane during the time of the Dutch.

Madison Court was formerly in the rear of No. 219 Madison St.

Maiden Lane was a country lane in the block now bounded by Broadway, Amsterdam Ave., 160th and 161st Sts.

Magazine Street was the former name of Pearl St. between Park Row and Broadway.

Manhattan Avenue was the former name of 5th Ave.

Manhattan Road was a country road which commenced at the Kingsbridge Road; about the present Lexington Ave. and 121st St. and ran southwesterly to a point in the block bounded by Park and Madison Aves., 118th and 119th Sts., then northwesterly, crossing 5th Ave. at 119th St., 6th Ave. between 120th and 121st Sts., 7th Ave. between 121st and 122nd Sts. to a point on the north side of 122nd St. about 200 ft. east of 8th Ave., then southwesterly to 8th Ave. about one-half way between 121st and 122nd Sts.

Mansfield Place was the former name of West 51st St. between 8th and 9th Aves.

Margaret Street was the former name of Cherry St.

Margaret Street was the former name of Willett St.

Maria Street, Kips Bay Farm, was the name of a country road which started from a point in the block bounded by 2nd and 3rd Aves., 29th and 30th Sts., and ran southeasterly to the East River between 28th and 29th Sts.

Marion Street was the former name of Cleveland Place and Lafayette St. between Broome and Prince Sts.

Market Street was the former name of South William St.

Marketfield Street was the former name of Battery Place between Broadway and Hudson River.

**Martha St.*

Martin Terrace was the former name of East 30th St. between 2nd and 3rd Aves.

Mary Street was the former name of Christopher St. between Greenwich Ave. and Waverly Pl.

Mary Street was the former name of Baxter St. between Leonard and Grand Sts.

Mary Street was the former name of Cleveland Pl. and Lafayette St. between Broome and Prince Sts.

Meadow Street was the former name of Grand St. between Broadway and Sullivan St.

Mechanics Alley formerly ran from No. 72 Monroe St. south to Cherry St. between Market and Pike Sts.; now the site of the Brooklyn Bridge approach.

Mechanics Place formerly ran from the east side of Avenue A, between 2nd and 3rd Sts.

Mechanics Place formerly ran in the rear of Rivington St. between Lewis and Goerck Sts.

Meek's Court was formerly in the rear of 55 Broad St.

Merchants Court was in the rear of No. 48 Broad St.

Merchants Place formerly ran in the rear of No. 28 Avenue A between 2nd and 3rd Sts.

Merchant Street was the former name of Beaver St.

Mesier's Alley was the former name of Cuyler's Alley.

Middle Road was a country road which started at the Eastern Post Road, about the present 4th Ave., between 28th and 29th Sts., and ran northwesterly, crossing Madison Ave. at 35th St. At 5th Ave. and 42nd St. (Burr's Corners) it turned northerly along the line of 5th Ave. to 90th St. where it terminated at the Eastern Post Road.

Middle Street was the former name of Monroe St. from Montgomery to Corlaer Sts.

Mill Lane runs from South William St. to Stone St. Opened 1657.

Mill Street was the former name of Stone St.

Mill Street was the former name of South William St. between Broad and Mill Lane.

Miller Place was formerly in the rear of No. 4 Macdougal St.

Milligan Place was formerly in the rear of No. 139 6th Ave. between 10th and 11th Sts.

Millward Place was formerly the name of West 31st St. between 8th and 9th Aves.

Mission Place, formerly Little Water St.

Mitchell Place was the former name of the north side of East 49th St. between 1st Ave. and Beekman Place.

Monroe Place was the former name of Monroe St. between Montgomery and Gouverneur Sts.

Monument Lane was a country road leading to Greenwich Village. It started at the Bowery and Astor Place and ran easterly, then northeasterly, following the present Greenwich Ave. which is a part of the old road; and ended at Gansevoort St.

Moore's Row was formerly between Catherine and Market Sts. and ran from Henry to Madison Sts.

Mortkile Street (Moord Kuyl Straat) was the former name of Barclay St.

Morton Street was the former name of Clarkson Street between Varick and Hudson Sts.

Mott's Lane; see Hopper's Lane.

Muddy Lane, former name of Mill St.

Mustary Street was the former name of Mulberry St. between Park Row and Park St.

Near Burke's, Spring St. near Hudson St.

Neilson Place was the former name of Mercer St. between Waverly Place and 8th St.

New Bowery, The, from the southerly side of Chatham Square to Franklin Square (Pearl St.).

New Dutch Church Street, name sometimes given to Crown St.

New English Church Street, a former name of Beekman St.

New Street was the former name of Nassau St.

New Street was the former name of Staple St.

Nicholas Street was the former name of Walker St. between Canal St. and West Broadway.

Nieuw Straat was the Dutch name of New Street.

North Street was the former name of East Houston St. between the Bowery and the East River; name was changed in 1833.

N.R. Furnace, foot of Hubert St.

Nyack Place was formerly in the rear of No. 31 Bethune St.

Oak Street, former name of Rutgers St.

Oblique Road, now Marketfield St.

Ogden Street was the former name of Perry St.

Old Dutch Church Street, now Exchange Place.

Old Street was former name of Mott St. between Park Row and Park St.

Old Kill Road; see Southampton Road.

Old Windmill Lane; see Windmill Lane.

Oliver Street was the former name of Spring St. between the Bowery and Broadway.

Orange Street was the former name of Cliff St.

Orchard Street was the former name of Broome St. west of Broadway.

Otters Alley formerly ran from Thompson to Sullivan Sts. between Broome and Grand Sts.

Oyster Pasty Alley (*Lane*) was the former name of Exchange Alley; was also known as Tin Pot Alley.

Pacific Place was formerly in the rear of No. 133 West 39th St.

Paisley Place, row of workshops of Scottish weavers in West 17th St.

Park Street was the former name of Park Row between Ann and Beekman Sts.

Partition Street was the former name of Fulton St. between Broadway to the Hudson River; east of Broadway this street was called Fair St.

Passage Place was the former name of Peck Slip.

Patchin Place was an alley in the rear of No. 111 West 10th St.

Penn Street was the former name of Pell St.

Perry Street, formerly Ogden St.

**Peter St.*

Petersfield Street was a country road on the Stuyvesant Farm; it started about the present 4th Ave. between 11th and 12th Sts.; crossing 3rd Ave. between 12th and 13th Sts., 2nd Ave. between 13th and 14th Sts., 1st Ave. at 15th St. and ended in the center of the block bounded by 1st Ave., Ave. A, 15th and 16th Sts.

Petticoat Lane was the former name of Marketfield St.; it was known by this name in 1791.

Pieter Jansen's Lane, Old Windmill Lane.

Pitt Street was the former name of Elm St. (now Lafayette St.) between Hester and Spring Sts.; known by this name in 1797.

Pleasant Avenue, former name of Avenue A.

Potters Hill, former name of Park St.

Prince Street was the former name of Rose St.; known by this name in 1766.

Princess Street was the former name of Beaver St. between William and Wall Sts. During the time of the Dutch it was known as Prinsen Straet.

Prospect Street was the former name of Thompson St.

Provost Street was the former name of Franklin St.; known by this name in 1797; known as Sugar Loaf St. in 1807; name changed to Franklin St. in 1833.

Pump Street was the former name of Canal St. It was known by this name in 1797.

Pye Woman's Lane, Pie Woman's Lane, was the former name of Nassau St. between Wall St. and Maiden Lane.

Quay Street was the former name of Water St. between Whitehall St. and Coenties Slip.

Queen Street was the former name of Pearl St. between Wall St. and Park Row. This street was known by various names at different periods; known in 1657 as Pearl St.; and in part Hoogh Straet and the Waal; in 1691 as Dock St.; and Great Queen St.; in 1728 as Queen St. and since 1797 as Pearl St.

Queen Street was the former name of Pine St.; was known by this name in 1691; known as King St. in 1728; name changed to Pine St. in 1794.

Queene Street was the former name of *Cedar St.* between William and West Sts.

Quick Street was the former name of East Broadway.

Raisan Street; see Reason St.

Randall Place was the former name of West 9th St. between Broadway and University Pl.

Reason Street was the former name of Barrow St. between Bleecker and Bedford Sts.; name changed in 1828.

Renwick Street was the former name of Baxter St. between Canal and Grand Sts.

Republican Alley, former name of Manhattan Place, from Elm (Lafayette) to Reade St.

Rhinelander Alley formerly ran from Greenwich to Washington Sts. between Beach and Hubert Sts.

Rhinelander Lane was a country road which ran from the Hell Gate Ferry Road at the present 2nd Ave. between 86th and 87th Sts. northeasterly to the south side of 90th St. between 1st Ave. and Ave. A.

Rider Street, *Ridder Street*, was the former name of Ryder's Alley.

Riker's Lane was a country road which ran from the Eastern Post Road, about the present 3rd Ave. and 76th St. and ran southeasterly, ending at the East River between 74th and 75th Sts.

Rivington Place was formerly in the rear of No. 316 Rivington St.

Robinson Street, was the former name of Park Place.

Roosevelt Lane was a country road which ran from the Old Harlem Road, about the present Lexington Ave. between 116th and 117th Sts. southeasterly, crossing 3rd Ave. at 115th St., 2nd Ave. at 112th St., then northwesterly to a point in the middle of the block bounded by 1st and 2nd Aves., 114th and 115th Sts., then southeasterly to the Harlem River between 110th and 111th Sts.

Rose Hill Lane, formerly ran from 8th Ave., between 21st and 22nd Sts., easterly between 21st and 22nd Sts., then northeasterly to 23rd St. and 3rd Ave.

**Rotterdam St.*

Rotten Row, *Rough Street*, *Ruff Street*, were the former names of Henry St.

Roy Road, *Fitzroy Road*, was a country road which ran north from Greenwich Village; it started at the Southampton Road about the present 14th St., between 7th and 8th Aves., and ran northwesterly crossing 8th Ave. at 22nd St., then north, parallel with and a little west of 8th Ave. and ending at a cross road about the present 42nd St. midway between 8th and 9th Aves. 9th Ave. was closed from 23rd to 42nd Sts. on Oct. 26, 1832.

Rosylyn Place was the former name of Greene St. between West 3rd and West 4th Sts.

Rudder Street was the former name of Ryder's Alley.

Russell Place was the former name of Greenwich Avenue between Charles and Perry Sts.

Rutgers' Hill was the former name of Gold St. between Maiden Lane and John St.

Rutgers' Place was the former name of Monroe St. between Clinton and Jefferson Sts.

Rutgers' Street was the former name of Oak St.; known by this name in 1755.

Ryder's Alley formerly called Eden's Alley.

Ryndert Street, Rindert Street, was the former name of Centre St. between Canal and Broome St.; known in 1797 as Potters Hill; known in 1807 as Collect St.; known in 1817 as Ryndert St.

Sackett Street was the former name of Cherry St.

St. Clamment's Place was the former name of Macdougal St. between Houston and Bleecker Sts.

St. David Street was one of the former names of Bleecker St.

St. Hevins Street was the former name of Broome St. between Broadway and Hudson St.; was known in 1755 as St. Hevins St.; known in 1766 as Bullock St.; known in 1797 in part as Bullock St. and in part as William St. and in part as Orchard St.; known since 1807 as Broome St.

St. Johns Street was the former name of John St.

St. Marks Place was the former name of East 8th St.

St. Nicholas Street was the former name of Walker St. between Canal St. and West Broadway.

St. Nicholas Street was the former name of Canal St. between Walker St. and the Bowery.

St. Peters Place was the former name of Church St. between Vesey and Barclay Sts.

Sand Hill Road, formerly called the "old highway" and a later name for Gerritsen's Wagon Way.

Schaape Waytie (*The Sheep Pasture*) was the Dutch name of Broad St. between Beaver and Wall Sts.

Scott Street was the former name of West 12th St. between Greenwich Ave. and Hudson St.; was also known as Troy St. and Abingdon Place.

Scott's Alley formerly ran south from No. 71 Franklin St. to White St.

Second Street was the former name of Greene St.

Second Street was the former name of Forsythe St.

Seventh Street was the former name of Macdougal St.

Seventh Street Place was a short alley, seven houses long, in the rear of No. 185 7th St.

Sheera Street, between Bowling Green and Wall Streets.

Shinbone Alley was the former name of Washington Mews; was also known as Washington Alley.

Shinbone Alley, between Lafayette & Bond Streets.

Sixth Street was the former name of Sullivan St.

Sixth Street was the former name of Waverly Place, between Broadway and Macdougal St.

Sixth Street was the former name of Ludlow St. It was known by this name in 1797.

Skinner Road was the former name of Christopher St.

Skinner Street was the former name of Cliff St. between Ferry and Hague Sts.,known by this name in 1755; known since 1791 as Cliff St.

Slaughter House Lane, *Slaughter House Street*, *Sloat Lane*, were the former names of Beaver St. between William and Pearl Sts.; name changed to Beaver St. Dec. 25, 1825.

Sloat, south of Wall, fromWilliam St. to Hanover Sq. Now obliterated.

Slyck Steegh ("*Dirty Lane*") was the Dutch name of a lane which was afterwards widened and is now South William St. In 1657 known as Slyck Steegh; in 1674, Mill Street Lane; in 1691, Mill Lane.

Smell Street Lane was the former name of Broad St. between Exchange Place and Wall St.

Smith Court was a short alley which formerly ran from Congress St.

Smith Street, *Smee Straet*, *Smeedes Straet*, *Smit Street*, were the former names of William St. between Wall and Pearl Sts.

Smith Street was the former name of Cedar St. between William and West Sts.; known in 1691 as Smith St.; known in 1728 as Little Queen St.; known since 1794 as Cedar St.

Smith Street was the former name of East Broadway.

Smith Street Lane was the former name of Beaver St. between William and Broad Sts.

Smith Street Valley, *Smith's Vall*, *Smith's Valley*, *Smith's Vly*, were the former names of Pearl St. between Wall St. and Peck Slip.

Southampton Road, *Great Kill Road*, was the principal road leading north from Greenwich Village. It started at Gansevoort St., this street being part of the original road; from the present easterly end of Gansevoort St. it ran northeasterly, crossing 8th Ave. at 14th St., 7th Ave. between 15th and 16th Sts., 6th Ave. at 17th St., then running northerly, just east of 6th Ave. and ending at Love Lane, about the present 21st St. a little east of 6th Ave.

South Fifth Avenue was the former name of West Broadway between Canal St. and Washington Sq.

Spencer Place was the former name of West 4th St. between Christopher and West 10th Sts.

Spingler Place was the former name of East 15th St. between Broadway and 5th Ave.

Stadt Huy Lane was the Dutch name of Coenties Alley.

Staggtown formed an independent community in the vicinity of Delancey and Attorney Streets in the first part of the nineteenth century.

Stanton Place was an alley formerly in the rear of No. 6 Stanton St.

Stewart Street formerly ran from Broadway between 30th and 31st Sts. southwesterly to a point in the block bounded by 6th Ave. and 7th Ave., 28th and 29th Sts.

Stillwell's Lane was a country road which started at the Bloomingdale Road (the present Broadway) and 87th St. and ran easterly, about 150 feet east of Amsterdam Ave.; it turned southerly, turning again easterly between 85th and 86th Sts. and ended in the present Central Park on a line with 7th Ave. and 86th St.

Stone Bridge Street was one of the former names of Broadway.

Stone Street was the former name of Pearl St.

Stone Street was the former name of Thames St.

Strand, The, was the name of the north side of Pearl St. between Broad St. and Old Slip; was known by this name when Pearl St. was fronting on the East River.

Striker's Lane; see Hopper's Lane.

Steuben Street formerly ran from the Eastern Post Road and 41st St. northwesterly to the Albany Road between 43rd and 44th Sts.

Stuyvesant Place was the former name of 2nd Ave. between 7th and 10th Sts.

Stuyvesant Street. The present street of this name, which now ends at 2nd Ave., formerly continued northeasterly, crossing 1st Ave. between 12th and 13th Sts., Ave. A at 14th St., and ended at the East River about the present 15th St. between Avenues A and B.

Sugar Loaf Street was the former name of Franklin St. between Broadway and Baxter St.; was known by this name in 1807.

Suice Straat was the Dutch name of William St.

Susan Street was a country road in the Kips Bay Farm. It ran from the Eastern Post Road, the present Lexington Avenue, between 38th and 39th Sts. southeasterly, crossing 38th St. between 2nd and 3rd Aves. and ending at the East River between 38th and 39th Sts.

Theatre Alley, between Ann and Beekman Streets.

Third Street was the former name of Wooster St.

Third Street was the former name of Eldridge St.

Thomas Street was the former name of Duane St. between Elm and Rose Sts.

Thomas Street was the former name of Pearl St. between Broadway and Park Row.

Thomas Street was the former name of William St. between Frankfort and Pearl Sts.

Thomas Street was the former name of Thames St.

Thompson's Court was an alley which formerly ran from No. 363 Rivington St.

Tienhoven Street was the former name of Liberty St.; known in 1691 as Crown St.; name changed to Liberty St. in 1794.

Tienhoven Street was the former name of Pine St.

Tin Pot Alley was the former name of Exchange Alley; was also known as Oyster Pasty Alley.

Thompkin's Place was the former name of East 10th St. between Greenwich Ave. and the Hudson River.

Torbet Street was a country road on the Rutgers' Farm; it ran from Henry to Madison Sts. between Catherine and Market Sts.

Troy Street was the former name of West 12th St. between Greenwich Ave. and the Hudson River.

Tulip Street was a country road on the Glass House Farm. It ran from 34th St. between 10th and 11th Aves. southerly to a point in the block bounded by 9th and 10th Aves. between 32nd and 33rd Sts.

Turin Lane was a country road which ran from the Bloomingdale Road (Broadway) between 93rd and 94th Sts., and ran easterly, ending at the Eastern Post Road, about the present 96th St.

Tuyn Straat was the name given to the present Exchange Place by the Dutch.

Tryon Row formerly ran from Centre St. to Park Row on the ground now occupied by the south end of the Municipal Building.

Union Court was formerly on University Place between 12th and 13th Sts.

Union Furnace, S.E. corner Broadway and Howard St.

Union Place was the former name of the west side of 4th Ave., and the east side of Broadway between 14th and 17th Sts.

Union Road formerly ran from the Skinner Road, in the block bounded by 5th and 6th Aves., 11th and 12th Sts., northwesterly to the Southampton Road at 7th Ave. and 15th Sts.

Union Street was the former name of Greene St.

Van Bruggen Street was the former name of Pine St.

Van Clyff's Street, former name of east part of John St.

Vandercliffe's Street, now Gold St.

Van Nest Place was the former name of Charles St. between 4th and Bleecker Sts.

Varick Place was the former name of Sullivan St. between Houston and Bleecker Sts.

Vauxhall, Broadway and Bowery, from 4th St. to Astor Place.

Verdant Lane; also called Feitner's Lane; was a country road which started at the Bloomingdale Road (Broadway) between 45th and 46th Sts., and ran northwesterly, crossing 8th Ave. between 46th and 47th Sts., 9th Ave. between 47th and 48th Sts., 10th Ave. between 48th and 49th Sts., 11th Ave. between 49th and 50th Sts., and ended at the Hudson River between 49th and 50th Sts.

Village Street was the former name of West Houston St. between Macdougal St. and the East River.

Walker Street was the former name of Canal St. between Baxter and Ludlow Sts.

Walnut Street was the former name of Jackson St.

Warren Place was the former name of Charles St. between Greenwich Ave. and Waverly Place.

Warren Road was a country road in Greenwich Village which ran from the Southampton Road to Love Lane, from the present 16th to 21st Sts. between 6th and 7th Aves.

Warren Street was the former name of Clinton St.

Washington Alley was the former name of Washington Mews.

Washington Place, formerly 5th St.

Washington Street was the former name of Jefferson St.

Weasver Street was the former name of Vesey St.

Weehawken Street, from Christopher to Amos Street.

Weigh House Street, now Moore St.

Wendel Street was the former name of Oak St.

Wesley Place was the former name of Mulberry St. between Houston and Bleecker Sts.

West Avenue; see Albany Road.

West Broadway Place was the former name of West Broadway between Canal and Grand Sts.

West Court was formerly in the rear of No. 66 West 22nd St.

West Road, was once the name of 6th Ave.

White Place was formerly in the rear of No. 134 West 18th St.

White Street was the former name of Ann St.

William Street was the former name of Broome St. between the Bowery and Sullivan St.; was known by this name in 1797.

William Street was the former name of West 4th St. between Christopher and West 13th Sts.; known by this name in 1807.

William Street was the former name of Madison St. between Catherine and Montgomery Sts.

Willow Street was the former name of Macdougal St.

Winckel Straat was a short street running north from Bridge St. just east of Whitehall St. It was closed in 1680.

Windmill Lane was a former name of Cortlandt St.; known by this name in 1782.

Winne (or Wynne) Street was the former name of Mott St. between Pell and Bleecker Sts. Known by this name in 1755.

Winthrop Place was the former name of Greene St. between Waverly Place and West 8th St.

Wooster Street was the former name of West Houston St. between Broadway and Macdougal St.

Wooster Street was the former name of University Pl. between Waverly Pl. and West 14th St.

Wynkoop Street was the former name of Bridge St.

**These streets ran north and south adjoining Houston Street on the extreme eastern side of the island. Goerck and Mangin—Map 183.*

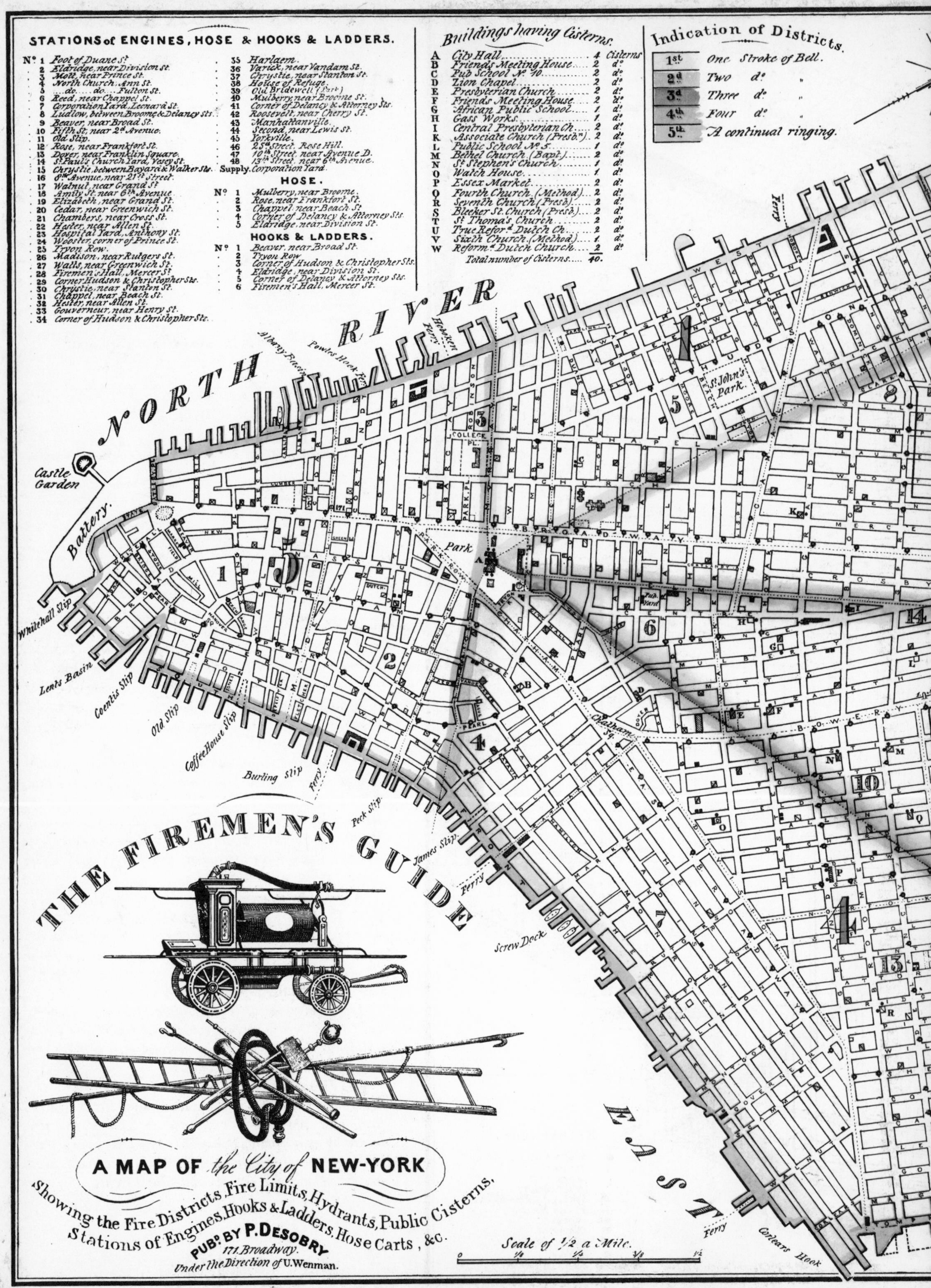
STATIONS of ENGINES, HOSE & HOOKS & LADDERS.
Nº 1 Foot of Duane St.
2 Eldridge, near Division St.
3 Mott, near Prince St.
4 North Church, Ann St.
5 ...do.....do... Fulton St.
6 Reed, near Chappel St.
7 Corporation Yard, Leonard St.
8 Ludlow, between Broome & Delancy Sts.
9 Beaver, near Broad St.
10 Fifth St. near 2d Avenue.
11 Old Slip.
12 Rose, near Frankfort St.
13 Dover, near Franklin Square.
14 St Paul's Church Yard, Vesey St.
15 Chrystie, between Bayard & Walker Sts.
16 8th Avenue, near 21st Street.
17 Walnut, near Grand St.
18 Amity St. near 6th Avenue.
19 Elizabeth, near Grand St.
20 Cedar, near Greenwich St.
21 Chambers, near Cross St.
22 Hester, near Allen St.
23 Hospital Yard, Anthony St.
24 Wooster, corner of Prince St.
25 Tryon Row.
26 Madison, near Rutgers St.
27 Watts, near Greenwich St.
28 Firemen's Hall, Mercer St.
29 Corner Hudson & Christopher Sts.
30 Chrystie, near Stanton St.
31 Chappel, near Beach St.
32 Hester, near Allen St.
33 Gouverneur, near Henry St.
34 Corner of Hudson & Christopher Sts.
35 Harlaem.
36 Varick, near Vandam St.
37 Chrystie, near Stanton St.
38 House of Refuge.
39 Old Bridewell (Park)
40 Mulberry, near Broome St.
41 Corner of Delancy & Attorney Sts.
42 Roosevelt, near Cherry St.
43 Manhattanville.
44 Second, near Lewis St.
45 Yorkville.
46 25th Street, Rose Hill.
47 10th Street, near Avenue D.
48 13th Street, near 6th Avenue.
Supply. Corporation Yard.
HOSE.
Nº 1 Mulberry, near Broome.
2 Rose, near Frankfort St.
3 Chappel, near Beach St.
4 Corner of Delancy & Attorney Sts.
5 Eldridge, near Division St.
HOOKS & LADDERS.
Nº 1 Beaver, near Broad St.
2 Tryon Row.
3 Corner of Hudson & Christopher Sts.
4 Eldridge, near Division St.
5 Corner of Delancy & Attorney Sts.
6 Firemen's Hall, Mercer St.
Buildings having Cisterns.
A City Hall ... 4 Cisterns
B Friends Meeting House ... 2 do
C Pub School Nº 10 ... 2 do
D Zion Chapel ... 2 do
E Presbyterian Church ... 2 do
F Friends Meeting House ... 2 do
G African Public School ... 1 do
H Gass Works ... 1 do
I Central Presbyterian Ch. ... 2 do
K Associate Church (Presbn.) ... 2 do
L Public School Nº 5 ... 1 do
M Bethel Church (Bapt.) ... 2 do
N St Stephen's Church ... 1 do
O Watch House ... 1 do
P Essex Market ... 2 do
Q Fourth Church (Method.) ... 2 do
R Seventh Church (Presb.) ... 2 do
S Bleeker St. Church (Presb.) ... 2 do
T St Thomas's Church ... 2 do
U True Refor.d Dutch Ch. ... 2 do
V Sixth Church (Method.) ... 1 do
W Reform.d Dutch Church ... 2 do
Total number of Cisterns ... 40.
Indication of Districts.
1st One Stroke of Bell.
2d Two do
3d Three do
4th Four do
5th A continual ringing.
NORTH RIVER
EAST
Castle Garden
Battery
Whitehall Slip
Lents Basin
Coentis Slip
Old Slip
Coffee House Slip
Burling Slip
Peck Slip
James Slip
Ferry
Screw Dock
Corlears Hook
Albany Ferry
Powles Hook Ferry
Hoboken Ferry
St. John's Park
College Pl.
Park
Chatham Sq.
THE FIREMEN'S GUIDE
A MAP OF the City of NEW-YORK
Showing the Fire Districts, Fire Limits, Hydrants, Public Cisterns,
Stations of Engines, Hooks & Ladders, Hose Carts, &c.
PUBD. BY P. DESOBRY
171, Broadway.
Under the Direction of U. Wenman.
Scale of 1/2 a Mile.
0 1/8 1/4 3/8 1/2